4 Crystals

Growth, Properties, and Applications

Managing Editor: H. C. Freyhardt

Editors: T. Arizumi, W. Bardsley, H. Bethge
A. A. Chernov, H. C. Freyhardt, J. Grabmeier
S. Haussühl, R. Hoppe, R. Kern, R. A. Laudise
R. Nitsche, A. Rabenau, C. J. M. Rooijmans
W. B. White, A. F. Witt, F. W. Young, Jr.

Organic Crystals Germanates Semiconductors

With Contributions by
L. N. Demianets, G. A. Emelchenko, J. Hesse,
N. Karl, A. N. Lobachev and H. Maier

Springer-Verlag
Berlin Heidelberg New York 1980

ISBN-13: 978-3-642-67766-3 e-ISBN-13: 978-3-642-67764-9
DOI: 10.1007/978-3-642-67764-9

Library of Congress Cataloging in Publication Data: Main entry under title: Organic crystals, germanates, semiconductors. (Crystals – Growth, Properties and Applications; 4) Includes bibliographical references. Contents: Karl, N. High purity organic molecular crystals. – Demianets, L. N., Lobachev, A. N., Emelchenko, G. A. Rare-earth germanates. – Maier, H. and Hesse, J. Growth, properties, and applications of narrow-gap semiconductors. 1. Molecular crystals. 2. Chemistry, Organic. 3. Germanates. 4. Narrow gap semiconductors. I. Series. QD921.064 548 80-20989

Softcover reprint of the hardcover 1st edition 1980

2152/3140-543210

Table of Contents

High Purity Organic Molecular Crystals

Norbert Karl

Physikalisches Institut and Kristallabor, Universität Stuttgart, Pfaffenwaldring 57,
D-7000 Stuttgart 80

The article reviews the preparation of highest purity organic crystals which are of rapidly growing importance for the understanding of basic processes and for the innovation of practical applications. High purity can be obtained by zone refining in combination with other methods. For the characterization of the crystals, grown from the melt or by sublimation, sensitive methods have been developed, being able to detect impurities down to well below 1 ppm. The author's extensive practical experience is condensed into reliable technical prescriptions and valuable advices and hints.

Table of Contents

1 Introduction

1.1 Preface

Organic chemistry had its origin in chemicals which are synthesized by living cells. These chemicals consist of *molecules* whose skeletons are built up of carbon atoms. The remaining valences are connected with ligands such as hydrogen, halogens, $-OH$,$=O$, $-NH_2$. Some of the skeletal carbon atoms can be replaced by nonmetals such as oxygen, nitrogen, or sulfur ("heteroatoms").

It is characteristic for the living world, *not* to be in a crystalline state.

However it is possible to obtain single crystals from many organic compounds both of natural and synthetic origin. For a number of years the physics and chemistry of these crystals have stimulated fundamental research on a rapidly growing scale.

The great variety of possible organic structures (as compared to inorganics) opens up a large field of new materials and of novel material properties; for previous literature reviews and data compilations see [1–40] and Chap. 6.

The art of producing good and pure organic single crystals has developed hand-in-hand with the ever growing requirements of basic research, arising from its interest in fundamental interactions in the solid state.

Interactions manifest themselves in a very detailed way by energy transfer. Energy transfer is very sensitively influenced by impurities acting on the transport properties as scattering centers or traps: Interfering effects by impurities have been recognized early on such elementary excitations as singlet excitons, cf.[41, 42], charge carriers, cf.[43–46], and triplet excitons, cf.[47], which are adversely affected even by concentrations as low as 10^{-6} mol/mol (1 ppm) and less. This means that highest purity is of prime importance; the requirements are comparable to those which are typical for inorganic semiconductors such as Ge, Si or some group III–V compounds. Therefore the emphasis of this review will lie on *high purity*.

The author wants to address all who wish to inform themselves on the general state of the art of growing and characterizing high purity organic crystals. In addition he wants to address also all those readers who seek information and references on special problems. As much as possible, personal practical experience is included in order to be helpful to anyone who is responsible for a crystal growth laboratory or who is doing the actual tricky work.

1.2 Scope of the Field Covered

The achievement of high purity implies the applicability of iterative methods of purification which are efficient in a nonspecific way and thus eliminate a broad spectrum of impurities. There is no doubt that *zone refining* meets these requirements in an optimal way. Zone refining is not restricted to special impurities because it is based essentially on the noncoincidence of solidus and liquidus curves in the great majority of phase diagrams. Repeated zone refining, if properly done, in principle can lead to any practically desirable level of purity. Therefore this contribution will concentrate on those organic materials which can be melted without decomposition.

Compounds for which one must confine oneself to other less efficient purification methods, such as fractional distillation, sublimation, recrystallization or chromatography will only be included where it is implied by general aspects. These less efficient purification methods, however, will deserve special attention as indispensable steps of overall prepurification, or in combination with additional impurity-specific chemical reactions which must be applied for some persistent impurities.

Crystal growth techniques are selected from the aspect of being able to maintain the high purity of the starting material. The Bridgman technique – growth from the melt by unidirectional cooling – offers itself in a quite natural way for compounds which remain stable at their melting point. It can be performed in a sealed ampoule which garantees optimal exclusion of any contaminants.

Sublimation growth in evacuated sealed ampoules can be a valuable alternative for compounds which undergo a phase transition at elevated temperatures, which destroys single crystallinity. Clearly it is also a useful method for compounds which are not stable at their melting point, but which are stable in lower temperature regimes, where their vapor pressure is sufficiently high for sublimation.

Pulling techniques like Nacken-Kyropoulos or Czochralski growth have not been attempted very often because of the high vapor pressure of most organic molecular crystals at their melting point and because of the relative weakness of the hot suspended crystal. Nevertheless, these methods seem to be a promising alternative[48, 49].

Solution and flux growth obviously have to battle against solvent impurities and solvent inclusions. Under the aspect of high purity it is advisable to rely on these methods only where the above-mentioned techniques fail or where it is implied by special aspects (such as growth, handling and use at the same temperature in order to avoid thermal stress). In general, solvents must be carefully purified as well.

The demand for undecomposed melting restricts the molecular weight even for "stable" compounds roughly to values below 300 [g/mol]; an example for a high-melting stable compound is picene, $C_{22}H_{14}$, M. W. = 278; m. p. = 364 °C (structure formula I, Table 1).

Too low melting points of small molecules on the other hand preclude convenient preparation and handling of crystals; therefore experience is only rarely available for such compounds; examples: methane, CH_4[50] (II); CD_4[51], triple point T_{tp} = 89,78 K at p_{tp} = 79.1 torr; n-heptane C_7H_{16}[52, 53] (III), m. p. = 182.55 K.

Concerning the chemical nature of material from which high-purity macroscopic single crystals have been obtained, the literature is dominated by unsubstituted parent hydrocarbons; less informations is available for molecules with small substituents and nearly none for those with large substituents. There are several reasons for this. First, many functional groups reduce the thermal stability of the parent molecule, e. g. $-C\begin{smallmatrix}\nearrow O\\ \searrow OH\end{smallmatrix}$, $-I$. Second, most functional groups introduce librational and rotational degrees of freedom whose activation at certain intermediate temperatures often leads to phase transitions which can be more or less destructive for single crystals. Third, large functional groups with many conformational degrees of freedom may lead to very slow crystallization with substantial tendency to supercooling.

Table 1. Chemical structures of molecules

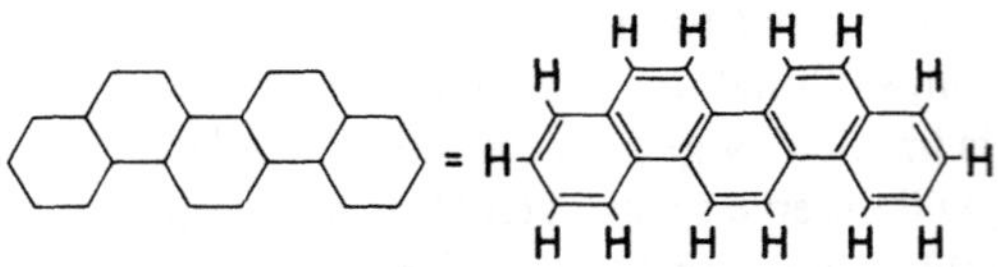

(I)

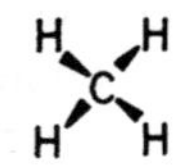

(II)

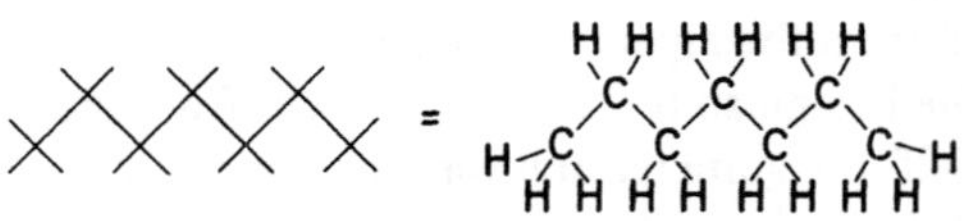

(III)

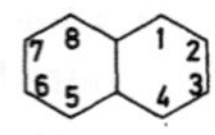

(IV)

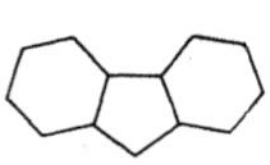

(V)

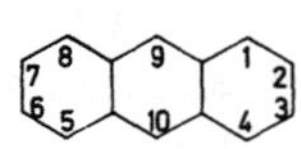

(VI)

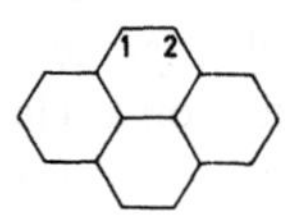

(VII)

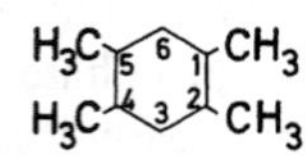

(VIII)

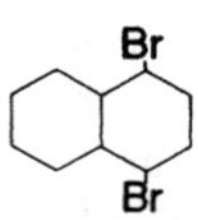

(IX)

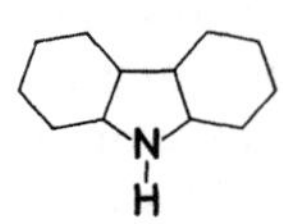

(X)

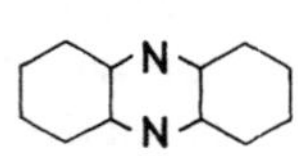

(XI)

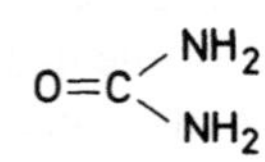

(XII)

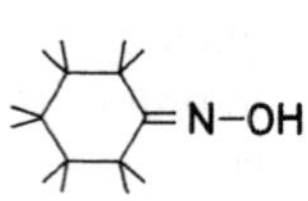

(XIII)

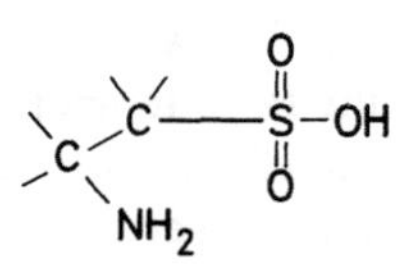

(XIV)

(XV)

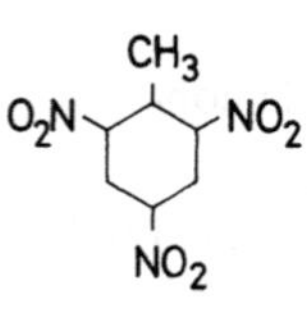

(XVI)

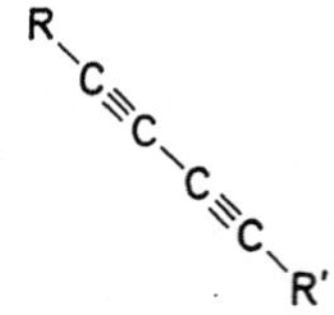

(XVII)

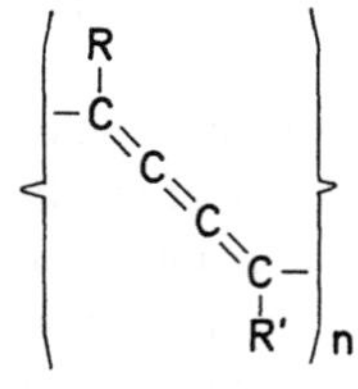

(XVIII)

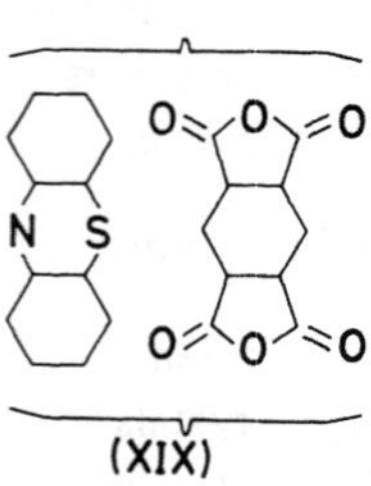

(XIX)

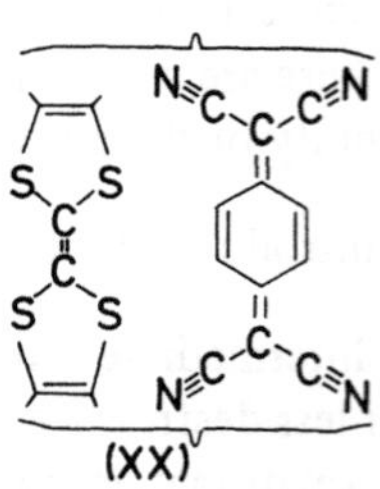

(XX)

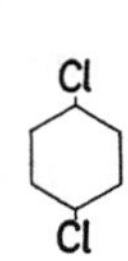

(XXI)

Table 1 (continued)

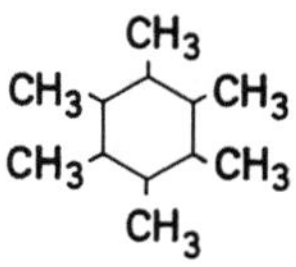
(XXII)

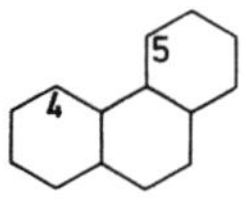
(XXIII)

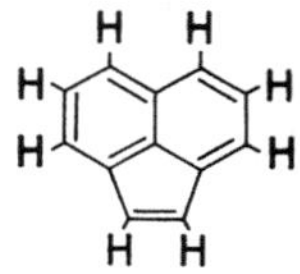
(XXIV)

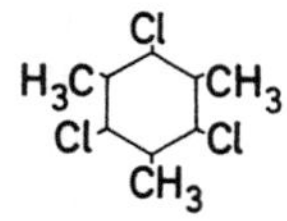
(XXV)

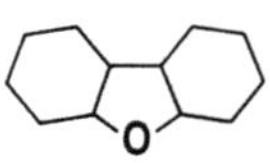
(XXVI)

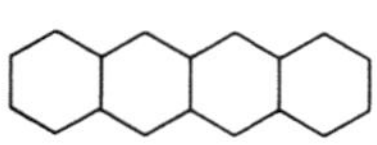
(XXVII)

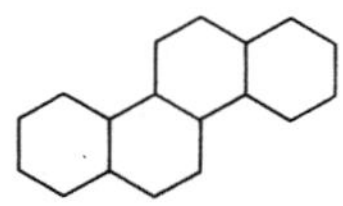
(XXVIII)

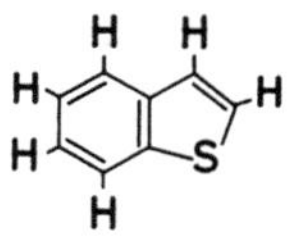
(XXIX)

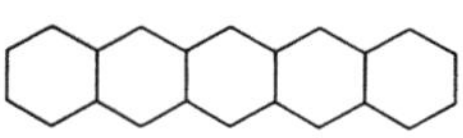
(XXX)

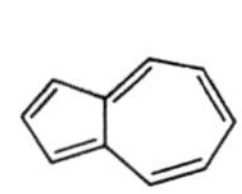
(XXXI)

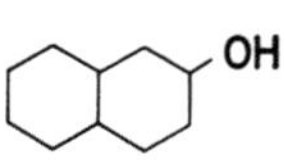
(XXXII)

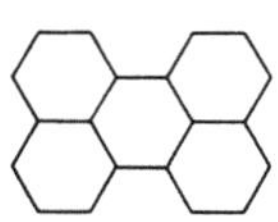
(XXXIII)

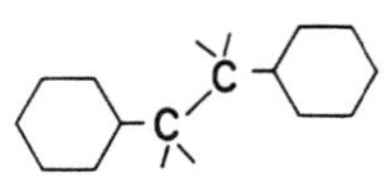
(XXXIV)

The most thoroughly investigated class of high purity organic crystals is that of rigid aromatic molecules such as simple hydrocarbons like naphthalene $C_{10}H_8$ (IV); fluorene, $C_{13}H_{10}$ (V); anthracene, $C_{14}H_{10}$ (VI); pyrene, $C_{16}H_{10}$ (VII); substituted aromatic hydrocarbons like 1,2,4,5-tetramethylbenzene ("durene"), $C_{10}H_{14}$ (VIII); 1,4-dibromonaphthalene, $C_{10}H_6Br_2$ (IX); and unsubstituted heteroaromatics such as carbazole, $C_{12}H_9N$ (X); and phenazine, $C_{12}H_8N_2$ (XI).

In view of the above qualifications we will not concentrate our attention on crystals of organic acids nor on ionic organic crystals; these compounds are mostly unstable at their melting points. Problems may also arise for such compounds as urea, CH_4N_2O (XII), cyclohexanone oxime, $C_6H_{11}NO$ (XIII), taurin, $C_2H_7NO_3S$ (XIV), dioctadecyladipate, $C_{42}H_{82}O_4$ (XV), and 2,4,6-trinitrotoluene $C_7H_5N_3O_6$ (XVI), to give some examples of compounds from which large single crystals have been successfully obtained by solution growth[54)].

Organic polymers are excluded because usually no macroscopic homogeneous single crystals can be obtained. The only exception is the class of poly-diacetylenes (XVIII) obtained as large single crystals by solid state polymerisation of solution grown monomer crystals (XVII)[55] see also[56]. However, crystals of this class of material have not yet been prepared in high purity.

Donor-acceptor (D:A) complexes between at least two molecular species (differing in their ionization energy and electron affinity) will be included inasmuch as the individual components or the combined complex can be zone refined. These complexes have a more or less neutral ground state lattice interaction but possess ionic excited states due to electron transfer from the donor to the acceptor, e. g. by light absorption. An example is the black D:A-crystal phenothiazine:pyromellitic dianhydride (PTZ:PMDA)[57] (XIX).

Closely related is the very exciting class of molecular radical ion salts like tetrathiafulvalene: tetracyanoquinodimethane ($TTF^{\cdot +}$: $TCNQ^{\cdot -}$)[58, 59] (XX), where partial or full electron transfer already occurs thermally, leading in some cases to quasimetallic conductivities (as in the example given). Unfortunately these complexes are presently not obtainable in very high purity, because of thermal instability at their melting points. However, the diffusional solution growth technique will be mentioned for this kind of material because it is of general interest for the preparation of donor-acceptor crystals from the individual components[60, 61].

1.3 Some General Properties of Organic Molecular Crystals

The molecular packing in pure van-der-Waals crystals is governed by atom-atom potentials. Lattice energy is minimized by a packing with as many as possible van der Waals atom-atom contacts[62, 63]. This is usually reached by highest density packing, often under creation of lattice symmetries such as glide reflection planes and/or screw axes. In fact, the great majority of organic molecular crystals belongs to the space groups $P2_1$, $P2_1/c$, $P2_1 2_1 2_1$, or to one of the space groups of the point group mmm with at least 2 or 3 glide reflection planes and 2 or 3 screw axes cf.[17, 64].

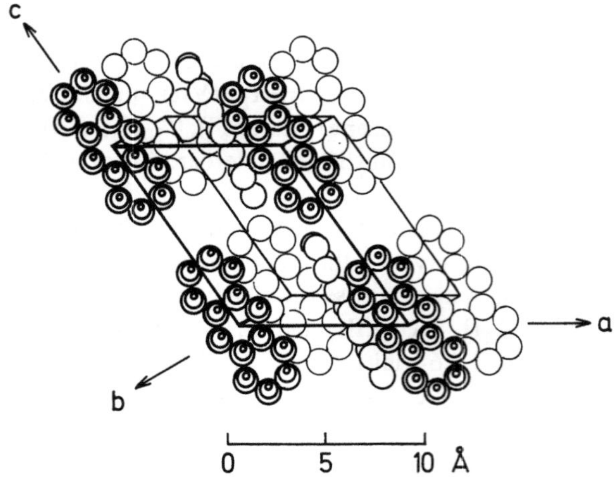

Fig. 1. The crystal and molecular structure of anthracene. Drawn is the carbon skeleton; the hydrogen atoms are not shown. After Sinclair, J. M. Robertson and Mathieson[65]

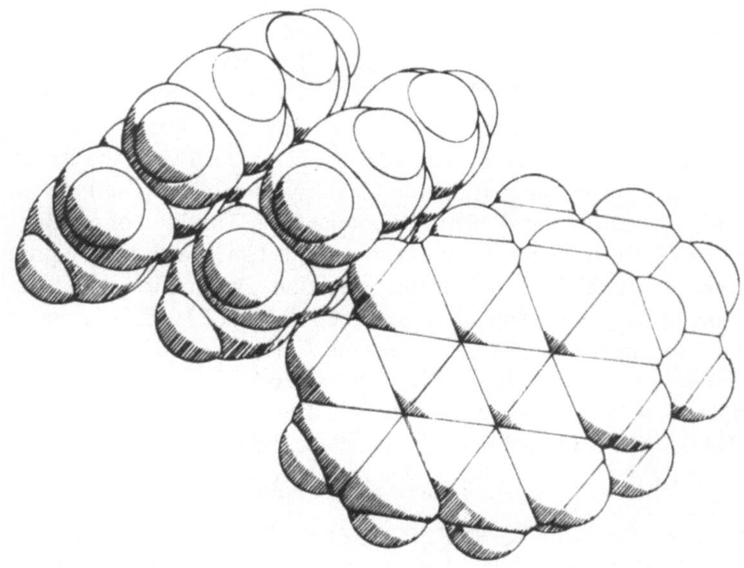

Fig. 2. The pairwise packing of pyrene in the room temperature phase. The pyrene molecules are represented here by their van der Waals contact surfaces. From[76], drawn after the data of Camerman and Trotter[66]

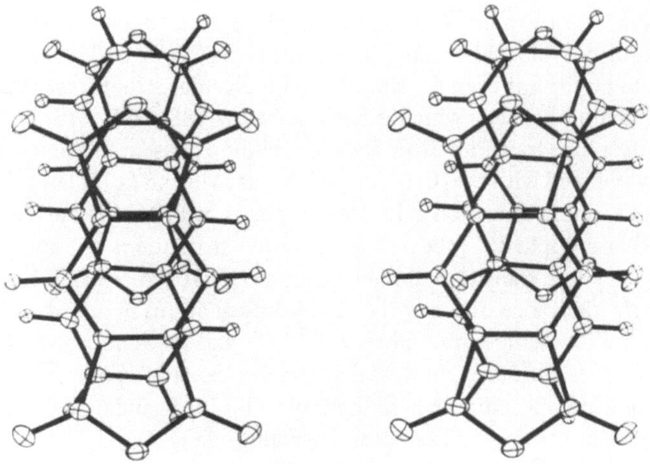

Fig. 3. A stereoscopic representation of the molecular stacking in the donor-acceptor complex anthracene: pyromelliticdianhydride viewed normal to the molecular planes[a]. Form B. E. Robertson and Stezowski[71].[b]

---

[a] The molecular skeleton (valance bonds) and the vibrational ellipsoids of the atoms are drawn

[b] A stereoscopic 3-dimensional impression can be obtained after some excercise without a pair of stereoscopic viewing lenses by adjusting the eye parallax to infinity while focussing on the figure from a not too close distance

Two different realizations of such symmetric packing can be found: *Individual molecules* are related by screw axes and/or glide reflection planes (example: anthracene, see Fig. 1[65]); *sandwich pairs* of molecules are so related. The latter form occurs for some planar aromatic molecules, (example: pyrene[66], see Fig. 2). Depending on the inclination of the molecular planes relative to the shortest translational unit vector, planar aromatic molecules may form distinct infinite linear stacks with predominantly one-dimensional interactions. This kind of packing usually manifests itself by a strong tendency towards needle-like growth and by pronounced cleavage planes parallel to the distinct direction. The simplest molecular packing of this sort belongs to the triclinic space groups P1 and P$\bar{1}$ with only *one molecule* per unit cell (examples: β-1,4-dichlorobenzene, $C_6H_4Cl_2$ (XXI), stable above 30.8 °C[67–69]; hexamethylbenzene, $C_{12}H_{18}$ (XXII), above 116 K cf.[70]); or, in the case of D:A complexes, with *one DA pair* per unit cell. The stack sequence is then either ... DADADA ... (mixed stacks), example anthracene: PMDA[71], see Fig. 3, or $\left\{\begin{array}{l}\ldots AAAAA \ldots\\ \ldots DDDDD \ldots\end{array}\right\}$ (separate donor and acceptor stacks) as realized in several TCNQ radical-ion salts like TTF:TCNQ (XX)[58]. In α-9,10-dichloroanthracene ($P2_1/a$, Z = 4) the molecular planes are almost exactly perpendicular to [100] which is the pronounced growth direction[72], (see also Fig. 45).

Packing calculations, using appropriate atom-atom potentials, show for many systems that the lowest minimum of the lattice energy is often joined by other side minima which are only slightly shallower and which belong to different relative arrangements of the molecules cf.[62]. This is a consequence of the rather unspecific character of van der Waals interactions. It is quite obvious from this, that the problem of polymorphism and phase transitions (cf.[73] and ref. given therein) becomes quite important even for simple rigid molecular structures (examples: α-1,2,4,5-tetrachlorobenzene which is triclinic below 188 ± 2 K with Z = 2; as a consequence of the phase transition from the monoclinic β phase stable above 188 K, the α-phase is usually twinned with the (010) normal, b*, as twin axis (albit law)[74] cf.[73]; pyrene shows a drastic phase transition in its optical spectra at 110 K (0 kbar) or at 300 K if a pressure of 4 kbar is applied[75]; the low temperature structure is not completely known[76].

A phase transition in phenanthrene (XXIII) at ~343 K has lead to some confusion. Steric hindrance of the two endo hydrogen atoms in pos. 4 and 5 may be responsible for the low temperature phase (cf.[77, 78] and references given therein). Crystals of molecules with near rotational symmetry can undergo order-disorder phase transitions at temperatures where the molecules (or ligands) become free to rotate or librate in the lattice; examples: acenaphthylene, $C_{12}H_8$(XXIV)[79], dichlorodurene $C_{10}H_{12}Cl_2$[80], anthracene: TCNB[310].

For reaching close packing, the necessary number of molecules per unit cell, Z, can become rather high. Z = 1 seldom occurs (p-dichlorobenzene, hexamethylbenzene, 9,10-dinitroanthracene[81], anthracene: PMDA). Z = 2 (naphthalene, anthracene) and Z = 4 (fluorene[82], dibenzofuran, $C_{12}H_8O$ (XXVI)[83, 84] are often realized; but higher numbers can occur (Z = 8: 1,4-dibromonaphthalene[85]). Such high Z-numbers may lead to an increased tendency for twinning, mosaic-structures and local structural defects. In fact it is rather difficult to obtain good 1,4-dibromonaphthalene single crystals.

All experience available so far has proven that a simple molecular crystal can be regarded as a close-packed arrangement of unspecifically interacting molecules. (Specific interactions with structural relevance occur in more complicated systems with hydrogen bonding or ionic forces; to some extend also in D:A crystals). Therefore in simple molecular crystals all those foreign molecules which are geometrically similar to the host molecule may be readily incorporated as impurities; they replace the host by closely fitting into the empty lattice site and with near conservation of lattice energy)[86–88]. Thus, depending on the material source or the method of synthesis, quite a number of geometrically similar impurities may be found in a crystal (examples: tetracene, $C_{18}H_{12}$ (XXVII), in anthracene and chrysene, $C_{18}H_{12}$ (XXVIII); 2-methylnaphthalene, $C_{11}H_{10}$, and thionaphthene, C_8H_6S (XXIX), in naphthalene). On the other hand, if the fit is not good, a molecule selected as dopant may be rejected even at a concentration as low as 10^{-5} mol/mol (example: tetracene in naphthalene), making intentional doping at higher levels a problem requiring careful selection. Double replacement seems to be possible (pentacene, $C_{22}H_{14}$ (XXX) goes into the naphthalene (IV) lattice; however the microscopic structure is not yet established for this example.

Further characteristics of organic molecular crystals are the smallness of the lattice energies[89–91], e. g. 1 eV (23 kcal/mol) for anthracene[92] which means only 40 meV per atom! [c] The resulting high sublimation vapor pressure well *below the melting* point (~0.1 torr at 115 °C, ~1.0 torr at 145 °C for anthracene, ~40 torr at the melting point (218 °C)[92]) calls for special precautions during zone refining in order to keep the removed impurities away from the purified part of the ingot. Difficulties may also arise for crystal growth techniques where the crystal is not at the coldest place of the growth chamber (e. g. Czochralski). For many compounds the vapor pressure is considerable even at room temperature, precluding, among other things, the preparation of well polished faces. Surface investigations may be adversely affected (or even made impossible if vacuum is needed) whereas self-cleaning may be helpful in other cases.

1.4 Outline of the Paper

These introductory remarks illustrating some of the peculiarities of organic crystals were meant to lead the reader immediately into the actual problems which are associated with the purification of such materials by different methods (Chap. 2) and with the growth of single crystals (Chap. 3). Many solutions have been proposed and tested in the literature and in our laboratory (Stuttgarter Kristallabor). Emphasis will be concentrated on standard techniques for standard high purity compounds such as naphthalene and anthracene on the one hand. On the other, simple reliable techniques will be treated which are universal enough to be promising as a first trial on the way to pure single crystals of novel compounds.

A chapter on characterization will follow (Chap. 4). Here the determination of trace amounts of chemical impurities ("chemical defects") down to and well below

[c] Typical values for inorganic ionic crystals are 8–10 eV

1 ppm will be of prime interest (part 4.1). But structural ("physical") defects also deserve appropriate attention, although much is still left to be discovered and investigated in this field, including the development of new experimental techniques (part 4.2).

Handling and machining of organic material require special precaution and techniques. Organic material is mostly more or less chemically unstable against oxygen, water vapor or heat, photochemically reactive, volatile and mechanically soft. Oriented cutting needs simple and reliable methods for orienting crystals of low symmetry. These points are treated in Chap. 5.

Finally the reader may feel the necessity for justification of this kind of research. The author will try to show (Chap. 6) that growing organic crystals with a fantastic purity of, say 99.999999% (10^{-8} mol/mol) with regard to certain interfering impurities is not just *l'art pour l'art*. Fundamental questions of basic research have found unforeseen answers; novel material properties and new physical effects have been discovered. Applications of such ideas pertain even to fields such as electrophotography, organic metals and photosynthesis.

2 Purification

2.1 Zone Refining

2.1.1 Theoretical Aspects

Zone refining (Pfann[93, 94]) is based on the fact that a binary liquid mixture, say of the component under consideration, A, and of a second component, B, usually forms crystals of a different relative composition on cooling (cf. Fig. 7 b), i.e. with

$$k_B = \left[\frac{B}{A+B}\right]_{\text{solid}} \Big/ \left[\frac{B}{A+B}\right]_{\text{liquid}} = \frac{c_B^s}{c_B^l} \neq 1 \tag{1}$$

where c_B^s and c_B^l are the concentration of the second component B in the solid (s) and liquid (l) phase, respectively, and k is called the distribution (or segregation) coefficient of B as a component of the system A (B). In typical binary phase diagrams, (naphthalene/azulene (XXXI), Fig. 4, is given as an example), liquidus (1) and solidus (s) curves are not straight lines, thus making k dependent on the actual concentration, i. e. on the position in the phase diagram. In what follows we shall consider B being an impurity which is present in the system under consideration only in a small ($\lesssim$ 1%) concentration (indicated by brackets around B in the symbol A (B)). For this case the liquidus and solidus curves can be approximated reasonably well by straight tangential lines at the point B = 0 and therefore k can be considered to be a constant (cf. Fig. 7b); (zone refining with a composition-dependent k has been treated by Matz[96]).

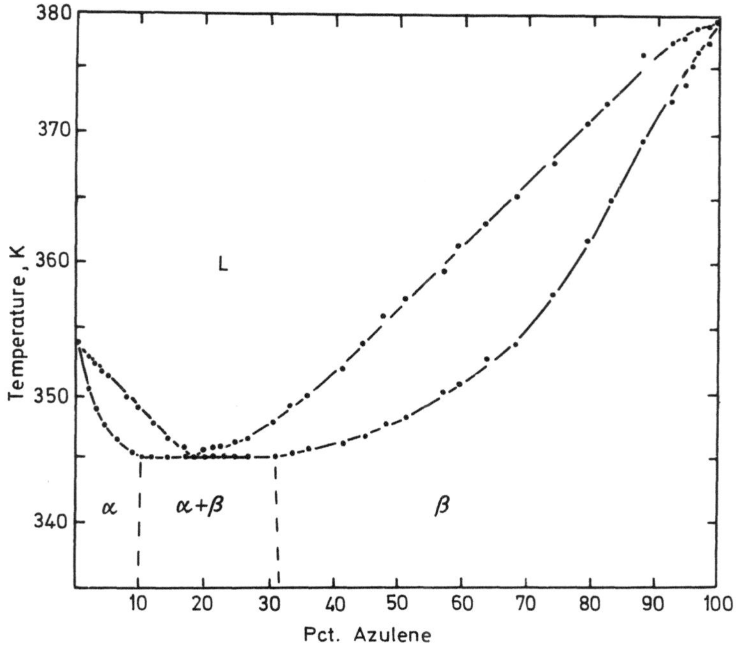

Fig. 4. The naphthalene/azulene phase diagram[95)]

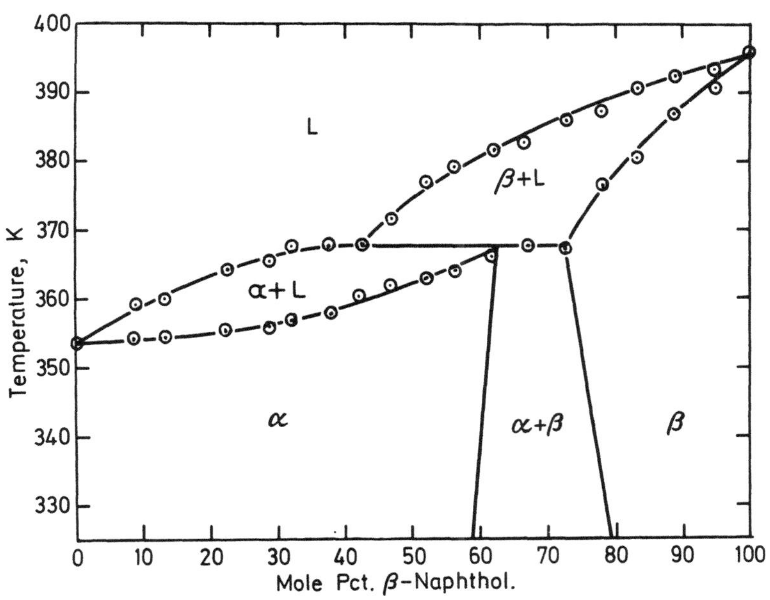

Fig. 5. The naphthalene/β-naphthol phase diagram after P. M. Robertson et al.[97)] (A later investigation by P. M. Robertson and Scott[98)] revealed the existence of two β-naphthol-rich solid phases. For clarity the initial phase diagram is shown here.)

The distribution coefficient k can be smaller than one, as e.g. in the right part of the phase diagram of naphthalene/β-naphthol (XXXII), (Fig. 5). $k < 1$ is connected with an impurity that is not readily incorporated into the crystal lattice and therefore depresses the melting point of the pure compound A. – In systems with a wide range of homogeneous miscibility in the solid state, i. e. with high dopant acceptibility (which is only rarely found), there may be a part of the phase diagram resembling the left side of Fig. 5, with $k > 1$, connected with impurities that raise the melting point. This case occurs relatively seldom. – For geometrically rather different molecules A and B, the critical selection rules mentioned in the introduction mostly lead to phase diagrams of the type given in Fig. 6 with a nearly complete miscibility gap and k_A as well as $k_B \ll 1$.

In systems with more than one impurity, k_B may depend on the concentration of the other impurities: $k_B = k_B\ (c_C, c_D \ldots)$. Complicated and mostly unknown ternary and higher phase diagrams must be considered. However, in dilute systems the interactions among the impurities themselves are frequently negligible in comparison to the interactions of each individual impurity molecule with the "host"-A-molecules. In this case segregation can be described to a good approximation by independent binary distribution coefficients $k_B, k_C, k_D \ldots$ for each impurity. However, especially for k_B not very different from 1, it may even happen that a second impurity switches the distribution coefficient of the first one from $k > 1$ to $k < 1$ or vice versa.

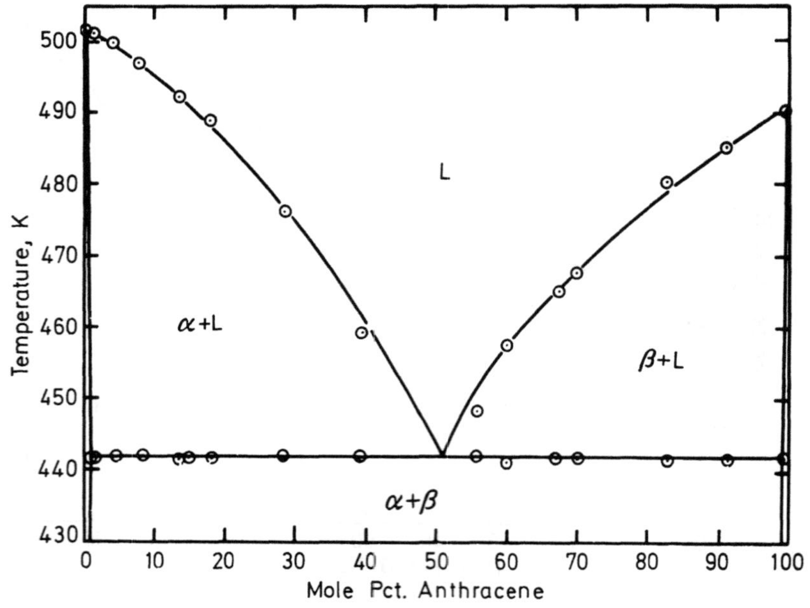

Fig. 6. The hexachlorobenzene/anthracene phase diagram[97] with a wide miscibility gap, that is typical for the majority of binary systems of geometrically different molecules

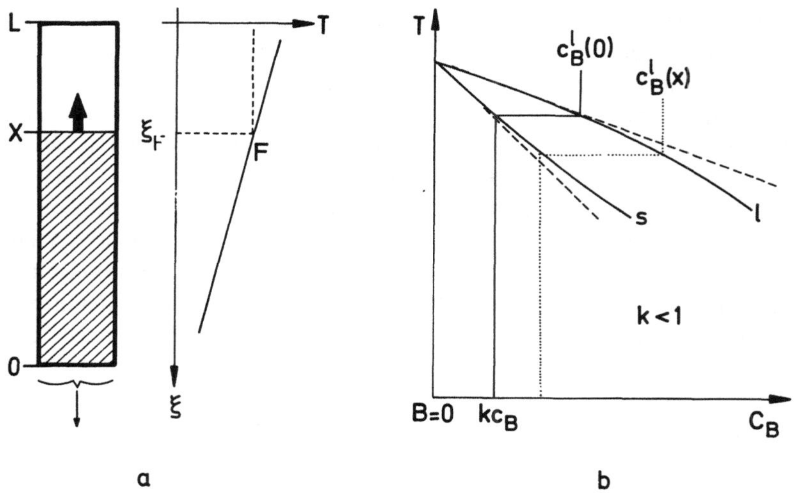

Fig. 7. a Normal freezing, schematic diagram. Pathways in the corresponding phase diagram are indicated in **b**

It is convenient to consider unidirectional cooling ("normal freezing") first. This can be realized experimentally by lowering a cylindrical container with the molten material down a temperature gradient in direction ξ (Fig. 7a). Suppose that supercooling is avoided by some means. The material starts to crystallize at $x = 0$ in the ampoule $0 \leqslant x \leqslant L$. Let the initial homogeneous concentration of the impurity in the completely molten material (i. e. with the crystallization front at $x = 0$) be c_B^l (0). According to the section $0 < [B] \leqslant [A]$ of a phase diagram, for which $k_B < 1$ holds (represented in Fig. 7b), the first crystals will contain $c_B^s = kc_B^l$. The amount of impurity which is not incorporated into the crystal (s) will be distributed in the melt which thus becomes more concentrated in B. A more concentrated melt in turn will deposit more concentrated crystals (dotted lines in Fig. 7b) on progress of the crystallization front, and so on.

The final impurity distribution in the solid, c_B^s, after normal freezing is completed, will be a function of x, or, more generally, of the crystallized fraction $g = \frac{x}{L}$ or of the uncrystallized remainder $r = 1 - g$.

Taking into account the volume and concentration changes for a progress dx of the crystallizing front one obtains the differential equation for normal freezing which has the solution[94].

$$c(x) = k\,c_o \left(1 - \frac{x}{L}\right)^{k-1}; \quad \log\left(\frac{c}{c_o}\right) = \log k + (k-1)\log\left(1 - \frac{x}{L}\right) \tag{2a}$$

Obviously for $k < 1$ the impurity concentrates toward the end of the ingot at $x = L$; this is easily seen for $k \ll 1$ where $c(x) \approx k\,c_0\,\frac{1}{r}$. Equation (2a) is also valid for $k > 1$ where the impurity concentrates at the starting end at $x = 0$. Notice that the concentration curve contains the distribution coefficient twice, a fact to which we shall return in point (2.1.4c): If the simple model is correct the slope and the intercept of the log/log-plot of Eq. (2a) should lead to identical k-values.

So far we have made a fundamental simplication: We have considered the system always to be in thermodynamic equilibrium. In fact the rejected impurity ($k_B < 1$) has to diffuse away from the growing crystal front (and vice versa for $k > 1$). Diffusion is a slow process. Convection and intentional stirring will assist or take over the impurity transport only beyond a diffusionally controlled boundary layer, Fig. 8. In this boundary layer impurity enrichment ($k_B < 1$) or depletion ($k_B > 1$) relative to the main part of the melt occurs. This fact has been taken into account by Burton, Prim and Slichter[99] who introduced a modified, kinetically controlled effective distribution coefficient k_{eff},

$$k_{eff} = k_0 \frac{1}{k_0 + (1-k_0)\exp(-fd/D)}, \qquad (3)$$

where k_0 is the equilibrium distribution coefficient and the exponent is the ratio of the travel velocity f of the crystallization front, to a diffusional velocity D/d of the impurity in the liquid with the diffusion constant D and a characteristic boundary layer width d (which itself is a function of stirring and/or convectional parameters). As long as the solidification speed f is small compared to the diffusion speed D/d, or as long as the boundary layer thickness can be made small compared to D/f, the distribution coefficient is essentially unaffected (k_0). For faster and faster speed and/or larger d, k_{eff} approaches unity, which means no segregation at all.

Normal freezing is not very efficient for purification because the purest fraction obtainable is only purer by a factor of k (considering the most commonly encountered case $k < 1$). Its significance lies in the fact that it can easily be used to determine effective distribution coefficients under real conditions, especially as a function of the crystallization speed[8]: Zone refining is governed by the effective distribution coefficient of the impurities but it does not give quantitative numbers for k in a straight forward way.

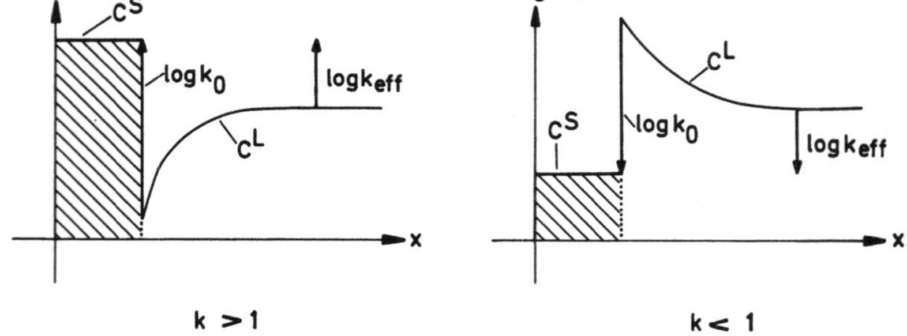

Fig. 8. The purification efficiency is reduced because of transport restrictions for the impurity at a diffusion-controlled boundary layer in front of the solidified material (hatched). The resulting effective distribution coefficient k_{eff}, is always closer to one than the equilibrium distribution coefficient k_0

The fact that normal freezing causes segregation and an inhomogeneous distribution should always be kept in mind when considering the dopant level in doped Bridgman crystals: For instance an anthracene crystal with 10^{-5} mol/mol tetracene ("added to the melt") will turn out to contain in reality only $\sim 10^{-6}$ mol/mol at the starting end of growth ($k_{eff} \sim 0.1$ [8]). The concentration will increase in the growth direction and may be 10 or even 100 times higher even well before the end at $x = L$. Here as well as for optimizing zone refining, knowledge of the distribution coefficient as a function of the growth or transport speed is highly desirable.

For obtaining information on the equilibrium distribution coefficient, it is necessary to extrapolate results obtained for different growth speeds f to $f = 0$. (Solid state diffusion, tending to equalize the concentration profile obtainable by normal freezing, will be excluded for the moment). A very constant growth speed of a uniform and smooth phase boundary in conjunction with efficient mixing of the melt is necessary. A separate apparatus may be useful for the determination of distribution coefficients[8, 100]. However it is often more realistic to use segregation coefficients k_{eff} obtained with the actual zone refining or crystal growth device, because convection and stirring (if applied) may be more or less effective under different experimental conditions. Moreover the segregation coefficient may depend on the degree of single crystallinity of the ingot, on details of void formation by thermal contraction of the crystal (see 2.1.4c), and on the kind and distribution of physical defects produced under the actual experimental situation. For single crystal growth k_{eff} may even depend on the crystallographic growth direction. (The latter has been observed by the author for twin anthracene crystals grown in parallel from a common tetracene-doped melt).

Therefore a modification of the normal freezing equation for the typical shape of a Bridgman crystal will also be given here. The crystal is assumed to consist of a cylindrical seeding part (1) of radius r_1, followed by a conical part (2) and a wider cylinderical part (3) with the volumes V_1, V_2, V_3 respectively, (Fig. 9). With $V = V_1 + V_2 + V_3$ the normal freezing equation reads

$$c_s^{(1)}(x) = kc_0 \left(\frac{V - \pi r_1^2 x}{V}\right)^{k-1} \qquad 0 \leqslant x \leqslant l_1 \tag{4a}$$

$$c_s^{(2)}(x) = kc_0 \left(\frac{V_2 + V_3 - \frac{\pi}{3}\left[r_x^2 \cdot (x - l_2 + h) - r_1^2 \cdot (l_1 - l_2 + h)\right]}{V}\right)^{k-1} \qquad l_1 \leqslant x \leqslant l_2 \tag{4b}$$

$$c_x^{(3)}(x) = kc_0 \left(\frac{V_3 - \pi r_2^2 (x - l_2)}{V}\right)^{k-1} \qquad l_2 \leqslant x \leqslant L \tag{4c}$$

where $r_x = r_1 + \frac{(x - l_1)}{h} r_2$ and k is the effective distribution coefficient.

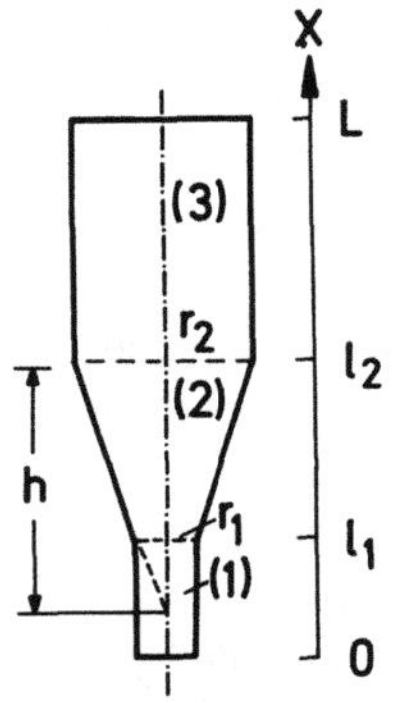

Fig. 9. Symbols used for the description of the dopant profile in a Bridgman crystal of the shape indicated

Equations 4a–c allow the dopant distribution to be calculated. There are discontinuities in the slope of the concentration curve c_s (x). They are removed if the concentration is plotted versus the fraction r of unsolidified volume, which is the generalization of Eq. (2a):

$$c(r) = kc_0 r^{k-1} \tag{2b}$$

and which allows one to obtain k from the slope and intercept of a log c (r) versus log r plot of points of the entire Bridgman ingot.

Zone refining applies *repeated* melting and recrystallization steps effected by the repeated passage of narrow molten zones along the ingot, (see Fig. 10). In contrast to normal freezing, fractional melting allows one to avoid complete equalization of the concentration profile obtained from the first (or previous) passage. Nevertheless, equalization occurs across the length of the molten zone. It is obvious already from this that a final distribution will be established, for which segregation and equalization keep in balance. A smaller zone length will yield a steeper gradient and a purer depleted end; however, narrower molten zones with a smaller transport volume will remove impurities less efficiently, making more zone passages necessary.

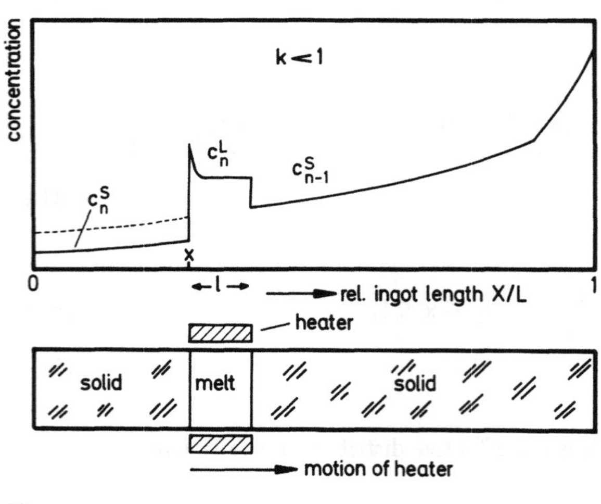

Fig. 10. Principle of zone refining. For the case k < 1 the molten zone transports the impurity to the end of the ingot

Zone refining mathematics[101,102] cf.[94] is a complicated subject. Let us start with the following simplifying assumptions:

1) The distribution coefficient is constant and independent of all other impurities. Then it is sufficient to do the calculation for one impurity.
2) Diffusion of the impurity in the solid phase can be neglected.
3) The zone has constant width and constant cross-section.
4) Macroscopic zone travel velocity and microscopic crystallization velocities are constant.
5) Melt and solid have the same specific density, independent of temperature.
6) No new impurities are created by isomerisation or decomposition during zone refining.

Following Braun et al.[102] it is convenient to normalize the cross-section to unity and to restrict the calculation to a (pseudo) one-dimensional model. The concentration is given in units of the initial (homogeneous) concentration c_0. Thus $c_0(x) = c_0 = 1$. The space coordinate x along the ingot is measured in units of the length ℓ of the molten zone. The ingot length is taken to be an integer multiple N of the zone length

$N = L/\ell$.

By this normalization the ingot is divided into N sections of unit length, (which is the normalized length of the molten zone), see Fig. 11.

Zone refining starts with equalization of the concentration in the section $0 \leqslant x \leqslant 1$. With the zone proceeding, crystallization occurs at the left side (Figs. 10, 11) whereas more material is melted at the right. Let $c_n(x)$ be the impurity concentration in the solid crystallizing at the point x at the nth zone passage. The impurity concentration in the melt is then $(^1/_k) \cdot c_n(x)$. With the above normalization this is at the same time the absolute quantity of the impurity in the melt. On progress of the zone by dx, an amount of impurity $c_n dx$ is withdrawn from the melt by crystallization into the volume dx at the point x, whereas a quantity $c_{n-1}(x+1)$ is added to the melt at the other side of the zone at x + 1.

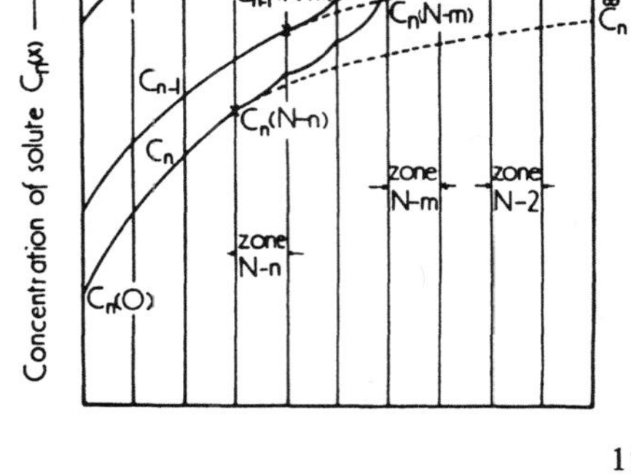

Fig. 11. Designations used in zone refining theory; from[102] (by permission of the Institute of Physics)

Therefore the concentration change in the liquid is

$$\frac{1}{k} \cdot dc_n(x) = \{c_{n-1}(x+1) - c_n(x)\}dx \quad (0 \leqslant x \leqslant N-1). \tag{5}$$

When the front of the molten zone reaches the end of the ingot (this corresponds to the coordinate $x = N-1$ for the crystallization front) no further material is melted and the zone length decreases. For this region of normal freezing the differential equation reads

$$dc_n(x) = \frac{1-k}{N-x} \cdot c_n(x)\, dx \quad (N-1 \leqslant x \leqslant N). \tag{6}$$

The solutions of Eqs. (5) and (6) after one zone passage are

$$c_1(x) = 1 - (1-k)e^{-kx} \quad (0 \leqslant x \leqslant N-1) \tag{7}$$

$$c_1(x) = c_1(N-1) \cdot (N-x)^{k-1} \quad (N-1 \leqslant x \leqslant N). \tag{8}$$

These equations were first derived independently by Lord[103] and Reiss[104].

A salient point occurs at $x = N-1$ after one zone passage, see Fig. 11, because from there on normal freezing takes over. This point is reflected back to $x = N-2$ by the second zone when its front reaches $x = N-1$. This game continues and after $n = N$ zone passages the whole ingot is mathematically divided into N sections separated by $N-1$ salient points of the distribution function of the impurity (see Fig. 11). For each individual section the concentration profile c_n can be derived from the knowledge of the distribution c_{n-1} after the previous passage, using a recursion formula[102]. Only for $n < N$ zone passages can a closed analytical expression be given which, however, holds only for the region $0 \leqslant x \leqslant N-n$. We will have to apply more than N zone passages usually for the attainment of highest purity organic material as we shall see later. Therefore prediction and evaluation of zone refining profiles generally requires computer calculations.

With a semi-infinite ingot, starting at $x = 0$ the salient point problem does not exist: no end is reached and therefore no back reflections occur. Analytical solutions have been derived; but as the impurity is not concentrated at a finite end, this procedure is a good approximation for a real ingot of length N only, if less than N zones have passed, and then only for the region $x < N-n$. For $x > N-n$ unrealistically low concentrations are obtained (dashed curves in Fig. 11).

For our purpose it is useful to consider the *ultimate distribution* which theoretically can be reached after an infinite number of zone passages. This case can be represented by an analytical solution to a reasonable approximation.

After Pfann[94] the final distribution is an exponential

$$c_\infty(x) = k' e^{k'' x} \tag{9}$$

where k' and k'' are implicit functions of the distribution coefficient

$$k = k''/(e^{k''} - 1) \tag{10}$$

$$k' = Nk''/(e^{Nk''} - 1). \tag{11}$$

Equations. (10) and (11) can be solved by using tables of the Einstein function $x/(e^x-1)$. (For a further, more rigorous treatment of the final distribution see[105]).

A number of zone refining distributions have been calculated by numerical treatment[101] of the zone refining differential Eqs. (5) and (6) and by using the analytical solutions[102].

Schildknecht, in his book on zone refining[106], has given a very clear representation of the numerical results.

In what follows a comprehensive summary will be given and conclusions for the laboratory will be drawn.

2.1.2 Conclusions for the Laboratory

Some basic features of zone refining as obtained from the elementary zone refining theory outlined above shall be discussed before going into practical details. In Fig. 12 impurity distribution functions are plotted for 10 zone passages with a zone length ℓ of 10% of the total ingot length L, i. e. $\ell/L = 0.1$. The parameter is the distribution coefficient. Impurity depletion is better the more k differs from one. Impurities with $k > 1$ concentrate at the starting end ($x = 0$), those with $k < 1$ at the opposite end of the ingot. If both types of impurities are present only a central medium purity fraction of the ingot can be gained at the expense of the end pieces where one impurity is well down but the other is accumulated.

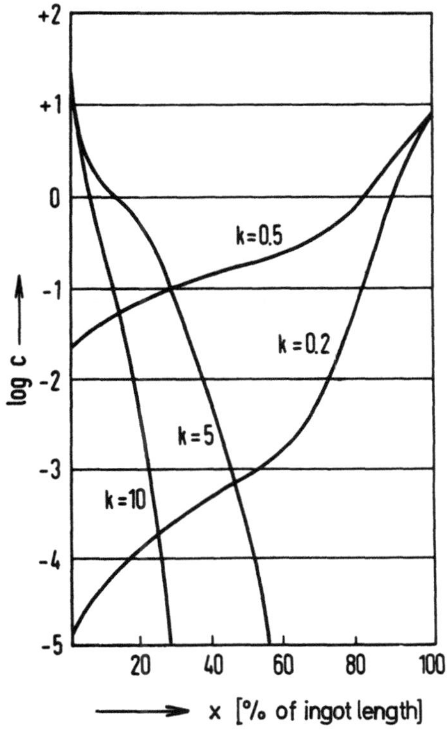

Fig. 12. Calculated impurity distribution functions for different distribution coefficients k after 10 zone passages with a zone length of 10% of the ingot length. Plotted on the abscissa is the impurity concentration relative to the starting concentration. From[106] (after[101])

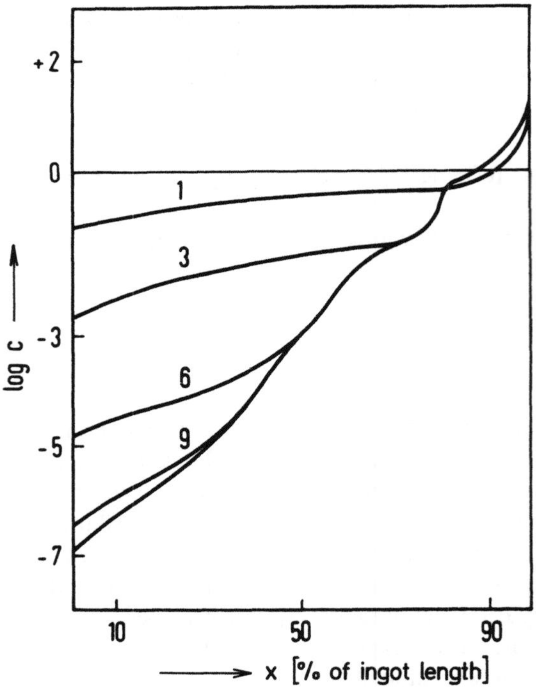

Fig. 13. Calculated impurity distributions for k = 0.1 and ℓ/L = 0.2 after 1, 3, 6, 9 and ∞ zone passages. From[106] (after[101])

Let us now consider an effective distribution coefficient $k_{eff} = 0.1$ and a zone length of 0.2 L. Tetracene as an impurity in anthracene could serve as a practical example for k = 0.1[8]. With this k-value, by definition, the starting end of the ingot will be one order of magnitude purer after one zone passage. From Fig. 13 it can be seen that after 3 passages nearly k^3, i. e. 3 orders of magnitude are reached at x = 0. n = 6 passages yield only ~5 orders of magnitude; since an effect of the piling up of the impurity (by normal freezing) at the end of the ingot is reflected back to the starting end after $n \geqslant N$ passages (where N was the ingot length in number of zone lengths; N ~ 5 in the example given).

With a few further passages the ultimate distribution is approached at n = 9. The highest purity obtainable for $k < 1$ at the purest point of the distribution function (x = 0) can be crudely estimated to amount to $c_{min} \lesssim k^N \cdot c_0$[1]. After the finite number of zone passes n = 2N a purity is already reached which is close to the ultimate distribution and which is characterizable at x = 0 by $(c_{min}/c_0)^{3/4}$. It follows, that for $k < 1$ more than n ~ 2N zone passages no longer improve the distribution significantly[1]. It should be realized that for the parameters of the example the impurity is depleted only 3 orders of magnitude on an average in (only) 60% of the ingot; the other 40% would have to be discarded.

In Fig. 14 the result for ℓ = 0.2 L (previous case) is compared with that for a shorter zone length ℓ = 0.1 L. The result of using a shorter zone length is striking: After n = 10 passages 60% of the material is already purer by 5 orders of magnitude! Further improvement would be possible by applying further passages up to n ~ 2 N = 20. Clearly the shortest zone length ℓ realizable and an ingot length L as long as

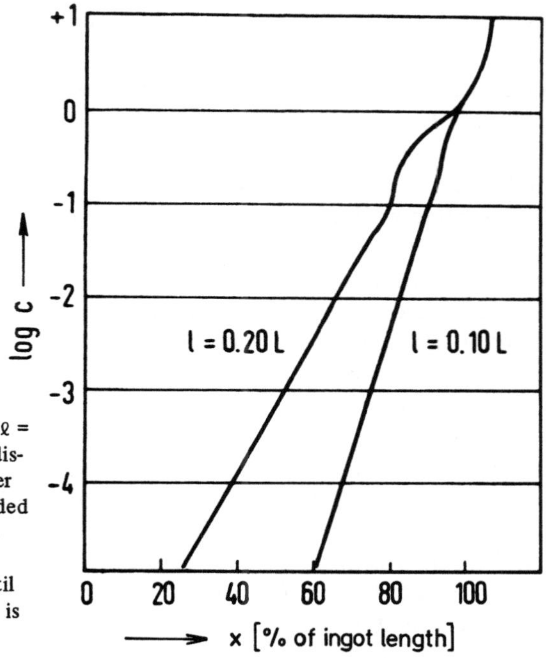

Fig. 14. The distribution function for ℓ = 0.2 L is already close to the ultimate distribution after 10 zone passes. However this distribution is considerably exceeded with the shorter zone length ℓ = 0.1 L for the same number of passes, and further purification is still possible until the ultimate distribution for ℓ/L = 0.1 is approximated. From[106] (after[101])

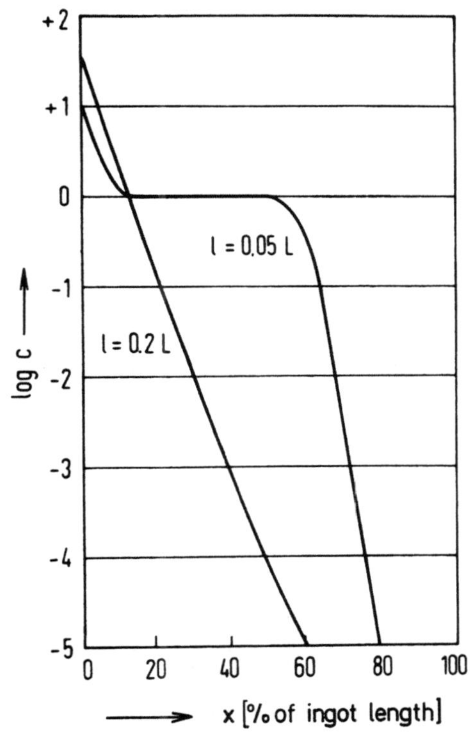

Fig. 15. Especially for k > 1 a longer zone is superior for the first few zone passes (n = 10; k = 5). From[106] (after[101])

possible should be used. Alternatively, as only the ratio $\ell/L = 1/N$ counts, expansion of the length (of a "pure" part) of a zone refined ingot under reduction of the diameter, will allow increasing the ultimate purity during further zone passages.

It is difficult for an impurity with $k > 1$ to reach the starting end of the ingot than for an impurity with $k < 1$ to reach the rear end, because it must travel opposite to the direction of motion of the molten zones. For each complete zone passage from $x = 0$ to $x = L$ a $k > 1$ impurity can go only a distance $\ell = L/N$, whereas a $k < 1$ impurity can travel through to the very end. Therefore for $k > 1$ a greater zone length is advantageous for the first passes, cf. Fig. 15 which compares $\ell = 0.05$ L to $\ell = 0.2$ L for $k = 5$ and $n = 10$ zone passages. However, for smaller ℓ/L the final distribution, although reached only after many more passages, is substantially better, cf. Fig. 16. The approximation of the ultimately obtainable distribution needs many more zone passages for $k > 1$ than for $k < 1$.

Finally the dependence on minor differences in the effective k-value will be considered. This point is of great practical importance for k close to one: According to zone refining theory, for k close to one, purity can be improved only by a comparatively small factor; the ultimate distribution is not too different. This can be seen from Fig. 17, which has been calculated for $\ell = 0.02$ L. Clearly all means should be applied in such cases to get the effective distribution coefficient as far away from one as possible. Reduction of the thickness of the diffusion boundary layer by some appropriate kind of stirring will be favourable.

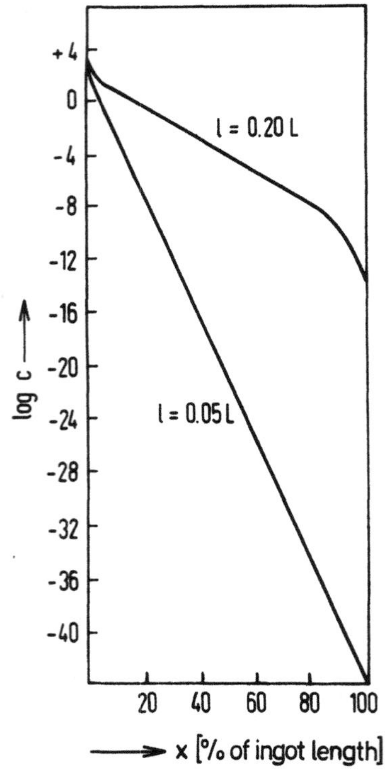

Fig. 16. The final distribution is again much better with the shorter zone length. From[106], (after[101])

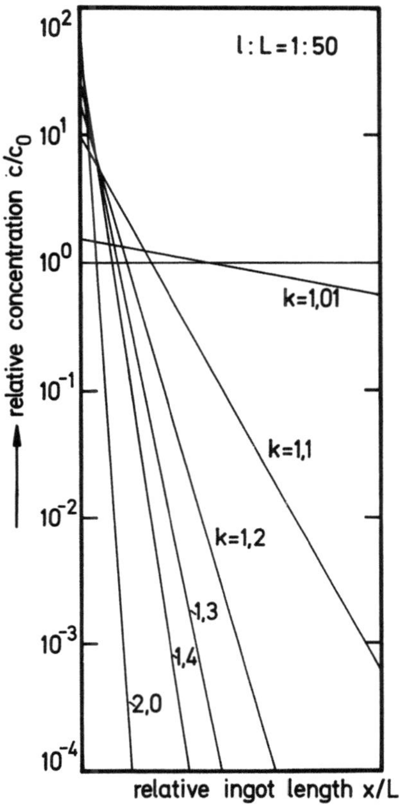

Fig. 17. Calculated ultimate distribution functions for different values of the (effective) distribution coefficient k (close to one) and with $\ell = 0.02$ L[107])

A few remarks should be made on the dependence of the distribution coefficient k on the speed with which the zones are pulled through the ingot (cf. Eq. (3)). For a given diffusion boundary layer of thickness d, the optimum zone refining speed will have to be adapted to the diffusion speed D/d in the liquid. Estimating d under convectional stirring to be 0.1 mm and taking $D = 10^{-5}$ cm^2/s, a zone speed of f = 3 cm/h follows for the exponent to be 1. For getting the distribution coefficient close to its equilibrium value, an exponent of 0.1 is desirable; this consideration suggests a crystallization speed of a few mm h^{-1}.

It is important that this speed be free of fluctuations. Fluctuations could be caused by uneven progress of the driving mechanism or by unstable heating, cooling, or surrounding temperature conditions, shifting the phase boundary liquid/solid. In short time intervalls with faster crystallization rates; the whole impurity content of the diffusion boundary layer ($k < 1$) may be trapped in the growing solid before being able to diffuse away. Thus a momentary $k_{eff} \sim 1$ may result.

It can be anticipated that the formation of a single crystal which is as perfect as equilibrium thermodynamics allow for, will be in favor of segregation. Defects in the crystallizing front such as dislocations and grain boundaries may trap additional impurities. Any cracks or voids may suck in the concentrated ($k < 1$) or depleted ($k > 1$) diffusion boundary layer, thereby reducing the average segregation power.

2.1.3 Practical Realization[d]

For organic materials, it is convenient to use sealed glass tubes as zone-refining containers; this is possible, because of the comparatively low melting points and the relative chemical inertness of most organic compounds. Their considerable vapor pressure usually prohibits application of "open-boat" geometries. First of all it is advisable to zone-refine under ambient inert gas. In addition, back sublimation can be hindered by filling the whole cross-section of the zone-refining tube with the material to be purified. For this purpose a vertical suspension and motion offers itself in a quite natural way (see Fig. 18); however, it is not optimal for all points under consideration: At the introduction of a new molten zone, space must be provided for material expansion on melting, amounting typically to between 10 and 20%. Therefore zones are usually introduced from the top where a hollow space exists anyway. However, it is nearly impossible to avoid introduction of voids into the ingot which tend to separate liquid and solid phases just at the crystallizing

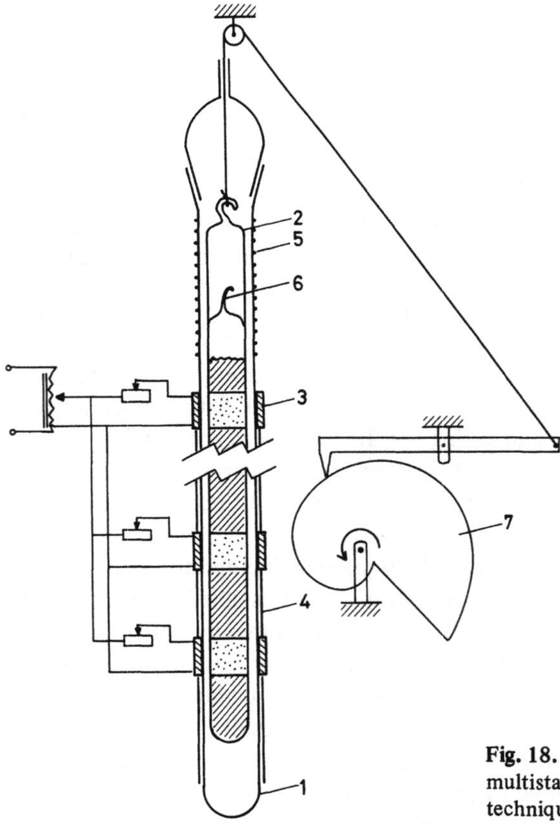

Fig. 18. A simple realization of a vertical multistage zone refiner using the intermittent technique[108)]

[d] Many additional practical aspects of zone refining which must be omitted here, may be found among others, in the books of Pfann[94)] and of Zief and Wilcox[130)]

front where phase contact should be maintained. (Segregation may also take place through the resulting gas phase but this is a different topic and shall not be considered for the moment). Bubbles are transported downwards, raising in turn the material to be zone-refined, until finally no expansion volume is left on top and cracking of the ampoule may result. Furthermore, aggregating bubbles usually restrict the heat flow from the zone heater to the rest of the ingot, thus causing the actual glass wall temperature to rise and eventually initiate thermal decomposition. Nevertheless this "normal-vertical (NV) zone refining" method is most widely used. Improvements by "reverse zone refining" or "horizontal zone refining" require increased apparatus prerequisites and more experience as we shall see below.

Because zone refining is a time consuming process, it is highly advisable to apply as many zones as possible at the same time by using a multistage zone refiner. Further, the "intermittent technique" is advantageous: With this technique the whole ingot is covered with zones all the time, in contrast to just pulling the ampoule out of the multistage heater assembly, where only half of the zone heaters are used on the average. (All apparatuses described in this section use the intermittent technique).

A simple useful assembly ("NV 1")[108] is shown in Fig. 18. Without stirring and with air cooling only, a minimal zone distance of 2 . . . 3 times the tube diameter (which is 14 mm in the example) and a minimum zone length of the order of the tube diameter can be realized. Zone heaters (3) are commercial 39 Ω wire-wound ceramic ring resistors (Draloric ZW 20/18 24 V), threaded on a supporting glass tube (1) and kept separated by sections of glass tubing (4). Linear motion is transmitted by a lever which glides on an Archimedes cam (7). The cam is driven by a synchronous motor equipped with a switchable gear (Halstrup, D 7815 Kirchzarten). The ampoule is slowly raised, which is equivalent to zones travelling slowly downwards, until the lever jumps back. At this instant the whole ampoule is quickly lowered by one zone distance; each zone is taken over by the subsequent heater. The basic idea is schematically demonstrated in Fig. 19, which, however, is drawn for the case of reverse movement.

Some additional features of the zone refining apparatus, Fig. 18, should be mentioned: The upper portion of supporting tube (1) is equipped with a heater winding (5) in order to avoid condensation from the gas phase which would retain impurities that should travel downwards ($k < 1$). In addition, sublimation – wherever it could occur – is kept down by addition of 53 kPa (400 torr) of an inert gas (N_2). For convenient handling under high purity conditions a break-seal (6) may be provided. The supporting tube is essentially closed in order to avoid contamination of the laboratory in case an ampoule cracks. Contamination may cause purity and health problems: some organic compounds, especially of aromatic nature, are known to be carcinogenic; the great majority are barely or not characterized, concerning this and other similar aspects.

With the NV-intermittent technique it is usually necessary to start by introducing a single zone first. An additional heater can be switched on whenever a new cycle of the cam begins. This needs readjustment of the current of the preceeding heater. Therefore each heater current must be adjustable individually. The melting temperature is reached by use of a variable transformer, fed by a stabilized power supply.

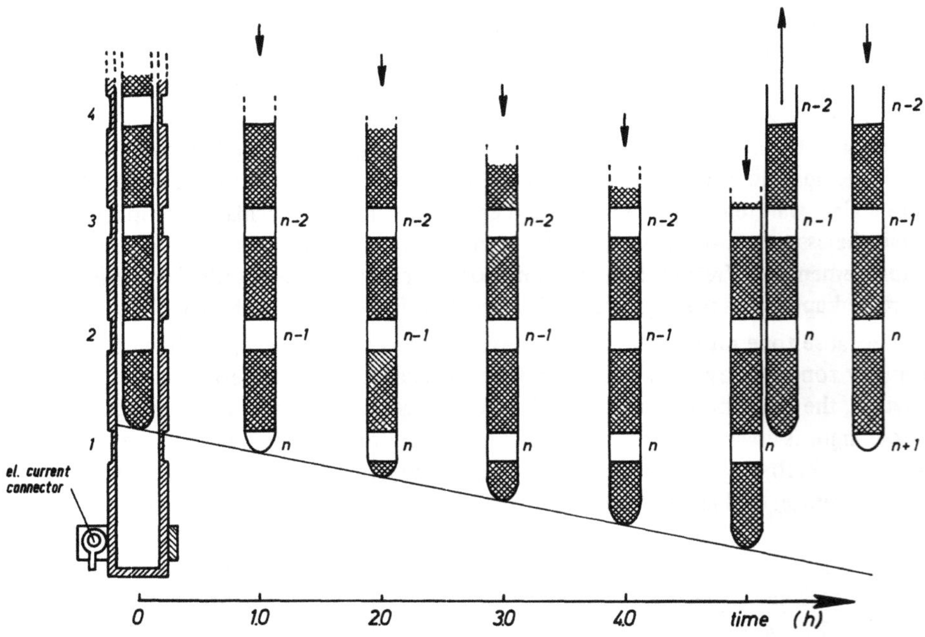

Fig. 19. Schematic representation of the intermittent zone refining technique. On the left side one version of the multistage zone refiner furnace of apparatus "RV 1" (see text) is indicated[112])

The room temperature is kept constant. The bubble problem is less serious, if the internal zone tube diameter is not smaller than ~10 mm. Compounds which melt at comparatively low temperatures with regard to their molecular weight cause less problems. Reuniting the ingot by fractional melting may be necessary from time to time for other compounds.

Finally an essential trick should be taken into account: If the lifting effected by the cam is made a little bit greater than the zone heater distance, minor remelting occurs on progress to the next cycle, making impossible any uncontrolled fast crystallization which would lead to the incorporation of increased ($k < 1$) impurity content into the solid.

There are some features which can be improved at the expense of simplicity of the equipment. To achieve short zones, short intermediate solid portions, well defined phase boundaries and zone lengths, relative independence from the actual room temperature and its constancy, it is necessary to provide distinct cooling elements. Heating and cooling elements should be coupled as closely as possible to the zone refining tube. Figure 20 gives an example[109)] cf.[118)], which will be referred to in what follows as "NV 2". Heating elements are self supporting coils of thick resistive wire, wound around sections of brass tubing. Heating and cooling elements have a slightly reduced bore-diameter over 1 mm length at their center in order to guide and center the zone refining tube. Cooling elements are connected in series to a thermostat which is usually operated at a temperature which lies between 10 and 100 deg. below the melting point. Tube diameter is either 1.2 or 0.6 cm; length of the molten zones is roughly equal to the tube diameter; zone distance is 5.5 cm.

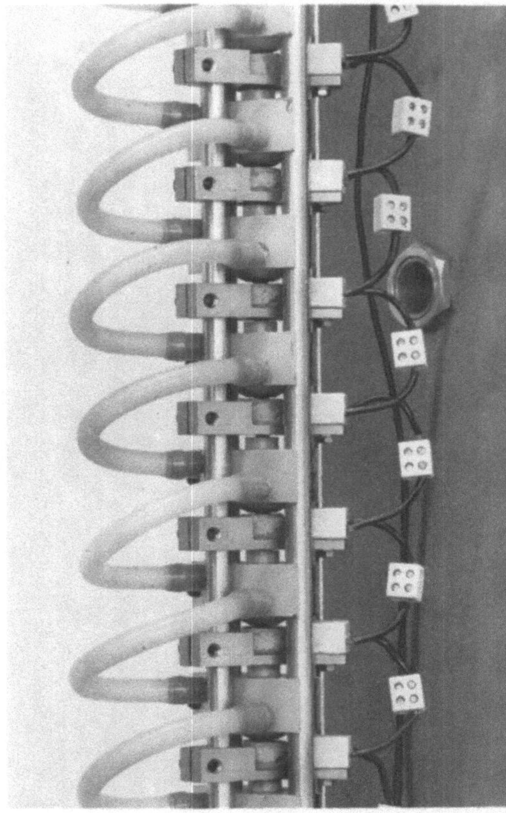

Fig. 20. Multistage zone refiner ("NV 2"), assembled with separate heating and cooling elements[109]. Zone distance is 5.5 cm

"Reverse zone refining" avoids some of the drawbacks of normal zone refining. With this technique zones are introduced from the closed (in the vertical case "lower") end of the ingot. Volume expansion at melting requires that the whole ingot move in the zone refining tube, otherwise the ampoule will inevitably crack. It was found that moving takes place if many molten zones with narrow separating solid portions are present in the ingot at the same time, and if the solid's temperature is not too far below its melting point. Obviously some sort of gliding is required.

It makes no major difference whether reverse zone refining is performed in a "vertical" or in a "horizontal" set up, if stirring is applied. (Stirring can be effected for instance by repetitively accelerating and stopping rotation of the ampoule around its long axis). Vertical reverse zone refining without external stirring does not seem to be very promising because of nearly vanishing driving forces for convectional mixing in this kind of geometry. In fact, diffusion boundary layers as thick as 5 mm have been observed under such conditions[8].

The advantages of reverse zone refining are that no free space is left in front of the ingot. Therefore neither impurity retention nor introduction of bubbles is a problem. Instead, the ingot becomes denser with time, approaching near single crystallinity. As mentioned above, segregation should be most efficient at the interface between melt and *single* crystal.

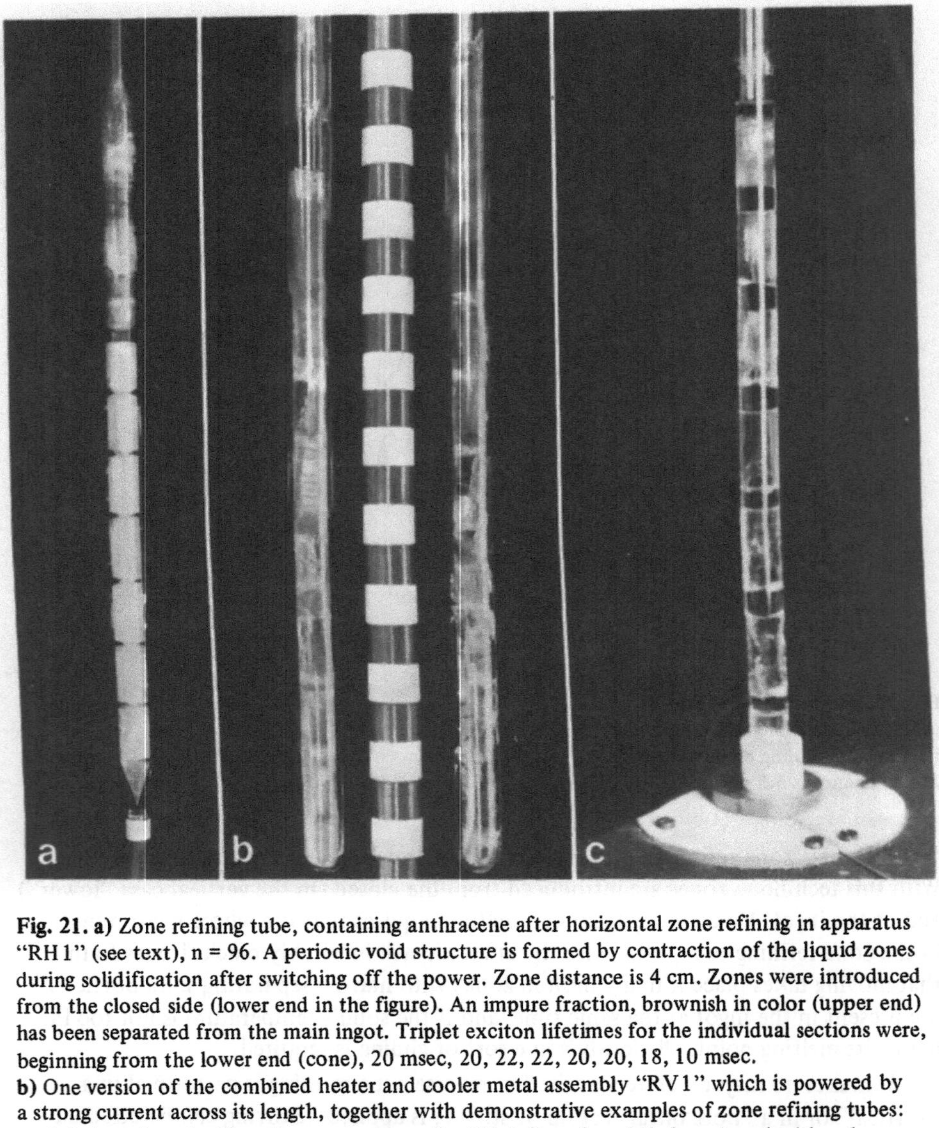

Fig. 21. a) Zone refining tube, containing anthracene after horizontal zone refining in apparatus "RH 1" (see text), n = 96. A periodic void structure is formed by contraction of the liquid zones during solidification after switching off the power. Zone distance is 4 cm. Zones were introduced from the closed side (lower end in the figure). An impure fraction, brownish in color (upper end) has been separated from the main ingot. Triplet exciton lifetimes for the individual sections were, beginning from the lower end (cone), 20 msec, 20, 22, 22, 20, 20, 18, 10 msec.
b) One version of the combined heater and cooler metal assembly "RV 1" which is powered by a strong current across its length, together with demonstrative examples of zone refining tubes: naphthalene doped with yellow azobenzene ($k < 1$), left, and naphthalene doped with red rhodamine B ($k < 1$), right. Zone refining was interrupted after 10 zone passage in order to make visible the impurity concentration in the previously molten zones and its transport towards the upper end. The lower ends are colourless.
c) A hot zone refining tube, containing naphtalene, pulled out for inspection from an other version of RV 1, having a greater zone distance than in Fig. 21 b. The uppermost solid ingot appears bluish fluorescing, caused by the impurity anthracene.

With the reverse technique it is necessary to start with *all* zones molten from the beginning. Therefore, in order to avoid cracking one must start with a less dense ingot; or with complete melting of the material and introduction of *solid* zones.

"Reverse horizontal" zone refining equipment developped by Fischer ("RH 1"), and by Benz and Gerdon ("RH 2") gave highly satisfactory results:

RH 1 is an apparatus for a tube diameter of 2.5 cm, using stirring by intermittent rotation. Fixed tube length is 50 cm, zone distance is 4 cm, zone length is 2 cm, giving $\ell/L \leq 0.05$ with 40 cm of filling. It is an improved version of the apparatus described in[110]. No individual cooling elements are provided. Instead the whole assembly consisting of heating coils plus zone refining tube is surrounded by an oil jacket, resembling a Liebig-cooler. The latter is connected to a thermostat which is operated at 80 °C for anthracene, for example. An actual zone refining tube containing anthracene is shown in Fig. 21 a.

RH 2 is a reverse horizontal apparatus[111] characterized by block integration of heating and cooling elements, see Fig. 22. Maximum ingot length is 35 cm, tube diameter is 1.5 cm, zone distance is 2.8 cm, zone length is 0.7 cm, giving $\ell/L = 0.02$. For possibly dangerous material a double wall zone refining tube is used. Small quantities of material can be expanded to fill the whole tube length by insertion of a central glass rod (or tube) into the normal ampoule. The body of the apparatus consists of two combined cast aluminium blocks (1) (each 20 cm long) provided with channels (2) for the coolant to flow through and with a central bore (3) for the zone refining tube. Into the block pockets (4) are milled for the heating coils (5), carried by individual supports (6). A stirring motor (7) is provided. For convenience the whole assembly is mounted on an optical bench (Zeiss system). The intermittent drive mechanism, using an Archimedes cam, has not been drawn.

A different development, operated in the reverse vertical version ("RV 1")[112] has already been presented in Fig. 19, (left side). A metal tube (e. g. stainless steel) with a periodic structure along its length is used to create a periodic temperature profile. The zone refining (glass) tube is inserted; it is transported in the usual

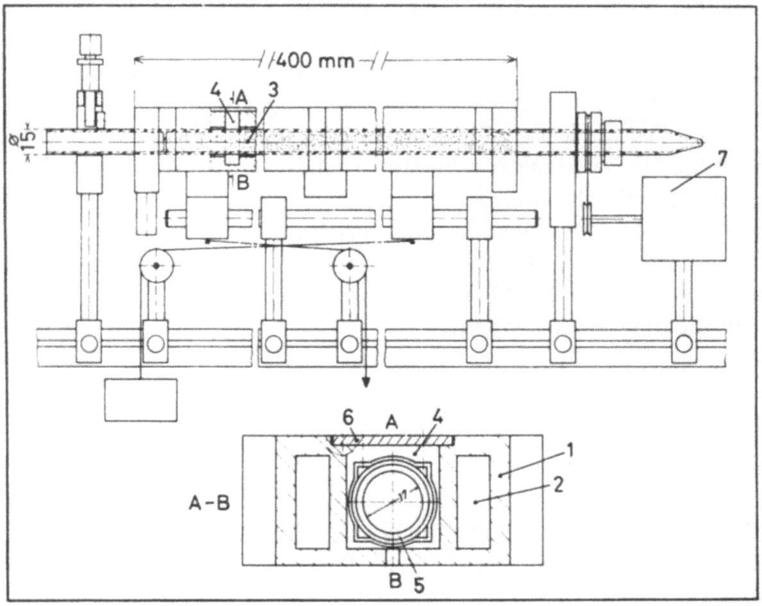

Fig. 22. Horizontal reverse multistage zone refiner; after Benz et al.[111]

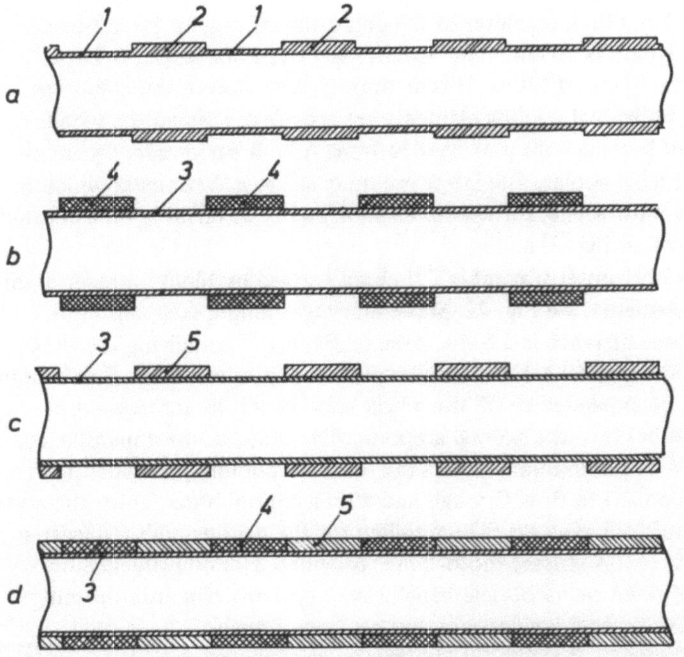

Fig. 23 a–d. Several versions of generating a periodic temperature profile for zone refining, see text[112)]

way as described above and should be stirred by intermittent rotation. The metal tube is heated as a whole by running a low voltage high current along it. Either sections (1), milled down from the full diameter (2) (Fig. 23 a), or a periodic heat insulation (4) on a constant thickness metal wall (3) (Fig. 23 b), or short-circuiting metal (e. g. Cu-) rings soldered onto the tube (Fig. 23 c) or a combination of a, b, and c create the heating and cooling zones respectively (see also photographs Fig. 21 b, c). The whole assembly is immersed into a tall thermostat of the kind available for thermometer calibration (Lauda, D-6980 Wertheim). This version is particularly simple, optimally isolated from variations of laboratory temperature, and well protected against fracture. However, the whole ampoule must be pulled out for inspection (Fig. 21 c) and the zones cannot be regulated individually. The last point requires a reasonable starting purity in order to keep the melting point and thus the zone length constant.

2.1.4 Results[e] and Improvements

a) Illustrative Example

Results obtained with zone refining of "technical purity" grade anthracene on apparatus "NV 2" will be given as an illustrative introductory example. Fig. 24

---

[e] Some organic compounds which have been zone refined successfully are listed in[113)]; [1)], p. 99; [2)]; [35)], p. 135, and [133)].

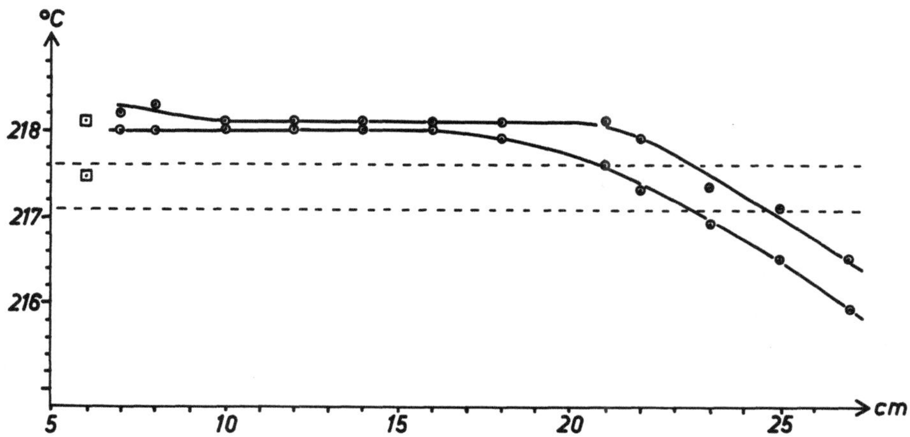

Fig. 24. Distribution of the melting interval along a zone refining tube of "technical grade" anthracene after n = 20 zone passages on equipment "NV 2"; $\ell/L = 0.025$, $f = 3\ cm\ h^{-1}$ [109]

illustrates the distribution of melting intervals along the ingot after n = 20 zone passages with $\ell/L = 0.025$, $f = 30\ mm\ h^{-1}$. It is observed that the melting point increases and the melting interval decreases for the main part of the ingot with respect to the initial values (dashed). Because the melting point decreases at the rear end of the bar, one can conclude that impurities with $k < 1$ are predominant. This situation is typical for anthracene and the great majority of other compounds. – Obviously impurities have been retained in the empty volume in front of the ingot (square symbols). The initially yellow-brownish material showed bright blue-violet fluorescence in the purified part; the end of the ingot had turned black (cm 23–27).

It is not advisable to start with technical grade substances if higher purity material is available. Selected "anthracene for scintillation" containing not more than 10^{-3} mol/mol carbazole would be a better choice as starting material.

b) Zone Refining Results with Easily Removable Impurities

Let us take the purification of naphthalene from the impurity anthracene ($k \sim 0.01$) as an example which has been studied in some detail[114]. Normal absorption spectroscopy allows the detection of anthracene down to about 10^{-7} mol/mol. With low-temperature fluorescence analysis 10^{-8} mol/mol is achieved. This (direct fluorescence)[f] detection limit has been reached for ≥80% of the ingot in a normal vertical zone refining apparatus after 200 zone passages with $\ell/L = 0.0067$ and $f = 9\ mm\ h^{-1}$. 50 zone passages were already sufficient for reaching 10^{-7} mol/mol; the starting concentration was in both cases $5 \cdot 10^{-4}$ mol/mol. Tetracene in anthracene behaves rather similarly[45].

[f] "Delayed fluorescence" reveals as little as $\sim 10^{-10}$ mol/mol anthracene, see paragraph 4.1.5

c) Experimental Distribution Coefficients

Knowledge of experimental distribution coefficients of impurities is highly desirable for an estimate of time and effort necessary for the attainment of a certain required purity. However, there are interpretational difficulties with measured distributions: Sloan[8)], Schmillen and Falter[45)], Probst[115)], Feederle[107)] and others have recognized that slope and intercept of log-log plots of normal freezing results do not give coinciding k-values as required by simple normal freezing theory (p. 15, 16). This contradiction can easily be removed if one takes into account that significant contraction with decreasing temperature is characteristic for most organic solids. A temperature gradient is necessarily maintained between the crystallizing front and other parts of the ingot. Thus on progress of the phase boundary successive parts of the ingot loosen from the glass wall; cracks occur, sucking in melt adjacent to the phase boundary. This impurity-enriched or depleted melt crystallizes quickly without essential segregation. By this mechanism the ingot will be surrounded by a less pure ($k < 1$) or purer ($k > 1$) jacket. This fact can easily be seen if one grows Bridgman crystals doped with a coloured impurity which is essentially rejected (such as anthracene (azobenzene)). It has also been studied by radioactive tracer techniques[116)].

Following Feederle[107)], the total amount of impurity in the melt is changed in addition to the normal freezing differential by an amount $-\bar{\alpha} c_\ell(x) d U dx$

$$dM_\ell = -c_s q dx - \bar{\alpha} c_\ell (x) d U dx \qquad (12)$$

where d is the mean distance between contracted ingot and glass wall, U is the circumference of the ingot, $\bar{\alpha} c_\ell(x)$ is the actual concentration in the boundary layer which is sucked in and which is different by an average factor $\bar{\alpha}$ from the concentration c_ℓ of the main part of the melt (notice that $k > 1$ gives $\bar{\alpha} < 0$); q is the cross-section of the melt. Combining the unknown factors $\bar{\alpha} dU = \alpha$ and neglecting in the total volume change,

$$dV_\ell = - q \, dx - U \, d \, dx, \qquad (13)$$

the small volume U d dx, which is filled for a second time with melt, one obtains with

$$dc_\ell = \frac{1}{V_\ell} dM_\ell - \frac{M_\ell}{V_\ell^2} dV_\ell \qquad (14)$$

the modified normal freezing equation

$$c_s = c_0 k (1-x)^{k + \alpha - 1} = c_0 k_b (1-x)^{k_m - 1} \qquad (15)$$

valid for the impurity distribution at the interior of the ingot. The prefactor $c_0 k$ was obtained by using the conservation of (impurity) mass, integrated over the length of the ingot. Obviously a modified slope $k_m = k + \alpha$ is expected in a log-log plot.

It follows that, for the determination of an effective distribution coefficient, one has to measure the *interior* impurity concentration, discarding the exterior layer from the analysis. The initially applied dopant concentration c_0 has to be used (and not a normalization with the integral of actually measured concentrations as was proposed elsewhere[115] under the assumption that thermal decomposition of the impurity was responsible for the observed discrepancies).

A quantitative estimate of α is difficult because it depends on factors such as relative concentration of ingot and container, strength of adhesion of the solidified material to the wall (which is also a function of wall pretreatment) eventual crack formation and concentration profile in the diffusion layer. Therefore the intercept will give an effective distribution coefficient more directly (k_b) than the slope.

A model taking into account this kind of solid state contraction during *zone refining* is not available. Therefore the numerical bearing of experimental distribution coefficients on actual zone refining impurity profiles or Bridgman crystal doping profiles must be considered with some caution.

In Fig. 25 normal freezing curves are given for fluorene containing 10^{-3} mol/mol anthracene[107]. The upper curve was obtained with a special normal freezing apparatus[100] with stirring; f = 9 mm h^{-1}. The lower curve pertains to a Bridgman crystal grown without stirring; f = 0.65 mm h^{-1}. In the latter case a plot of the normal freezing distributions, Eqs. 4a–c, versus the unsolidified remainder (according to Eq. 2b) was necessary; α comes out to be negative as postulated by the model

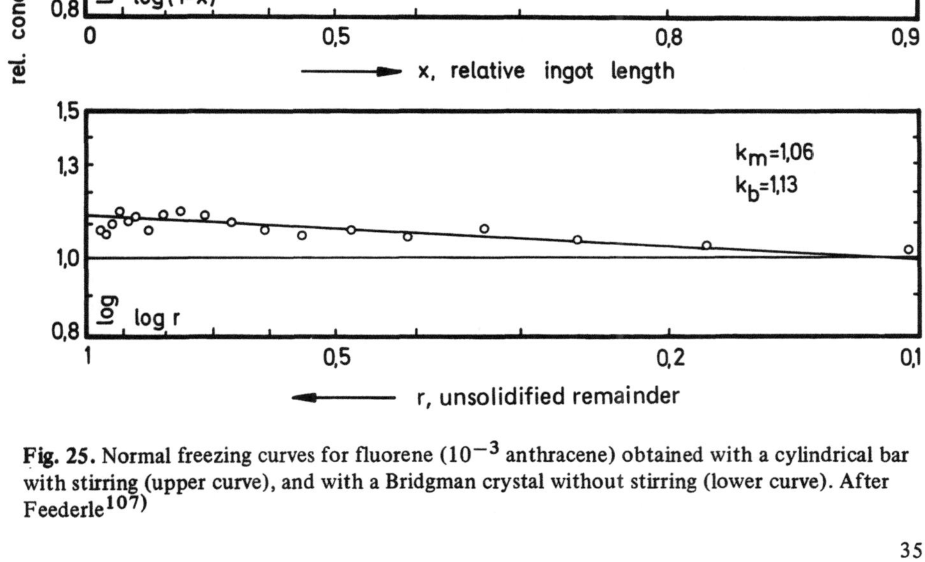

Fig. 25. Normal freezing curves for fluorene (10^{-3} anthracene) obtained with a cylindrical bar with stirring (upper curve), and with a Bridgman crystal without stirring (lower curve). After Feederle[107]

for k > 1. For the unstirred sample the distribution coefficient is closer to one ($k_{eff}^{(b)} = 1.13$) as compared to that of the stirred sample ($k_{eff}^{(b)} = 1.33$), in spite of the slower growth speed. (Note that the purer outer layer is omitted in the analysis, leading to $\int_0^L c(x)\,dx > c_0$ for k > 1).

A selection of effective distribution coefficients is collected in Table 2. In the light of the above discussion these numbers should be considered only as a quide for estimates of segregation problems.

Table 2. A collection of effective distribution coefficients

System host (guest)	k/f [mm h^{-1}] without stirring	with stirring	Ref.
naphthalene (β-naphthol)	2.3 /	–	131)
	1.85/	–	132)
fluorene (anthracene)	1.13/0.65	1.33 /9	107)
phenanthrene (anthracene)	–	1.6 /	119)
2,3-dimethylnaphthalene (anthracene)	2.1	–	311)
2,3-dimethylanthracene (tetracene)	> 1	–	134)
dibenzofuran (anthracene)	1.1–1.3	–	312)
brasan[a] (tetracene)	> 1	–	134)
p-terphenyl (tetracene)	> 1	–	134)
naphthalene (anthracene)	0.01 /7	–	114)
carbazole (anthracene)		0.5 /5	117)
anthracene (tetracene)	0.10 /1	0.05 /4	8)
	0.05 /1	–	45)
anthracene (anthraquinone)	0.02 /1	0.006/4	8)
anthracene (fluorene)	0.2 /1	–	8)
anthracene (phenanthrene)	0.1 /1	–	8)
anthracene (carbazole)	0.9 /1	0.84 /	8, 117)
anthracene (acridine)	–	0.14 /0.25	115, 118)
anthracene (phenazine)	0.1 /1–10	0.1 /1–10	115, 118)
anthracene (azobenzene)	0.04 /10	–	115)
phenanthrene (fluorene)	–	0.95 /	119)
phenanthrene (dibenzothiophene)	–	0.8 /	119)
phenanthrene (dihydrophenanthrene)	–	0.8 /	119)
	k	χ^b	
cyclododecane (acenaphthene)	0.08	0.26	86)
cyclododecane (naphthalene)	0.01	0.39	86)
cyclododecane (perylene)	$2 \cdot 10^{-3}$	0.49	86)
cyclododecane (anthracene)	$3 \cdot 10^{-5}$	0.74	86)
cyclododecane (chrysene)	$2 \cdot 10^{-6}$	0.92	86)

[a] 2,3-benzo-dibenzofuran
[b] χ is a misfit factor defined as V_{nol}/V_{ol}, the ratio of the nonoverlapping to the overlapping molecular volumes of host and guest molecules

d) Critical Zone Refining Tests

There are different zone refining techniques: normal vertical, reverse vertical, reverse horizontal, with and without stirring and/or intermittent transport technique.[g] In addition there are a great number of parameters: zone length relative to ingot length (ℓ/L), zone speed (f), temperature gradient, cooling temperature, free space above the ingot (retaining impurities in the vapour phase or in sublimed crystals), thermal decomposition, back-diffusion (see below), uneven zone lengths or insufficient jump-back distance (with the intermittent technique), fluctuations of the surrounding temperature or of the power supply etc. It was demonstrated by Karl and Probst[118)], and Fischer[110)] for $k < 1$, and by Feederle[107)] for $k > 1$ that different equipment gives substantially different purification results. Characterizing the success of zone refining by just refering to the number of zone passes has hardly any significance.

A comparative zone refining test has therefore been proposed by Karl and Probst[118)] using the system anthracene (phenazine) and gas chromatographic analysis: For a given ingot length ($L > (m + n)\,\ell$) the integral of the average relative concentration is determined over m zone lengths ($0 \leqslant x \leqslant m$) after n zone passes

$$E_m\,(4) = -\frac{1}{m}\int_0^m \log\frac{c\,(x)}{c_0}\,dx\,. \qquad (16)$$

$E_m\,(n)$ is a useful measure of the orders of magnitude of the average depletion obtained for the purest fraction of the ingot, m zone lengths long, after n zone passages. $n = 4$ proved appropriate for the proposed test system:

With an initial concentration of 10^{-2} mol/mol phenazine and a distribution coefficient of ~ 0.1 the purest parts are still in the region of gas chromatographic detectability for $n = 4$ zone passes[118)].

With the zone refiner RH 1 the test number so obtained is $E_8\,(4) = 3.1$, whereas RV 2 (without using the intermittent technique) gave only $E_8\,(4) = 1.9$ and equipment similar to RV 1 was characterized by $E_8\,(4) = 2.3$.

If purity is not the only consideration, then an output efficiency may be characterized by introduction of the quantity M of material of average purity $E_m(n)$ produced during the zone refining time t.

$$Q\,(n, M, t) = E_m\,(n)\ \ \{M/t\}\,. \qquad (17)$$

A zone refiner which is optimal for $k < 1$ is not necessarily also optimal for $k > 1$. Therefore a separate test has been developped for $k > 1$[107)], using fluorene and its persistent ("natural"[h]) impurity anthracene ($k = 1.3$) which is usually present at a concentration of $\sim 10^{-3}$ [mol/mol]. Other "natural" impurities were not found

---

g A set up with the zone refining tube inclined at $\sim 45°$ and stirred by rotation has been used by other authors[313, 314)]

h Many organic compounds, especially those of aromatic nature, are isolated from coal tar. They contain as "natural" impurities the whole family of related compounds formed during cracking of coal to interfere (by comparison with synthetic fluorene). 100 zone passes and an integration over 8 zone lengths from the purest point are suggested as test parameters.

A zone refiner similar to RV 1 gave E_8 (100) = 0.6 for f = 9 mm h^{-1}; using additional cooling elements at 42 ° C with a similar apparatus E_8 (100) = 1.0 was reached with f = 11 mm h^{-1}, whereas a horizontal zone refiner, RH 2, gave only E_8 (100) = 0.8 for f = 4.6 mm h^{-1}.

e) Purification with k Close to One

The closer the distribution coefficient approaches one, the less purification can be obtained even if the ultimate distribution can be approximated. In the actual purification problem fluorene (anthracene) with k $\gtrsim$ 1 it was important to learn how much depletion could be reached in a given time. The results are plotted in Fig. 26. Even after 300 zone passages (n = 6N) further improvement is possible. After 500 passages the theoretically expected final distribution is still far from being reached. Impurity retention in the empty part of the tube makes itself remarkable in a drastic way in the curve for n = 500. (Necessarily a separate zone refining tube belongs to each curve; L/ℓ was between 50 and 80 for the different ingots).

500 zone passages took nearly a quarter of a year. Nevertheless, only in a small fraction of the fluorene sample could the anthracene be brought down from 10^{-3} to $\sim 10^{-6}$ mol/mol. This is certainly a practical limit for useful zone refining. Any improvement of the efficiency would be highly desirable.

The purity can still be improved by zone refining if by some means the slope of the impurity concentration profile obtained can be reduced again. This is indeed possible by expanding the length of a pure part of the ingot, e. g. to the full length of the initial ingot. Either a smaller bore diameter tube must be used or the pure parts of several zone refining tubes must be united.

As an alternative, synthetic preparation of fluorene was checked[120], see point 2.2.6. No anthracene could be detected in the synthetic samples. Other impurities of the synthetic material can be easily removed by zone refining.

f) Back Diffusion

For k close to one, all tricks must be used to obtain a reasonable purity. Fischer[117] has studied carbazole (anthracene) with k_{eff} = 0.5 and anthracene (carbazole) with k_{eff} = 0.84 with respect to back diffusion of the impurity. Experimental concentration profiles are given and compared to results obtained with a technique where at the impure end after each zone passage a certain fraction of a zone length was excluded from further remelting at subsequent zone passages. The basic idea is that for k < 1 the concentration gradient, driving back-diffusion in the solid state, increases with x. Because of normal freezing in the last zone length, the main part of the impurity is concentrated at the very end in conjunction with the steepest gradient. Fischer has shown, both mathematically and experimentally, that order of magnitude ef-

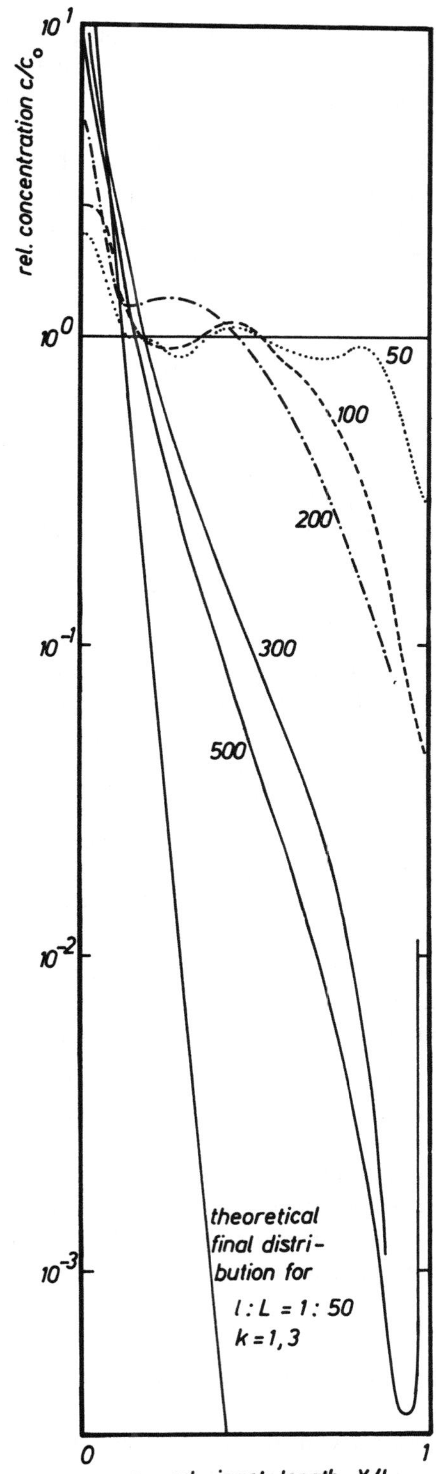

Fig. 26. Experimental distribution profiles of anthracene in fluorene after 50, 100, 200, 300, and 500 zone passages, compared to the theoretical final distribution for k = 1.3,[107)]

fects can be achieved by "cropping" the impure end successively. This success is due, among others, to the fact that in real crystals and close to the melting point non-negligible diffusion takes place along dislocations and grain boundaries. This part of the diffusion coefficient is orders of magnitude higher than diffusion in the bulk lattice[121–123]. Back sublimation along cracks is certainly another detrimental effect which is also reduced by reduction of the concentration gradient by the cropping method. In addition, back sublimation can be reduced by zone refining under inert gas pressure (~ 400 torr).

g) Other Difficult Systems

Commercial pyrene contains at least two impurities which are difficult to remove by zone refining. One with $k < 1$ has the molecular wight 208 and is probably a pyrene in which the carbon atoms 1 and 2 (VII) are replaced by one sulfur atom[i 124]. The other with $k > 1$ leads to a yellow-green fluorescence and affects the triplet exciton lifetime adversely (cf. Fig. 54). Pretreatment by "potassium fusion" (see below) is helpful[134].

Naphthalene contains among others[113] thionaphthene and β-methylnaphthalene. The latter is easily reduced to a content of between 10^{-5} and 10^{-6} mol/mol by zone refining alone, but hardly below. Additional potassium fusion (see point 2.2.5) can reduce both impurities below the limit of detectability, which for β-methylnaphthalene is about 10^{-8} mol/mol (by crystal fluorescence spectroscopy at liquid helium temperatures (see point 4.1.5)).

1,4-Dibromonaphthalene even after prolonged zone refining still contains 10^{-4} mol/mol naphthalene, 10^{-3} mol/mol monobromonaphthalene, probably other isomers of dibromonaphthalene, 10^{-5} mol/mol tribromonaphthalene and $\leqslant 10^{-5}$ mol/mol dibromothionaphthene[128]. This and other findings are indicative of a slow thermal isomerization and decomposition under cleavage and partial rebinding of bromine in the molten material. Chromatographic purification alone can yield purer material concerning these impurities.

In zone refined sym.-tetrachlorobenzene are found different mono-, di-, tri- and tetrachloro-homologs and isomers totalling $\geqslant 10^{-4}$ mol/mol[128]. Part of them are persistent side products from the synthesis, part of them seem to be continously forming during zone refining. Similar findings hold for other halogenated aromatic hydrocarbons such as 9,10-dichloroanthracene[134]. Sym.-octahydroanthracene can be zone-refined successfully. However, a strong phase transition usually explodes the ampoule during or after cooling, sometimes after days of apparent peace[134].

h) Normal or Reverse Zone Refining?

As already mentioned, provision of a hollow space is necessary in a closed glass zone refining tube containing organic material. This hollow space is at the entrance of the zones in the "normal" procedure and at the exit in the "reverse". Being always a source of retained impurities it makes "normal" zone refining more promising for impurities with $k > 1$ whereas the reverse variant should be prefered for $k < 1$. This is especially important if reverse *horizontal* zone refining is applied: With this method a portion at the end of the ingot must remain unmolten, and therefore unpurified, in order to avoid flowing away of the melt[g]. If both kinds of impurities are present a combination of both techniques in two zone refining steps is advisable.

i) Eutectic Zone Refining

A eutectic mixture such as e. g. the 50% point in the hexachlorobenzene/anthracene phase diagram (Fig. 6) has a sharp melting point. Therefore a eutectic mixture can be zone-refined like a pure compound. Fortunately, as is also illustrated by Fig. 6, the eutectic melting point can be much lower than that of the individual components. According to this concept tetracene has been zone-refined (much below its own melting point) with 2-naphthoic acid as an intentionally added component[125)]. Tetracene alone melts under decomposition. – Obviously a suitable method for the final separation of the eutectic mixture must be available if the procedure is to be of practical interest.

j) Complex Zone Refining

Stoichiometric adducts like donor-acceptor complexes usually posess a pronounced maximum in their phase diagram at the stoichiometric mixture (cf.[126)]). The adduct then behaves like a single chemical compound, showing a sharp congruent melting point. It can be zone-refined without being segregated into the individual constituents. Zone refining of the complex therefore can be successful even if one of the components (or both) are thermally unstable at their own higher melting points and therefore cannot be zone-refined individually[127)], cf.[57)].

2.2 Other Purification Techniques

2.2.1 Sublimation

Sublimation can be done on several levels of improvement. In simple high vacuum sublimation onto a collecting condenser cooled by water, volatile solvents are pumped away which may be present from preceeding steps of preparation or prepurification. Higher molecular wight impurities and decomposition products which do not

---

[i] phenanthro [4,5 – bcd] thiophene[129)]

sublime readily remain in the residue, together with possible mechanical contaminants. Therefore a sublimation step should always precede zone refining. A commercially available sublimation apparatus for laboratory purposes, like that fabricated by Schott und Gen., (Mainz) is sufficient for this purpose. Heating is conveniently done by embedding the beaker into heat conductive sand (SiC) on a hot plate. Alternatively a set-up as described by Sherwood[135] may be used.

"Fractional" or gradient sublimation in a temperature gradient is a more scrutinized method[136, 137]. It is important to use a temperature gradient which is extended over a sufficiently long tube and to work under an inert gas pressure which, however low, is high enough to preclude mean free paths of the subliming molecules from reaching apparatus dimensions (e. g. 1 torr). Among others, donors and acceptors for radical ion salts like tetrathiafulvalene (TTF), (see XX) an tetracyanoquiodimethane (TCNQ), (see XX) have been purified by this method[138].

Vapor zone refining has been proposed[125] as an interesting repetitive variant of fractionation, which is especially useful for compounds like tetracene, which cannot be zone-refined because of thermal decomposition at their melting points.

2.2.2 Distillation

Distillation is widely used for chemical separations. With suitable techniques of fractional distillation which are described in standard textbooks on chemical laboratory techniques, high efficiencies can be reached. Decomposition of higher molecular weight compounds can sometimes be prevented by distillation under reduced pressure. However, a high constancy of the pressure must be provided in order to maintain a high separation efficiency of distillation columns. Separation efficiency is usually quoted as "number of theoretical plates", that is the number of times liquid and vapor are in successive phase equilibrium.

For obtaining ultra-high purity compounds even fractional distillation is often not efficient enough for certain impurities. Addition of a further component forming an azeotropic mixture with the compound to be purified has led to satisfactory results for some systems. Azeotropic distillation resembles eutectic zone refining in that distribution coefficients are altered by the addition of a third component. Anthracene, e. g., can be purified by azeotropic distillation with ethylene glycol[139].

2.2.3 Recrystdllization

Recrystallization – in contrast to normal crystallization methods like normal freezing – uses at least a ternary phase system for the purification of a substance A from its impurity (B). For this purpose a "solvent" S is added. S, for convenience, is usually selected to be liquid at room temperature. The main condition, however, is that S is not incorporated into the crystal lattice of the compound, A, to be purified.

A suitable mixture A(B)S is transferred to the liquid state by heating. Subsequent cooling leads to the formation of crystals, usually of the component A. The first fractions have usually a higher purity than the initially added material. However, the starting point in the phase diagram could also be selected in such a way that the added component (S) crystallizes first, incorporating and thus withdrawing the impurity from the melt. Other alternatives are possible, depending on the corresponding phase diagram and on the selection of the starting concentrations.

Clearly a repetitive procedure would be highly desirable also for fractional recrystallization. One way of automatic realization could be to pull subsequent *solution zones* through a bar instead of just using *molten* zones as in zone refining. However, practical experience is scarce with this version. It should be mentioned that classically the idea of fractional crystallization has often successfully been applied to difficult separation problems by following a systematic scheme of repeated fractionation and combination (of recrystallized fractions).

In the actual situation with the preparation of high purity material usually more than one impurity is present; the underlying phase diagrams or even the chemical nature of the impurities are frequently not known. Practical experience has shown that in these cases a simple standard recrystallization with fractional collection is often not very efficient. Optimization by close analytical inspection is necessary.

2.2.4 Column Chromatography

Column chromatography has been widely used for the removal of persistent impurities from organic compounds. It is a counter-current technique where molecules distribute themselves between a flowing liquid phase (as dissolved molecules) and a stationary solid phase (as adsorbed molecules). The solid phase is usually a powder like Al_2O_3 with a high specific surface, filled tightly into a tube ("column"). The tube is filled up with a pure solvent flowing slowly and continuously through the assembly.

There are essentially two versions: Highest separation efficiency is reached if (at optimized parameters) a small length of solution of the compounds to be separated is brought onto the column at the solvent entrance. This solution fraction of the liquid column travels along the tube together with the other pure solvent. With inefficient adsorbent this section would appear at the other end essentially with the same geometrical shape, (actually diffusion and the radial velocity distribution change the spacial distribution to some extent). Assuming two different solutes, A and B, being present in the section containing the solution, and with the adsorbent being active, the solutes will appear as separate fractions at the outlet because of different trapping times at the adsorbent. In this version, liquid chromatography is essentially equal to gas chromatography, both being a delay time ("retention time") analysis.

Microscopically many successive steps of travel in the liquid and (reversible) adsorption on the powder follow each other. The trapping energy at the adsorption centers and hence the Boltzmann detrapping probability, or the average trap occupation times are a function of the chemical nature of the solute. These parameters are therefore in general different for the different molecules A and B. Strongly adsorbed components can sometimes be separated efficiently by using differential elution techniques, i. e. by gradually changing the composition and thereby the polarity of the solvent. Instead of being adsorbed on a powder the solute can also be distributed between the mobile liquid phase (solvent) and a (different) stationary liquid phase adsorbed on the powder.

The version described above is commercially available in the form of technically highly sophisticated instruments, generally working under a high liquid pressure in order to increase speed ("High Pressure Liquid Chromatography"). This technique is widely used for analytical purposes. For preparative purposes it has mostly the disadvantage of being capable of separating only small quantities per cycle at high separation power.

Separation of aromatic hydrocarbons is described in Refs.[140–145]. A second, older version of column chromatography allows the separation of much larger quantities, (on a laboratory scale usually between 10 and 50 g). However, this can only be achieved at the expense of separation power. For hard-to-dissolve compounds liquid chromatography is done near the boiling temperature of the solvent used. An improved version of the original "Sangster and Irvine" assembly[145] (see also Meier and Fletschinger[146]) has been described by McGhie (Fig. 27)[147]. Solvent in the lower reservoir G is heated to boiling by a heating mantle J. The vapor condenses in the condensers K; the liquid collects in the upper reservoir H, passes through glass wool C, dissolves the crude material A placed onto a coarse frit B and drops onto the column which is filled with the adsorbent D, kept by the lower frit B. E is a pressure relief line. Excess solvent overflows and returns to the lower reservoir. A

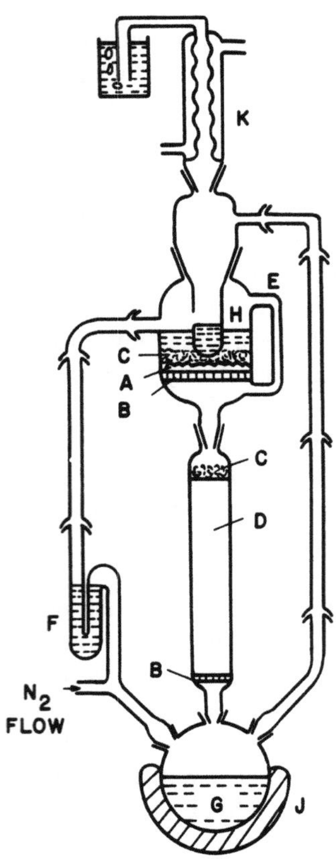

Fig. 27. Set-up for liquid chromatography on a preparative scale, for details see text. From McGhie[147]

liquid trap F prevents the inert ambient gas (N_2) from flowing through the return line. The design of the condensate-return below the condenser is such that mixing in the upper reservoir, due to falling droplets, is minimized in order to avoid overflow of impure solution. It is necessary to adjust the heating power to such a level that sufficient condensate is available in order that the column never sucks in gas. An automatic level control has been suggested for this purpose by Sloan[8]. For liquid chromatography near the boiling temperature of the solvent, the column is heated by the vapor, passing through a jacket which surrounds the column, (not shown in Fig. 27).

It is important to realize that in contrast to the first version, the process of dissolving the starting material on top of the column may take longer than the retention time for some fast impurities. Therefore only such impurities are separated which travel so slowly on the column that they are still on it when the main fraction appears at the outlet. The progress of the different fractions on the column often can be judged by observing coloured rings in normal or UV light. In addition the contents of the outcoming solution should be checked at regular intervals. – For further literature on liquid chromatography see[148, 149].

Prepurification by column chromatography for the preparation of very high purity organic crystals has been described in the literature among others for naphthalene[150], anthracene[8, 135], phenanthrene[119, 151], and chrysene[147, 152]. In the case of anthracene liquid chromatography mainly serves to remove its natural companion carbazole (usually present at 10^{-2}–10^{-3} mol/mol) which by zone refining alone is difficult to reduce below a level of 10^{-5} mol/mol.

2.2.5 *Purification via Chemical Reactions*

An impurity which withstands all purification efforts can sometimes be removed easily after a transformation into another chemical species by a selected chemical reaction. The condition however is that all added reactants plus eventually contained impurities as well as all other possible reaction products, especially with the host material, are also easily removable. Chemical reaction in a sense leads to a "chemically improved segregation coefficient".

Mainly two types of reactions have been used in the literature: Diels-Alder condensation and potassium fusion. As an example for Diels-Alder condensation the reaction of anthracene with maleic-anhydride may be considered which has been used e. g. for the purification of phenanthrene from anthracene[119, 151]:

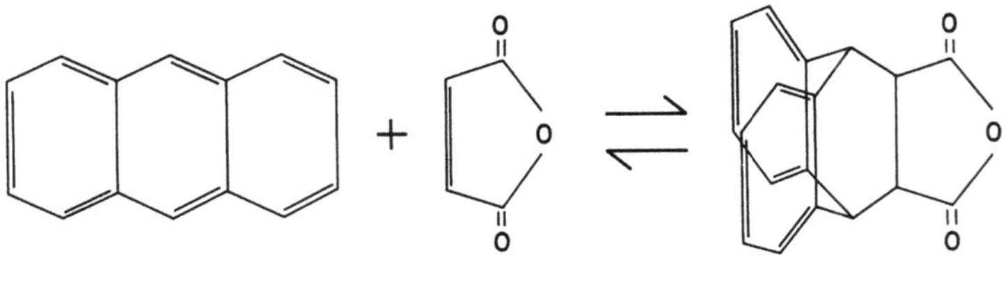

The above reaction can easily be performed by refluxing a solution of the reactants in benzene, toluene or xylene for a period of several hours. (Notice that the reaction is reversible at higher temperatures.) The anthracene Diels-Alder adduct is removed from the solution by extraction with aqueous potassium hydroxide solution.

"Potassium (or sodium) fusion" is a method where molten organic compounds react selectively with molten potassium metal under vacuum. It is useful for impurity molecules with sites or side groups which lead to the formation of non volatile potassium salts such as thionaphthene or β-methylnaphthalene. Both of these impurities are contained in "natural" naphthalene. Potassium fusion may also be useful where final reaction products are formed which have more favourable distribution coefficients at zone refining or where an impurity is selectively or at least preferentially decomposed.

After the potassium treatment purified naphthalene is collected by vacuum sublimation. Potassium fusion may be useful for many parent hydrocarbons; however care should be taken that no halogenated compounds are admitted to react with potassium metal; explosions may be the consequence!

2.2.6 Synthetic Preparation

In some cases synthetic preparation leads to pure material with less effort than prolonged purification. Fluorene is an example[163)]. Commercial fluorene contains a number of structurally related compounds[153)] which are not readily separated by the usual methods of purification, including zone refining (see paragraph 2.1.4 e).

Synthesis cf.[163)] was started from commercial 2-amino-diphenyl. By diazotation and reaction with KI, 2-iodo-diphenyl was obtained. Reaction with CO_2 of the Grignard product led to diphenyl-2-carboxylic-acid. The ring was closed with conc. H_2SO_4. Final reduction with hydrazine of the fluorenone so obtained, gave fluorene. The product was found to be essentially free of those impurities which are present in "natural" fluorene. The over-all reaction yield is high: triplet exciton lifetime (cf. paragraph 4.1.6) reached a value of 120 ms after zone refining[134)].

2.3 Practical Results of Combined Purification Methods

The highest over-all purity of any organic material seems to have been obtained for naphthalene. Persistent impurities such as β-methylnaphthalene and thionaphthene can be removed well below 10^{-7} mol/mol. Although it is analytically impossible to trace *all* impurities down to this level, it may safely be assumed that the over-all purity in best naphthalene samples is somewhere around 10^{-8} mol/mol (0.01 ppm). Therefore in what follows, ultrapurification will be exemplified by a prescription which has been successfully used for naphthalene[120)]. Naphthalene for scintillation is chromatographed at room temperature on basic Al_2O_3, Merck, activity I, using a 1 : 1 mixture of benzene and petroleum ether 40/60 °C. The naphthalene-containing fraction is concentrated by partial evaporation of the solvent. The naphthalene precipitate is collected and dried in a desiccator over paraffin flakes. In a subsequent step the material is zone-refined for a first time with some sodium metal added at the top of the zone refining tube. Next the main part of the zone-refined material is reacted at 170 °C for 12 h with a potassium mirror produced inside an evacuated

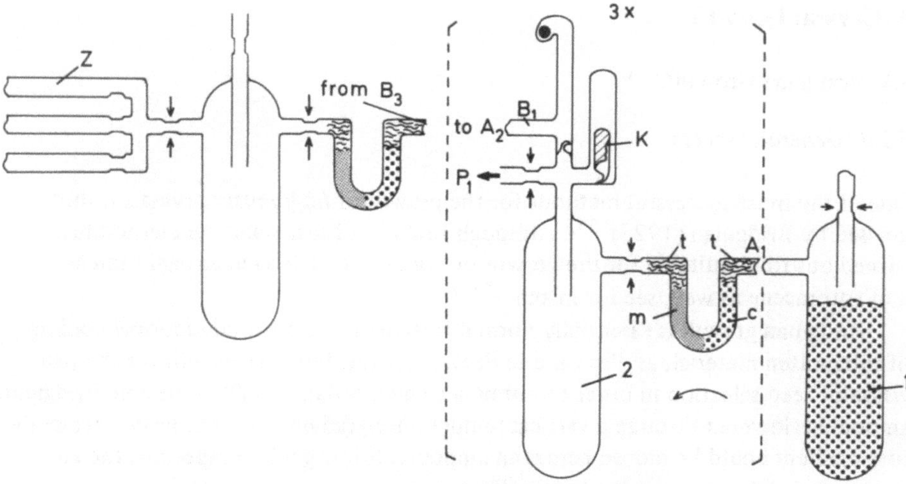

Fig. 28. Set-up for purification by potassium fusion and sublimation. Material from a first zone-refining step is collected in flask 1, sublimed through Teflon wool (*t*), charcoal (*c*) and 4 Å molecular sieve (*m*), and reacted in flask 2 with potassium metal. Sublimation through adsorbents and potassium fusion are repeated twice in two subsequent identical sections of the apparatus (not drawn) before the final product is sublimed into tubes (*z*) for a second step of zone refining. Several connectors are provided for pumping. Useless parts can be sealed off successively by a flame at restrictions provided for that purpose. Heating baths, heating tape and infrared radiators are not drawn. After W. Tuffentsammer[120]

flask. The apparatus is shown in Fig. 28. Sublimation through Teflon felt, molecular sieve and charcoal transfers the material into another flask from where a second similar potassium treatment and sublimation step is started. This step is repeated for a third time before final zone refining with 100–200 zone passages is applied. The whole procedure is done in an evacuated sealed set-up. Material transfer is steered by magnetically cracking breakseals which are provided from the very beginning.

Single crystals grown by the Bridgman technique in vacuo from naphthalene so purified show the following exceptional characteristics: They reach a lifetime of triplet excitons of 400–500 ms (cf. point 4.1.6). Microscopic charge carrier mobilities can be measured down to liquid nitrogen temperature and below[154]. In fluorescence spectra at 4.2 K no β-methylnaphthalene emission lines are found any longer.

Highest purity anthracene is obtained after liquid chromatography on a vapour heated column, cf.[8], using n-hexane as boiling solvent, vacuum sublimation and two steps of multistage zone refining. Oxygen and UV light must be excluded during handling of the ultrapure product. Details of purification and handling of anthracene have been described by Sloan[8] and by Sherwood[135]. Examples of typical levels of impurities are carbazole $\leq$ 1 ppm; tetracene $<$ 0.01 ppm. The lifetime of triplet excitons at room temperature reached for pure anthracene single crystals is 25 ms.

A detailed prescription for obtaining phenanthrene with a total impurity content $<<$ 1 ppm has been given by McArdle, Sherwood and Damask[119].

3 Crystal Growth

3.1 Bridgman Growth

3.1.1 General Aspects

One of the most successful methods for the growth of *high purity* crystals is that devised by Bridgman (1925)[155]. Although first used for metals, the method has turned out to be suitable for the growth of organic crystals such as naphthalene and anthracene as well, see for instance[156].

Bridgman growth is essentially normal freezing i. e. slow unidirectional cooling of the molten material, as discussed in Sect. 2.1.1 (cf. Fig. 7a), except for the provision of seed selection in order to obtain a single crystal. Usually a suitable Bridgman ampoule is lowered through a vertical temperature gradient. Alternatively a temperature gradient could be moved across an ampoule. For organic compounds the ampoule normally can be made of glass (Pyrex).

In principle Bridgman growth could be done directly in the zone refining tube after zone refining. However seed formation would have to take place somewhere at the impure ends. To avoid redistribution of the impurities, only remelting of one zone length is permissible and this zone should travel in the same direction as during zone refining. Therefore Bridgman growth immediately in the zone refining tube seems feasible only with the reverse zone refining technique. In fact promising results have been obtained with apparatus NH 1 and NH 2 (see paragraph 2.1.4) by using zone refining tubes provided with a cone (cf. Fig. 21a) at the "filled" end for seed formation and applying a reduced speed for the last zone. However, a separate crystal growth step is generally preferable.

3.1.2 Material Transfer

Transfer of zone-refined material under high purity conditions needs some attention. The pure fraction of a zone-refined ingot is either cut off, inserted (together with its surrounding glass envelope) into a stock chamber which is connected with the Bridgman ampoule, evacuated, and sealed off; material is sublimed or gently melted into the Bridgman ampoule; the ampoule is sealed and removed with a flame. Or, to avoid contamination by oxygen or water vapour, the zone refining tube could be opened in a suitable glove box or manipulated in a flexible vacuum line being composed of transparent glass sections and flexible metal bellows allowing inspection, cracking and moving of the cracked sections[134].

Alternatively high purity can be maintained by applying a break-seal technique for material transfer. A zone refining tube equipped with a break-seal is drawn as detail No. 6 in Fig. 18. After zone refining, the zone refining tube is sealed onto a glass apparatus containing several side arms which can be removed later by a flame. The glass apparatus is connected to a high vacuum system, provided with cold traps and an inlet system for inert gas. The break-seal can be opened by a magnetic ball and a permanent magnet. A pulsed high magnetic field, produced by discharging a capacitor bank through a low inductance coil, would be an alternative method, especially if higher accelerating forces are needed. After use, the magnetic ball may be transferred to a side arm and sealed off; (if necessary for sealing, an inert gas pressure close to the ambient air pressure may be admitted). The whole assembly is then evacuated and sealed off at a restriction which had been provided for that purpose.

Next, a potentially impure leading fraction of the zone-refined material is removed carefully into one side arm. The pure part is transferred to one (or more) Bridgman ampoules by sublimation or by melting. To avoid superheating this may be done by heating the whole assembly in a temperature controlled drying oven. Regions from where no material should sublime off or where it should condense are cooled[134].

Working under vacuum makes material transfer by sublimation or melting easier. Further, vacuum in the crystal growth tube ensures higher quality crystals since inert gases have been found to adversely affect crystal quality[157].

3.1.3 Practical Realization of Bridgman Growth

a) Equipment

A Bridgman furnace needs an upper portion which is hotter than the melting temperature of the material to be grown and a lower one which is colder. Several designs for organic compounds have appeared in the literature (cf.[8, 10, 156–159], and Refs. given therein) differing essentially in the technical solution of basic problems such as spatial homogeneity and temporal constancy of the temperatures, shape of the temperature gradient between the two halves of the furnace, heat flow and thermal coupling to the ampoule. Further points are thermal and vibrational shielding, realization of a constant speed drive and use of safety lights.

In what follows, four essentially different concepts[134] will be described (Fig. 29a–d) which have proven useful in the Stuttgarter Kristallabor over a long period and which seem to meet most practical requirements.

The furnace sketched in Fig. 29a essentially consists of a wire-wound heating glass tube. Two zones of different temperature are produced by a different pitch of the windings, or, for more flexibility in use, by bypassing a fraction of the current from a central connection to the lower connection over a variable shunt (not shown). Two additional outer glass tubes are provided for thermal shielding if a temperature stabilized room is available. Otherwise shielding by a passive water jacket with sufficient thermal inertia or some kind of homogeneous counterflow cooler connected to a thermostat may be used. To minimize convection, the furnace is closed on both sides (not shown in the figure). Bridgman crystals often stick to the wall of the ampoule. Different thermal expansion coefficients and expansion anisotropy may lead to considerable stress and stress-induced crystal defects. Therefore it may be desirable to melt the crystal free from the wall for its most part, before cooling down. This can be accomplished[161] if thin supporting woven wires are attached to small permanent magnets, (one upper magnet serving for coupling to the main transport mechanism, and one lower magnet serving as weight w during the growth period). After growth the ampoule position in the furnace is fixed (by attaching the magnets to suitable supports at the furnace) and the whole assembly, crystal ampoule plus furnace, is turned upside down. Partial melting is possible with carefully readjusted currents. A subsequent annealing phase close to the melting point will be useful. For this step, in order to avoid subliming-off, the coldest point in the ampoule must lie in the crystal.

Figure 29b depicts a "reflector furnace"[160], which is unformly heated. Generation and movement of a molten zone is effected by a gold-coated, semitransparent heat reflector r which is slowly pulled upwards. This furnace has the peculiarity that the crystal ampoule remains fixed. Besides being useful for observations and manipulations, such as applying an electric field during growth, this type of furnace has produced many good crystals even in its simplest version.

Intimate heat transfer is possible in the model of Fig. 29c where the ampoule is immersed in a cylindrical bath containing two immiscible liquids, l_1 and l_2, held at different temperatures. Surprisingly, a relatively steep temperature gradient is maintained at the boundary where heat transfer by convection is hindered and only diffusional heat conduction is possible.

The double Liebig-condenser version of Fig. 29d needs the connection of the two chambers th_1 and th_2 to two thermostats. Very uniform heating and high constancy of the temperatures are the main feature. Temperature programming of the thermostats can be used. Annealing of grown crystals is possible very close to their melting points ($\Delta T \leqslant 1$ °C). Immersion of the lower part of the Bridgman ampoule into silicone oil leads to an improvement of the heat flow and hence of the isothermals. The oil level is balanced by a large diameter container c, connected by a Viton rubber tube. This increased heat transfer has lead to satisfactory pyrene crystals which were otherwise rarely obtained free of cracks.

An alternative, cheaper but less flexible method of heating is found in the literature[119]: under reflux, vapor of two liquids, boiling at different temperatures, is passed through the two chambers. The double Liebig condenser version is also suitable for growth below 0 °C.

b) Seeding

A single crystalline seed could be placed in the lower part of a Bridgman ampoule from the very beginning. For destroying parasitic seeds, partial remelting would be necessary. This is difficult to control.

Another method has turned out to be more practicable: A small amount of polycrystalline material is left unmelted in the lower stem. This is useful for avoiding rapid uncontrolled growth after a period of supercooling which may otherwise be established. In general, several seeds will start to grow. The combined action of the capillary and the hollow sphere (cf. Fig. 29) leads to a seed selection: All seeds with an inclined direction of principal growth will die out and therefore not reach the main part of the ampoule. In the great majority of experiments single seeding has been observed. 2,3-Dimethyl-naphthalene, 2,3-dimethyl-anthracene, fluorene and p-terphenyl tend to come up in several parallel lamellae. With these compounds the ampoule forms proposed in[135] may lead to more satisfactory results.

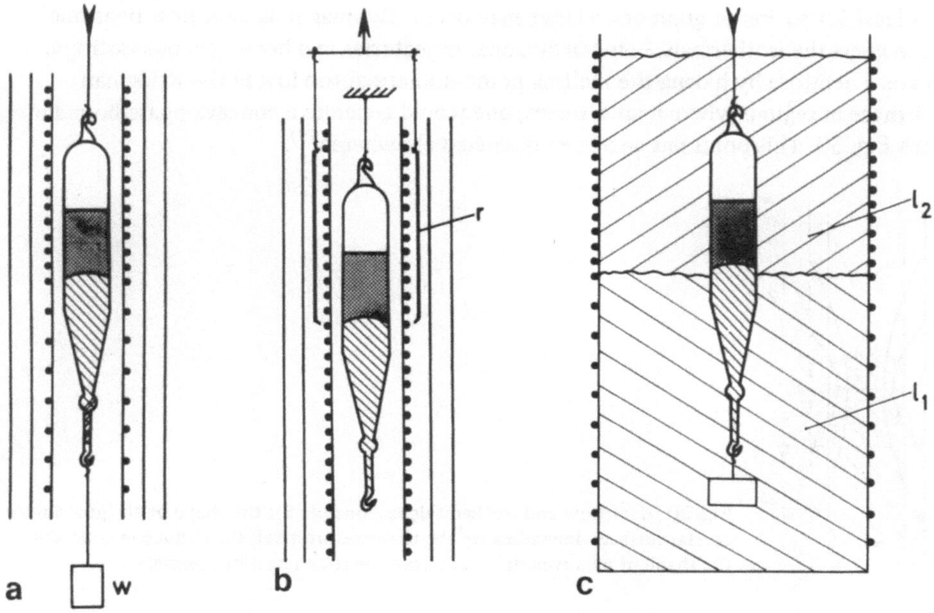

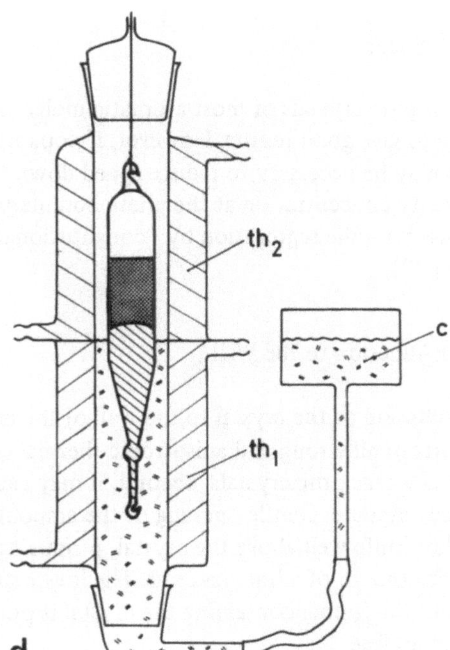

Fig. 29 a–d. Several versions of Bridgman furnaces: **a** conventional set-up; **b** travelling reflector; **c** two immiscible liquids; **d** double Liebig condensor. (Schematic drawings; all furnaces are actually longer in their vertical dimensions; the ampoule diameter is typically between 10 and 25 mm)

c) Phase Boundary

An (upwards) convex shape of the solid-liquid phase boundary is required for the formation of a good single crystal. With concave boundaries additional seeding or at least formation of grain boundaries may occur. By imagining heat flow lines one can guess the isothermals, being orthogonal trajectories in a homogeneous isotropic case. Therefore by having the melting point isothermal too low in the Bridgman furnace in regions where heat diverges, one would generate a concave phase boundary, see Fig. 30. This point has also been discussed elsewhere[159].

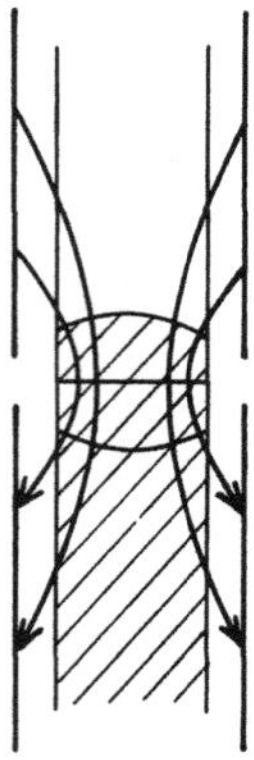

Fig. 30. Heat flow and isothermals responsible for the shape of the growing crystal surface; depending on the temperature level, the surface may accept the shape of a convex or of a concave melting point isothermal

d) Speed

For pure crystals of most aromatic molecules a growing speed of 1 mm h^{-1} has proven to give good results. However, if dopants are added which are partly segregated, it may be necessary to reduce speed down to 0,1 mmh^{-1} in order to avoid an impurity concentration at the phase boundary beyond the solubility. In the worst case periodic segregation by "constitutional supercooling"[162] may occur (cf.[7], Fig. 9).

e) Adhesion to the Wall

Adhesion of the crystal to the wall of the ampoule may be a problem for two reasons: First of all strong and anisotropic thermal contraction on cooling may strain and eventually crack the crystals. Second, it may cause trouble in removing crystals from the ampoule. Gentle opening of the ampoule can be done by scratching it with a glass knife well above the crystal, producing and pulling around a crack by touching with the tip of a hot glass rod. The lower stem can be broken off after scratching. Into the funnel containing the crystal is poured a pure solvent until the crystal becomes free.

It has been found that zone-refined anthracene does not stick in well purified Pyrex ampoules, if heating never exceeds the melting point (218 °C) at any place by more than a few degrees. These findings in turn justify the material transfer method by sublimation described under point 3.1.2. Growth in Teflon or Teflon-coated ampoules may be a useful alternative. Treating the ampoule with a silicone release agent and baking it before filling has been proposed elsewhere[135].

f) Doping

It is practically impossible to grow a homogeneously doped crystal by the Bridgman method. This is due to the unequal distribution of impurities or dopants (characterized by the distribution coefficient (see point 2.1.1)) between crystal and melt during normal freezing. For example in the main part of a Bridgman crystal the actual dopant concentration may be a factor of 10 lower than the "added" concentration, and it may vary by an order of magnitude along the crystal boule (example: tetracene in anthracene).

In addition there is often an inhomogeneous radial distribution as a consequence of void formation by crystal contraction upon cooling (cf. 2.1.4c). Therefore it may be necessary to determine the actual concentration (or concentration profile) by analytical methods (see Sect. 4.1) in each individual sample being used for measurements.

The preparation of highly (> 1%) doped mixed crystals is restricted to the rare case of phase diagrams with small (or even vanishing) miscibility gaps (such as naphthalene: azulene[95]), see Fig. 4; naphthalene : β-naphthol[97, 98]), see Fig. 5; fluorene : dibenzofuran[163]); anthracene : acridine[164]); anthracene : phenanthrene[165]); anthracene : carbazole[166, 167]); anthrone : anthraquinone[168])).

Efforts to find, grow and characterize highly doped mixed crystals of the highest purity are only now being made. Distribution coefficients close to one may be taken as an experimental indication of possible high miscibility (see Table 2).

Isotopically mixed crystals have attracted considerably more attention. These crystals are exceptional because they can be prepared at any composition; examples: naphthalene-h_8 : naphthalene-d_8 ($C_{10}H_8 : C_{10}D_8$); anthracene-h_{10} : d_{10}; pyrene-h_{10} : d_{10}; etc.). No considerable segregation occurs during crystal growth.

Direct handling and weighing of the dopant is possible down to ~0.01 mg, using an ultramicro balance. Smaller doping concentrations than reachable with this dopant quantity must be prepared by dilution. – For highest purity requirements dopants may also be added by a break-seal technique. This has led to highest quality doped anthracene crystals[46]).

3.1.4 Examples of Bridgman Crystals

Undecomposed melting is a condition for reaching high purity by zone refining. Therefore crystal growth from the melt by the Bridgman method suggests itself in all cases where the material does not undergo a destructive phase transition on cooling. We shall give a few illustrative examples of Bridgman crystals, rather than trying to list all results obtained in our laboratory or reported in the literature.

A naphthalene single crystal grown by the Bridgman method is shown in Fig. 31. The rough part of the surface is due to sublimation after (anisotropic) contraction of the crystal. Single crystallinity of a boule like the one depicted is not obvious. It often cannot be easily ascertained without close examination. Striations due to optical inhomogeneities of the optically highly anisotropic material are usually indicative of small angle grain boundaries. In such cases additional faint boundary lines may be developed by sublimation on top of the crystal.

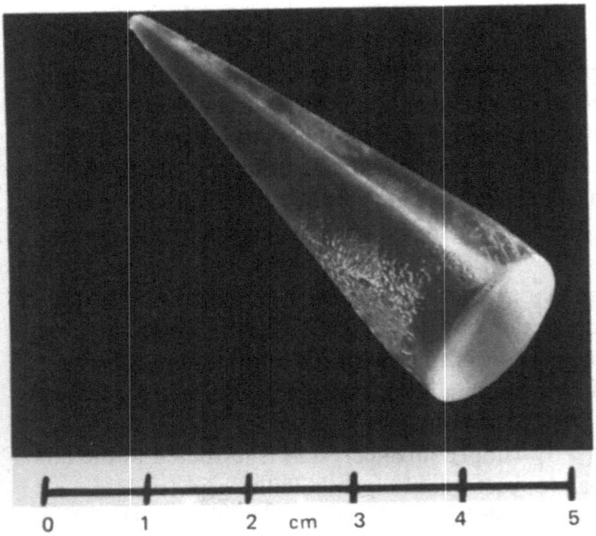

Fig. 31. A naphthalene single crystal grown by the Bridgman method[134]

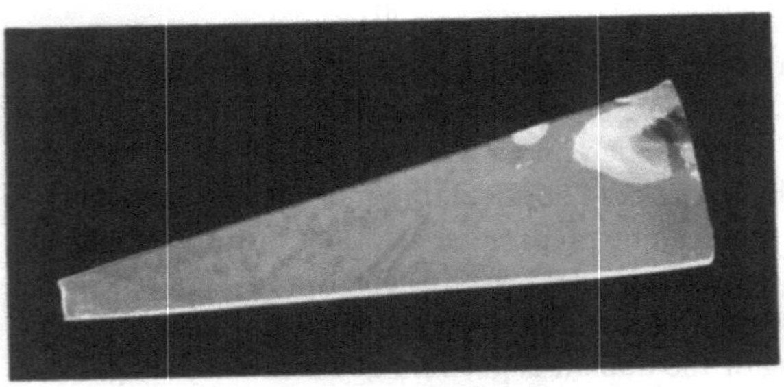

Fig. 32. A cleaved slice of 2,3-dimethyl-naphthalene. Same scale as in Fig. 31[134]

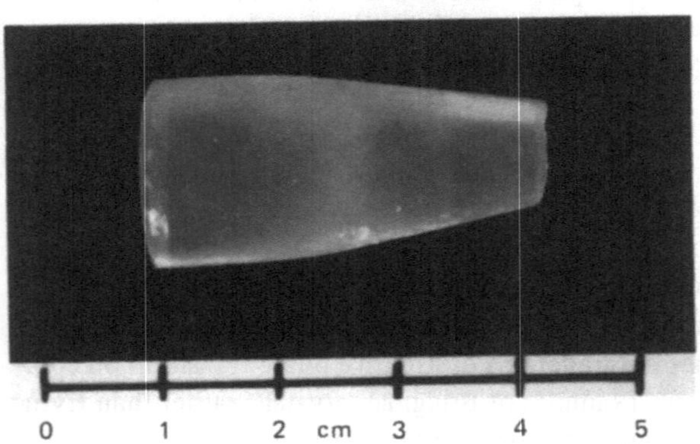

Fig. 33. A Bridgman crystal of dibenzyl[134]

If a given crystal shows well developed cleavage, the homogeneity of a cleavage plane may be a good indication of the degree of single crystallinity. A cleaved slice of 2,3-dimethylnaphthalene (doped with 10^{-3} mol/mol anthracene for laser emission) is given as an example in Fig. 32. With this crystal, cleavage is so pronounced that it is possible to cleave an optical resonator, plane parallel to a fraction of a wavelength, cf.[169].

Dibenzyl (XXXIV) is an example of the other extreme. A dibenzyl Bridgman crystal, (doped very uniformly with $\sim 10^{-4}$ mol/mol tetracene, which gives the otherwise colourless crystal a yellow colour and green fluorescence) is shown in Fig. 33, cleavage is completely lacking. Single crystallinity must be checked by polarization optics or X-ray methods, see point 5.2.

n-Heptane crystals, m. p. -90.6 °C, were prepared by slowly cooling the degassed liquid contained in a sealed pyrex tube, ending in a fine capillary[52, 53)].

3.2 Sublimation Growth

3.2.1 General Aspects

Sublimation growth, like melt growth, is a crystal growth method which is based simply on a phase transition. Therefore it can be performed while maintaining extremely clean conditions. Except for the container, no additional material is required.

The outstanding features of sublimation growth are: I. Growth at temperatures considerably below the melting point. II. Possibility of growth of several individual single crystals at the same time. III. Development of natural faces. IV. Possibility of homogeneous doping.

Lower temperature growth may be demanded by thermal instabilities such as decomposition near the melting point or destructive solid state phase transitions. Alternatively a lower temperature phase may be disired in cases where otherwise a metastable high temperature phase will survive. At lower temperatures the concentration of "unavoidable" lattice defects (originating from equilibrium thermodynamics) is smaller. Stress problems by inhomogeneous contraction will be reduced; this is especially true if the crystals are freely grown, i. e. with their natural faces developed. Natural faces may be desired for certain experiments such as reflection spectroscopy or epitaxy; they may be of great help for orienting crystals. Plane parallel samples obtained without any mechanical treatment are often useful, e. g. for optical spectroscopy, for crystal optics in polarized light, or for the detection by X-ray topography of intrinsic growth defects (i. e. defects not artifically introduced by handling, cutting or polishing).

Last but not least, perfect crystals of an otherwise inaccessible habit such as thin (1 μm), large area (1 cm^3) flakes may be obtained for some compounds (e. g. anthracene) by sublimation under appropriate conditions.

In contrast to melt growth, spatially constant doping is no problem with sublimation growth; a constant vapor partial pressure of the dopant is the only requirement. This is automatically established, either in equilibrium with solid dopant material, or in a kinetically controlled evaporization situation with homogeneously mixed source material. In the equilibrium situation the doping level is a function of the ratio of the vapour pressures of the host material and of the dopant, which can be controlled by the absolute temperature.

From a systematic point of view sublimation growth can be classified into methods where condensation occurs at a more or less one-dimensional (point-like) heat sink; at optimum conditions only one single crystal grows. Alternatively a two-dimensionally extended isothermal plane may be provided as a heat sink in order to obtain several individual crystals simultaneously.

3.2.2 Sublimation Growth with Point-Like Heat Sink

With this technique the temperature gradients converge to one coldest *point* where nucleation takes place. One single seed is required. Growth and selection of a good quality seed may be controlled manually by the experimenter: First, by reversing the temperature gradient, eventual parasitic nuclei are removed. In the next step a (weak) supersaturation in the Ostwald-Miers region is established, insufficient for nucleation to take place. After sufficient time has been provided for the temperature to stabilize, the temperature difference between source and sink is increased to such a value that the critical supersaturation is just exceeded until the first crystal eventually appears. (Observation is best done through a telescope). Now, depending on its appearent crystallographic quality, the seed is either allowed to continue growth at the quickly reestablished initial weak supersaturation, or the seeding cycle is started again.

Two versions of the controlled point leak technique shall be considered: One has been described by Honigmann[170–173] who has studied growth speed as a function of supersaturation for hexamethylenetetramine, and later by Sloan who has investigated the kinetics of crystallization of anthracene from the vapor[174]. This version uses an essentially isothermal closed chamber containing the material and its vapor. Temperature constancy is $\pm$ 0.1 °C. A point leak is realized by a metal rod outside the chamber, which is separately heated. The difference temperature is stabilized to ~0.02 °C. At 120 °C under 133 Pa (1 torr) N_2, Sloan found anthracene nucleation to occur in a few minutes at $\Delta T = 9.5 \pm 0.2°$. (This seems to be the externally measured temperature difference which does not necessarily correspond to the actual internal parameters). Crystals up to one cm diameter and a few mm thick were obtained, with ab basal plane, at supersaturations σ of a few percent. Growth speed was linear with σ, amounting to ~0.1 mm min^{-1} for $\sigma = 3\%$. The anthracene vapor pressure calculated with the equation given by Sears and Hopke[175] was 50 Pa (0.38 torr) in the experiment. (Slightly differing data on vapor pressures may be found elsewhere, cf.[92].)

The other version was introduced by Scholz[176,177] who has studied among other things the growth of HgI_2. The closed ampoule containing the starting material sublimed onto its wall, has a rotationally symmetric shape resembling a hanging pear. On the symmetry axis at the "blossom" a heat leak is provided. Supersaturation is either controlled by cooling (or heating) at this point from outside or by leaving its absolute temperature constant and changing the temperature of the whole remainder of the ampoule. The main characteristic of the Scholz technique, besides rotation of the ampoule, is a periodic oscillation of the supersaturation which is so adjusted that condensation and reevaporization follow each other periodically, however with condensation slightly prevailing on the average. The periodic reversal of the temperature gradient serves to remove parasitic seeds and to heal defects, and thus allows a considerably increased growth rate. The Scholz technique has yielded many large (on the order of 10^2 g) good quality HgI_2 crystals[178]. Efforts[179] to apply this interesting technique to organic compounds like anthracene and acenaphthene however have not yet given results superior to what is reached by the "plate sublimation technique" to be described below.

Sublimation growth with *automatic* seed formation and single seed selection using the point leak technique can be realized by adopting the Bridgman technique to sublimation growth. As in Bridgman growth, compact crystals accepting the shape of the ampoule are obtained. The form of the ampoule may be similar to that shown in Fig. 29. In a first step the material is sublimed away from the end provided with the capillary stem and the bulb (which may point down, as necessary for liquid-solid growth, or up). The next step is slow sublimation into that end. The temperature profile of a Bridgman furnace may be used for that purpose or a hotter zone may be introduced between source material and growth region[180]. Jones[181] has preferred to withdraw the ampoule from an oil bath, in order to establish the moving temperature gradient required. For seed formation to take place in the capillary stem and not where the sphere is opening towards the cone, a good heat coupling, such as provided by the furnaces of Figs. 29c and d is advantageous. The ampoule is slowly moved "down" the temperature gradient (which in space may be any direction, preferentially up or down). Speed must never exceed the growth speed in order to avoid dendritic growth and additional seeding. The process depends sensitively on such parameters as absolute growth temperature (absolute vapor pressure), vapor pressure relative to that of an added inert ambient gas, temperature gradient, convection and heat conduction (both being a function of the ampoule dimensions), heteronucleation at defects on the wall. Jones[181] has been able to grow anthracene crystals 8 mm diameter and several cm long, (vacuum, $T = 200\ ^\circ C$, $\Delta T \sim 0.8\ ^\circ C$, growth rate between 1 and 6 mm per day). However, triplet exciton lifetime (see paragraph 4.1.6) was short and cleavage planes were poorly defined. Vitovski et al.[180] have grown compact crystals up to several cm^3 of hexamethylenetetramine, anthracene and phenanthrene using a rotating ampoule. Unfortunately no temperature, inert gas pressure, growth speed and crystal quality data are given. Meuser[182] has grown large crystals of anthracene under 13 kPa (100 torr) He at 130 °C; triplet exciton lifetime was lower than for good Bridgman crystals and single crystallinity was imperfect. Our own results, after extended trials with different modifications of the Bridgman-geometry "compact sublimation" have confirmed these findings.

Low temperature compact sublimation with a modified geometry has been used by Zwerdling and Halford[183] to grow thin (0.1 mm) extended (300 mm^2) benzene crystals (m. p. 5.6 °C) in (010) orientation. Growth of cylindrical carbon tetrabromide, CBr_4, single crystals in the monoclinic phase II with 1–3 cm length and 0.6–1 cm diameter at $T = 32\ ^\circ C$ has been described by Ebisuzaki[184].

3.2.3 Extended Heat Sink

A two dimensionally extended isothermal heat sink may be preferable for sublimation growth. Isothermal surfaces can most conveniently be realized from an experimental point of view in plane, cylindrical or spherical geometries. The features are: I) The crystals grow under a more homogeneous temperature distribution. II) Several crystals can be grown in one run. In what follows, two versions will be discussed.

a) Capillary Sublimation

A cylindrical geometry with a tubular furnace and ampoule is often used for sublimation growth, cf. Aleksandrov et al.[9]. If only one single, freely grown crystal is desired instead of a compact boule (see above) growth by "capillary sublimation"[185], Fig. 34, with automatic single seed formation may be considered. By a suitable temperature profile the cylindrical wall of the condensation vessel is kept isothermal. With the temperature profile indicated in the figure, supersaturation is maintained in the condensation vessel as long as no seed has formed. After the first nucleation, supersaturation is kinetically restricted below its critical value by the capillary (c). Flow is a function of bore diameter and length, vapor viscosity, absolute pressures, pressure difference, and dilution by an inert gas. Because of limited flow, the method is suitable only for the growth of small crystals. The source temperature may be gradually raised to some extent during growth of the crystal in order to increase the flow rate.

A few remarks should be made about the furnace used for these experiments because its basic concepts may be of more general interest [cf. Fig. 34]. Two sections of thick-walled aluminium tubing are evenly heated by a distributed heat source, realized by winding Philips "Thermocoax" heaters tightly around into a spiral groove. The two halves of the furnace are insulated from one another by a thin asbestos sheet. Both ends are provided with heated caps (heat sources not drawn). The exact desired temperature profile, as indicated in the figure, is finally reached by painting a graphite suspension onto the bare metal at places where more heat

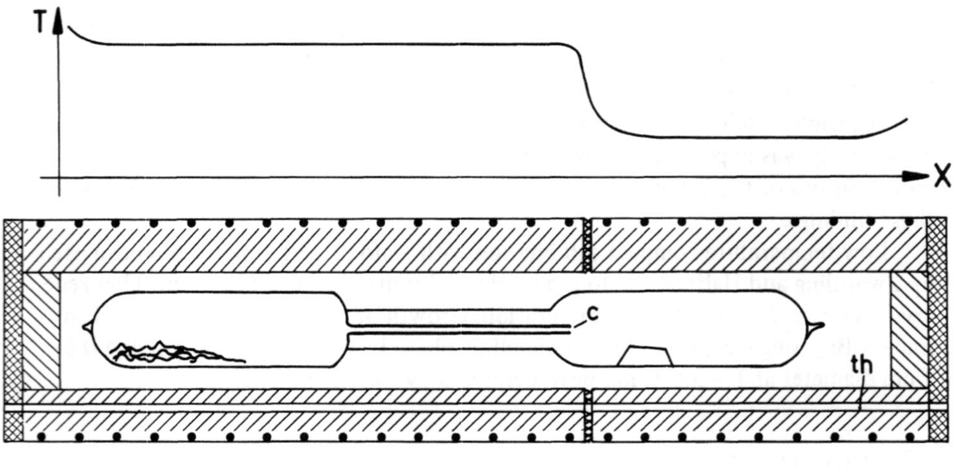

Fig. 34. Principle of crystal growth with automatic single seed selection by "capillary sublimation"[185]. The temperature profile used is indicated; the temperature difference between the two chambers is a few degrees

should be radiated off. Temperature is measured with Philips mantle thermocouples inserted into bores (th).

b) Plate Sublimation

Several crystals growing simultaneously are characteristic for a method[185, 186, 187], cf.[31], which has been given the name "plate sublimation". This technique has proven especially suitable for the vapor growth of organic molecular crystals, including two component donor-acceptor complexes.

Plate sublimation is based essentially on the sublimation between two extended isothermal plates (see Fig. 35), the distance between which is small as compared to the lateral extension. The material is enclosed in a sealed circular, cuvette-like glass vessel resembling a French cheese-box, (cf. Fig. 42 p. 65). The two large circular faces are in touch with two larger thick metal plates, containing distributed heat sources and temperature sensors. Evenly distributed heat sources are realized by soldering Philips "Thermocoax" wire in a meander-like way into the metal plates. The external faces of the plates are covered by thick asbestos sheets. To correct for a true one-dimensional temperature distribution (with position independent gradient normal to the cuvette), the latter is surrounded by a section of large diameter glass tubing of a length equal to the distance of the plates, heated by a few windings of resistive wire. With this ring heater the lateral temperature homogeneity obtainable is of the order of 0.1 °C. The perpendicular temperature gradient can be controlled by the temperature difference of the plates to within 0.1 °C. The equipment is powered by a stabilized power supply and operated in a constant temperature room (± 1 °C).

The advantage of the method is that only statistical control of seed formation is necessary thus avoiding tedious and time-consuming procedures. Owing to the quasi one-dimensional laterally extended thermal geometry, the simultaneous growth of a number of separate crystals (developing natural faces) is possible. The statistical probability of seed density per unit area is a function of magnitude and duration of an intentional, short duration increase of supersaturation, beyond its critical value, which can be controlled experimentally in a well reproducible way by the difference of the plate temperatures.

The growth procedure is as follows: The upper plate, the lower plate and the ring heater are powered in such a way, that the desired growth temperature is approximately reached everywhere. Both plates are brought to equal temperatures and the

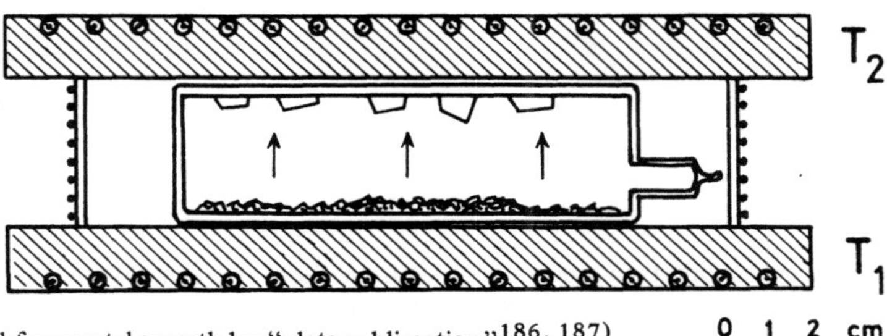

Fig. 35. Set-up used for crystal growth by "plate sublimation"[186, 187]

current through the ring heater is adjusted until the lateral circumference of the evacuated ampoule is free of condensed material but with the upper and lower plates remaining covered across ~80% of their area. After one has succeeded in maintaining this status in equilibrium without further adjustments for at least one day, the readings of the temperature sensors of the two plates are corrected for the true temperature difference which in equilibrium is zero. (Alternatively relative calibration can be done by interchanging the two sensors.) Then the temperature difference desired for growth is adjusted by a suitable electric shunt across the heater of one plate, say the upper one. Usually a ΔT of 0.5° is appropriate. Now by electric means or by placing a thermally insulating cover on top of the upper plate all material is transfered to the lower side and the inital conditions are restored.

For the growth of anthracene crystals (m. p. 218 °C) under vacuum, to give an example, at T = 80 °C and ΔT = 0.5° . . . 1.0 °C no seeds will form during several days, if the ampoule has been etched (cf.[7]) before to avoid heteronucleation at surface defects of the glass wall.

Seed formation is achieved by further increasing the difference Δj of the heater currents of the two plates, and thus the temperature difference, in a few minutes to a selected value (ΔT = 1 . . . 5 °C) and restoring the initial heater currents immediately after this new ΔT-value has been reached. (With the design described no thermal swing-over is observed.) A current and temperature program is shown in Fig. 36 for zone-refined pyromellitic dianhydride (PMDA) for which a high initial temperature difference (> 1.5 °C) can be maintained and a relatively high supersaturation is necessary for seed formation. Large crystals of this compound can easily be grown. An example for the other extreme is pyrene (dotted line).

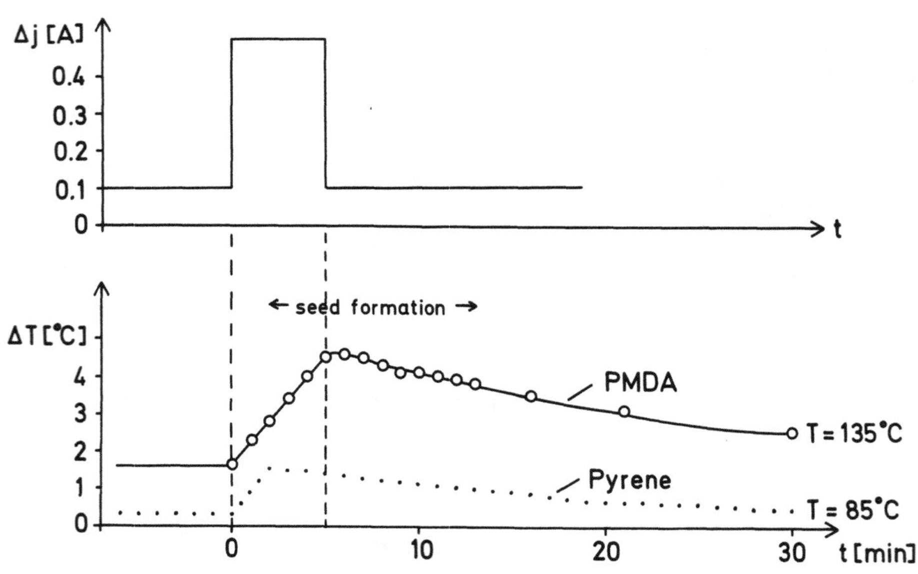

Fig. 36. Temperature-time program for seed formation and growth by the plate sublimation technique (*lower diagram*), as obtained by the timing of the plate difference current, represented in the *upper diagram*. (Plate currents are in the order of 2 A)[187]

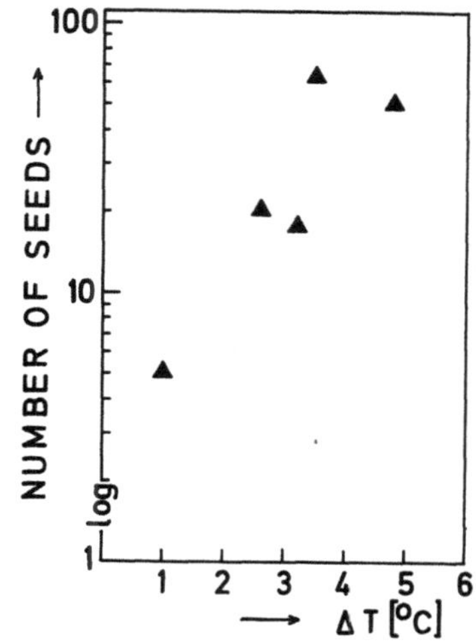

Fig. 37. Number of (anthracene) seeds obtained with the plate sublimation technique under vacuum, as a function of the maximum plate difference temperature ΔT during seed formation cycles similar to Fig. 36[187]

After a few hours any seeds formed are grown to such a size that they can be observed looking from the side through the ring heater. For obtaining crystals with well developed natural faces it is often necessary to interrupt the growth process before all polycrystalline material at the starting plate has sublimed off. Prolonged sublimation will cause the crystal surfaces to approach the isothermal planes. This effect is more pronounced the more elevated the temperature is. – Usually T and ΔT parameters leading to a growth of one gram of material during a period of between several days and one month have proved useful (for the dimensions of the set-up as described above). As already mentioned the number of seeds is statistically controlled by the amount and duration of excess supersaturation. This is shown in Fig. 37, where the number of anthracene seeds formed by the procedure described is given as a function of the maximum temperature difference during seed formation (with a thermal relaxation time of the plates similar to that in Fig. 36). – (The scatter in the data shown has its origin partly in nucleation statistics, to some extent it may be due to the fact that no special pretreatment of the glass surfaces had been applied in these experiments).

With anthracene and a number of other compounds grown under *high vacuum* conditions *bulky* crystals of a size of the order of $10 \times 10 \times 5\ mm^3$ have been obtained. – Naphthalene and pyrene examples are shown in Figs. 38a, b.

It was found that those individual crystals are most perfect which grew with only a small face in contact with the ampoule. The phenothiazine crystal shown in Fig. 39 is a typical example. Anthracene crystals of similar orientation and quality can be obtained, but anthracene grows statistically less frequently in this orientation.

The plate sublimation method is very suitable for the growth of weakly bonded stoichiometric complexes like donor-acceptor complexes. These systems allow only small temperature differences to be present in the whole set-up, otherwise they will segregate into the individual donor and acceptor phases. A number of complexes such as anthracene: PMDA[186] and phenothiazine: PMDA[57] have been grown successfully, cf. Table 3.

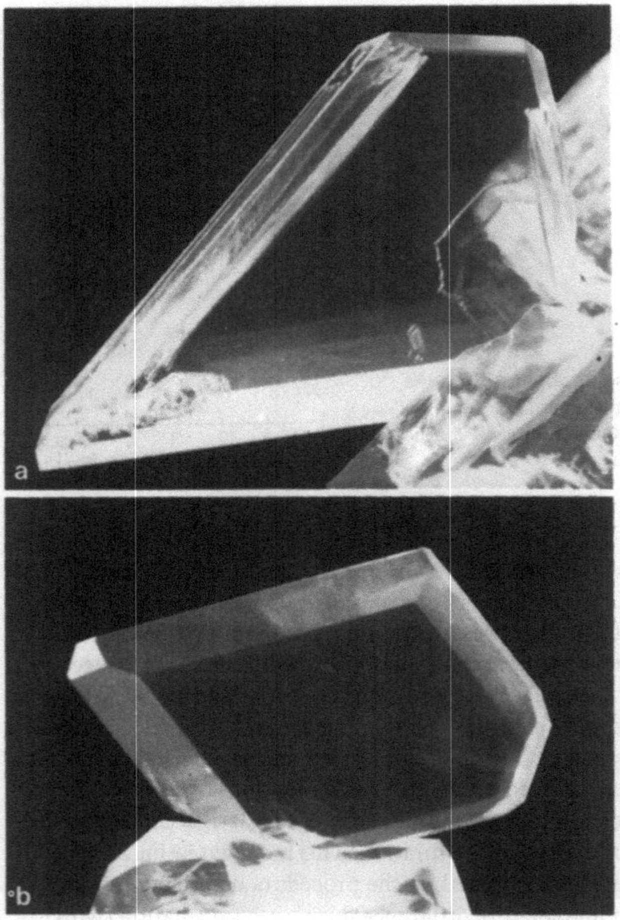

Fig. 38a, b. Examples of crystals obtained by plate sublimation: naphthalene (a) and pyrene (b). Horizontal dimensions ~ 10 mm; thickness ~ 2 mm[187])

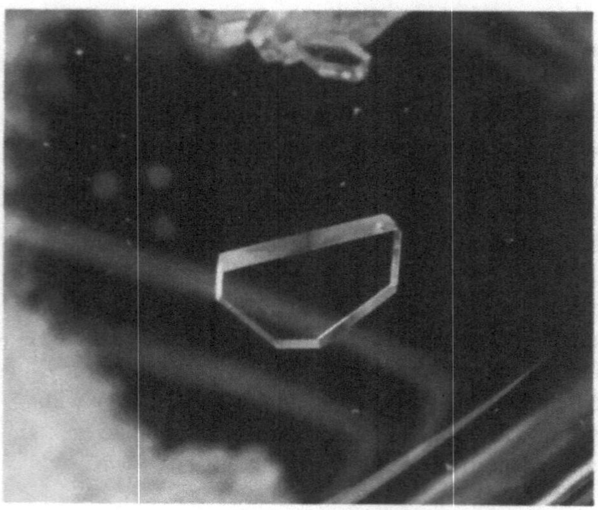

Fig. 39. A highly perfect sublimed phenothiazine crystal, grown ~ perpendicularly to the glass wall. The large face is parallel to the cleavage plane; ($6 \times 4 \times 1.3$ mm^3)[161])

Table 3. List of compounds and stoichiometric donor-acceptor complexes obtained as single crystals by "plate sublimation" in vacuo at the specified growth temperature T[161]

	T
naphthalene	25
1,4-dibromonaphthalene	75
tetrathiafulvalene (TTF)	80
anthracene and anthracene-h_{10}: -d_{10} mixed crystals	80
pyrene-h_{10} and -d_{10}	80
p-terphenyl (10^{-3} tetracene)	100
triphenylene	120
perylene	120
9,10-dichloroanthracene	130
pyromellitic dianhydride (PMDA)	140
tetracyanobenzene (TCNB)	140
picene (10^{-3} pentacene)	190
acridine: PMDA	108
carbazole: PMDA	115
anthracene: PMDA	135
anthracene-9,10-d_2: PMDA	135
anthracene: PMDA-d_2	135
phenazine: PMDA	150
phenothiazine: PMDA	135
pyrene: PMDA	125
naphthalene-d_8: TCNB	150
anthracene: TCNB	140

Under inert ambient gas (400 torr N_2) and with increased temperature gradient, crystals of a different habit have been obtained for naphthalene, anthracene and pyrene; these crystals are thin (0.05 . . . 0.5 mm) ab-platelets with an area up to several cm^2, cf. Fig. 40, and/or needles, elongated in the crystallographic c-direction. Growth in a moving temperature gradient has been studied by Mark[188]. – Extremely thin (0.1 . . . 10 μm) extended (0.5 . . . 25 mm^2) anthracene (001) sublimation crystals of high crystallographic and optical quality are obtained by another extended heat-sink method, cf.[156] and Refs. given therein. Under 200–230 torr (at 20 °C) argon, anthracene is heated from 160 to 185 °C. A few cm away a metal sheet is exposed to the highly supersaturated vapour. Crystals grow in a few minutes. A slightly modified version of the apparatus is shown in Fig. 41[189].

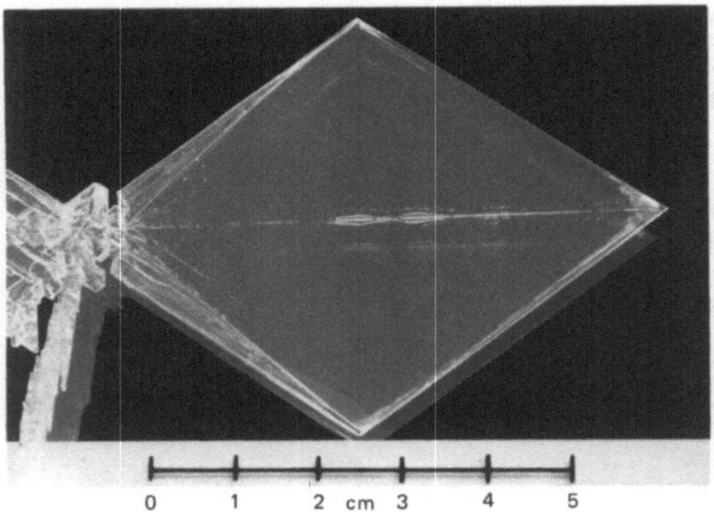

Fig. 40. Large anthracene crystal obtained by the plate sublimation technique under ambient inert gas. Thickness is 0.4 mm[185)]

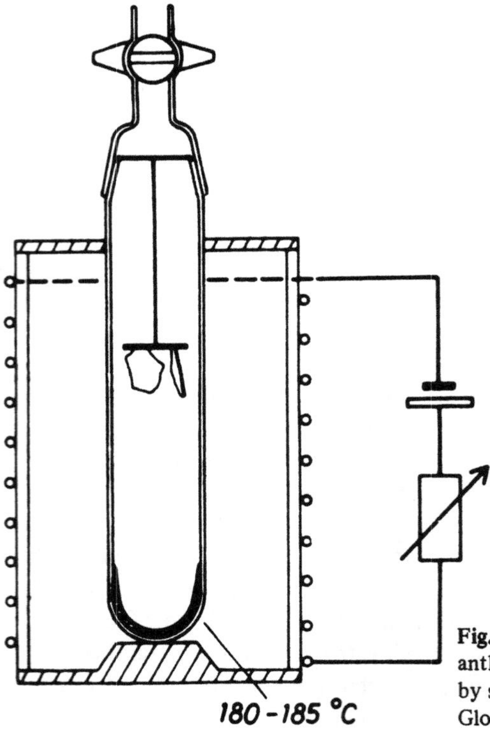

Fig. 41. Apparatus for the growth of large thin anthracene and 2,3-dimethylanthracene crystals by sublimation under inert atmosphere, from Glockner[189)]

3.2.4 Epitaxial Sublimation Growth

A few promising results with epitaxial growth of organic crystals by sublimation shall briefly be mentioned here. This field is just beginning to develop. Depending on the substrate, i. e. the material onto which a sublimation crystal shall grow, homo- and heteroepitaxy should be distinguished. The prime condition is some kind of close fitting of certain lattice parameters of substrate and epilayer. In heteroepitaxy the substrate is a different compound. The borderline to homoepitaxy is a substrate which is only isotopically different. Perdeuterated anthracene ($A-d_{10}$) grown on normal anthracene ($A-h_{10}$) may be given as an example.

Growth of a purer crystal onto a less pure substrate or growth of a doped layer onto an undoped substrate would be examples of homoepitaxy.

Plate sublimation (3.2.3b) can easily be modified for epitaxial growth in such a way that the growth parameters are not grossly distorted during addition of the second component[187]. A flexible metal bellows has been added for that purpose, allowing one to crack open a tubular stock ampoule containing the epi-material or the dopant to be added. Material addition is thus done under crystal growth conditions, maintaining vacuum, high purity conditions, and the existing temperature distribution. The ampoule used is shown in Fig. 42.

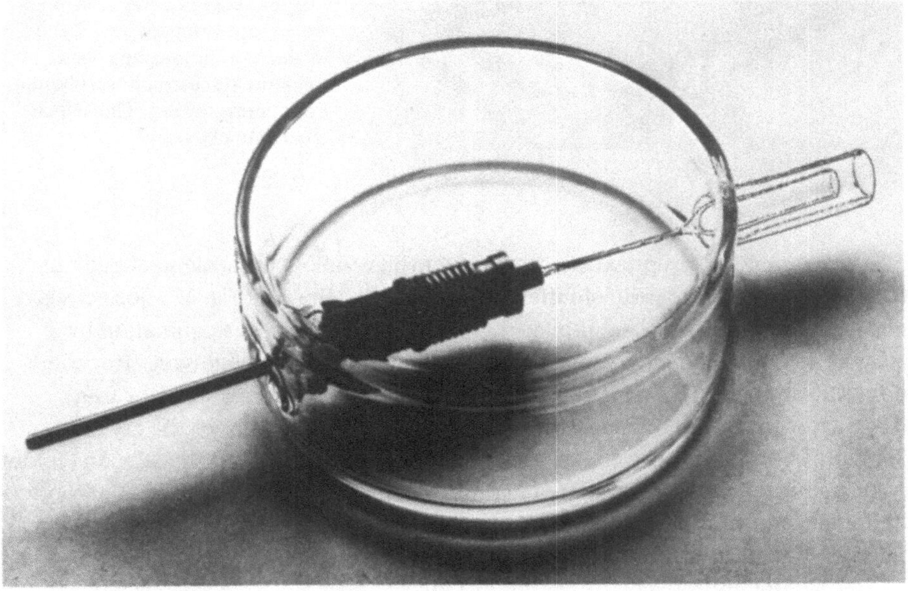

Fig. 42. Ampoule used for plate sublimation with modifications for adding a second component for epitaxial growth[187]. Ampoule diameter is 10 cm. After loading, connecting to a vacuum pump and evacuating, the main ampoule and the small tubular plug-in ampoule are sealed in common

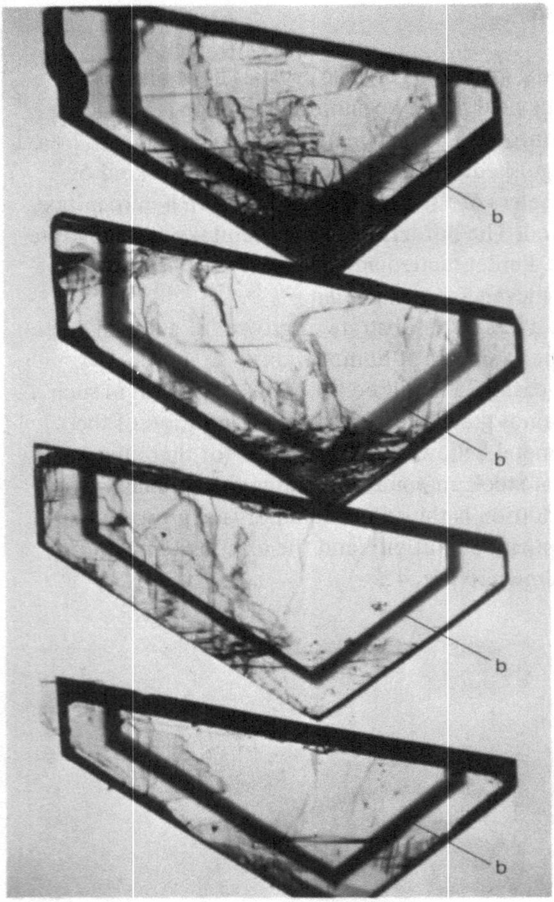

Fig. 43. Four cleaved slices of an anthracene crystal obtained by plate sublimation (in vacuum) with an epitaxial layer of benzazulene-doped anthracene, overgrown by undoped anthracene[190]. The long edge is 5 mm. The blue-violet doped ribbon, *b*, is slighly inclined to the cleavage plane; therefore its sharp internal border line appears diffuse. The defects are due to cleavage

An example of an epitaxially grown layer (blue-violet benzazulene-doped[j] anthracene between undoped colourless anthracene) is shown in Fig. 43. Four cleaved slices of the same crystal are displayed[190], revealing (on closer examination by a polarization microscope) I) undisturbed continuation of crystal growth after addition of the dopant, II) oriented incorporation of the dopant (strong dichroism), III) microscopic sharp boundaries between pure and doped parts of the crystals. Similarly epitaxial layers were grown of (blue) azulene-doped naphthalene on (colourless) naphthalene, of (green fluorescing) tetracene-doped anthracene on (blue-violet fluorescing) anthracene and of deuterated anthracene on undeuterated anthracene[187, 190]. The substrate can either be grown in situ before addition of the dopant or epi-material respectively, or placed into the chamber from the very beginning.

[j] Benz [f] azulene was obtained through the courtsy of Prof. Chr. Jutz, Technical University, Munich

3.3 Other Crystal Growth Techniques

3.3.1 Nacken-Kyropoulos and Czochralski Growth

Crystal pulling techniques like Nacken-Kyropoulos and Czochralski growth may be considered for those organic compounds which have comparatively low melting points with respect to their molecular weight. Sublimation problems are reduced with these compounds since the sublimation vapour pressure decreases exponentially with decreasing temperature. Thus e. g. acridine ($C_{13}H_9N$, mol. weight 179, m. p. 111 °C) would be a much better candidate than anthracene ($C_{14}H_{10}$, mol. wt. 178, m. p. 218 °C). However, comparatively small lattice energies of most organic crystals lead to high vapor pressure at the melting point; there is often a strong tendency towards gliding which is another detrimental feature that argues against pulling techniques. It is probably because of these aspects that pulling techniques have not been applied for a long time.

However it has been demonstrated quite recently that Nacken-Kyropoulos growth[48)] and Czochralski growth[49)] can be performed quite successfully for compounds like benzophenone and benzil. These compounds are compared to anthracene and acridine in Table 4.

The simplest way to obtain large (100 g) benzophenone and benzil crystals is to supercool their melt by 0.2–2 deg., to immerse a seed crystal while maintaining the supercooling of the melt and to withdraw the growing crystal from the residual melt before it reaches the wall of the container; a suitable growth speed is $f \leqslant 2$ mm h^{-1} [192)].

A Czochralski apparatus suitable for the growth of organic material has been described by Bleay et al.[49)], see Fig. 44. The material was melted in a Pyrex glass crucible (A) surrounded by a thick-walled aluminium block (B) in order to obtain an even temperature distribution in combination with high thermal inertia. A Pyrex pipe section (C), bolted and sealed to the furnace, provided a transparent enclosure for the growing crystal and permitted the control of the surrounding atmosphere via the ports (D). Undesired condensation was prevented by independently powered heaters wound around (C) and (E). The concentric water cooled crystal holder (F) could be slowly withdrawn by means of a motor-driven lead screw. A typical useful

Table 4. Melting points and vapor pressure data for compounds of similar molecular weights[a]

	formula	Mol. wt. [g/mol]	p at 200 °C [torr]	m. p. [° C]	p at m. p. [torr]
anthracene	$C_{14}H_{10}$	178	~20	218	~40
acridine	$C_{13}H_9N$	179	~20	111	~ 0.4
benzil	$C_{14}H_{10}O_2$	210	~20	96	~ 0.2
benzophenone	$C_{13}H_{10}O$	182	~20	48	~ 0.03

[a] Because of the great scatter of published experimental vapor pressure data, cf. anthracene[92)], these numbers cannot be assumed to have a greater precision than (roughly) –50/+100%. The melting point vapor pressures for acridine, benzil and benzophenone are extrapolated from the data collected in[191)]

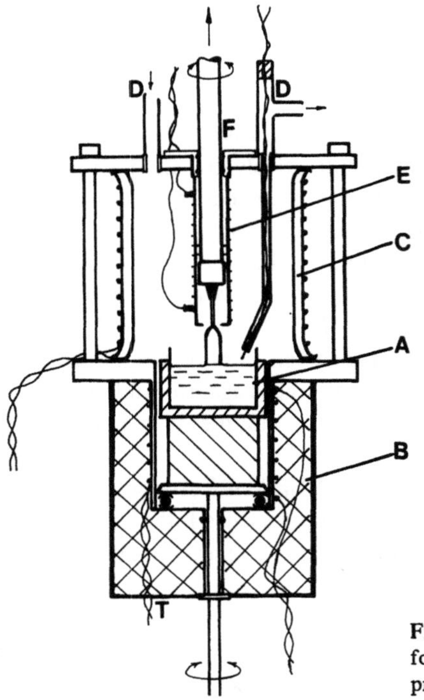

Fig. 44. Apparatus for Czochralski growth, useful for selected organic compounds with low vapour pressure at the melting point. From Bleay et al.[49]

speed was between 0.1 and 0.5 mm h^{-1}. Growth was initiated by dipping a Bridgman- or solution-grown seed crystal, attached to the crystal holder, into the melt, and lowering the temperature until the crystal just began to grow. The apparatus was then left to equilibrate for several hours before pulling was commenced. First results[49] obtained with benzophenone and benzil were encouraging. Crystals of dimensions up to 10 cm^3 were pulled in periods of 1–2 days. Analyses by X-ray topography revealed substantially fewer dislocations than generally contained in Bridgman crystals.

3.3.2 Solution Growth

Difficulties with the growth of high purity material from solution were already mentioned: Solvents must be purified to high standards as well or their residual impurities may be found in the crystal. Depending on the distribution coefficient between solution and crystal, impurity depletion or enrichment may occur. In addition solvent incorporation on a molecular and on a macroscopic scale (inclusions) may happen. Nevertheless, there exist situations where solution growth must be considered even for high purity material. A few examples will be given.

A great variety of different concepts for solution growth can be found in the literature, cf.[11]. These can be classified as isothermal evaporization methods, as techniques, where supersaturation is obtained by cooling, and as reactive condensation methods, driven by concentration gradients via diffusion, or by electrochemical potential gradients via ionic currents.

Isothermal evaporation of the solvent can best be done by placing beakers, containing the solution, in an (empty) desiccator and passing purified nitrogen through it at a rate of several cm^3 per min. Solvents can be purified by low temperature zone refining. Thorough degassing from oxygen etc. may be an important point. The donor : acceptor complexes naphthalene : tetracyanobenzene and anthracene : tetracyanobenzene can be obtained as large single crystals by this method, using acetone as solvent. α-9,10-dichloroanthracene crystals were obtained in the same way from benzene[193] see Fig. 45. (Tetracyanobenzene cannot be melted without decomposition; 9,10-dichloroanthracene is usually obtained in the β-modification from the melt; during cooling it mostly shatters upon transformation into the α-phase). – Slow stirring of the solvent and use of seed crystals is not necessary for the examples given, but it may be helpful in other cases in order to increase quality and size of the crystals and to avoid spurious nucleation.

Cooling methods have been used to grow thin extended crystals of anthracene, naphthalene, biphenyl and other compounds. In the version described by Kallmann and Pope[194], a temperature and density gradient is created within the solution so that the crystal will grow slowly and will remain floating during its growth. For that purpose 4 g of anthracene are dissolved in 80 g of ethylene dichloride in a 100 ml beaker by boiling the solvent (~ 1 min). After 3 min of partial cooling a 20 mm layer of (specifically lighter) xylene is carefully placed on top of the anthracene solution so that mixing is minimized. Further cooling leads to crystallization. Xylene, which is not a good solvent for anthracene, has a boiling point high enough to pro-

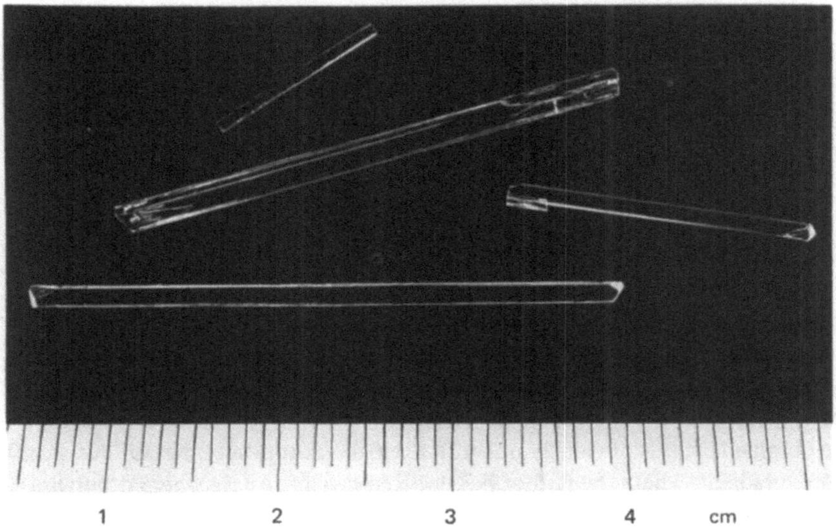

Fig. 45. Solution-grown crystals of α-9,10-dichloroanthracene. Thickness is a few tenths of a mm. Crystals were grown by M. Herb[193]

tect the hot ethylene dichloride solution from evaporation and hence eliminates surface nucleation. Further it serves as a washing solution during removal of the grown crystals. Anthracene crystals so produced range in thickness from 4 μm (thinner crystals cannot be removed safely from the solvents) to 80 μm and from 0.5 to 2 cm^2 in area. A slight modification of the method described may be found elsewhere[195].

A diffusion and reaction technique[196] has been used for the growth of the organic donor-acceptor radical-ion salt TTF : TCNQ from solutions of the neutral molecules. This complex belongs to a small family of organic crystals exhibiting exceptional metal-like conductivities. The problem of growing larger[60, 61] and purer[199] crystals has been a point of considerable attention since its discovery in 1973[200, cf. 201]; TCNQ and its complex with TTF can not be melted without decomposition.

The diffusion and reaction technique is generally useful for the growth of fairly insoluble complexes (or compounds) from their constituents. For that purpose a solvent-filled reaction chamber is separated by frits from two nutrient chambers, containing solutions of the donor and acceptor, respectively. Slow diffusion of the two components into the growth chamber leads to single crystalline precipitate. The only condition is that the complex be considerably less soluble than the individual components.

By a combination of gradient sublimation, use of a glove box, filled with 5N (99.999%) Ar, and multply distilled acetonitrile as solvent, an impurity level on the scale of 10 ppm was assumed to have been reached[199] for the components TTF and TCNQ. However, from low temperature magnetic susceptibility measurements 100 ppm spin 1/2 impurities were derived for the complex. The nature of these impurities remained unclear.

3.3.3 Multiple Crystals from the Melt

Melt growth between NaCl windows of a set of thin single crystal films with different orientations, emerging from one single common seed has been demonstrated by Kruse[202]. This technique deserves special mention because it may be of great practical help, even if maintaining high purity may be a problem with this concept.

3.4 Comparison of Growth Methods

Under the aspect of maintaining highest purity and avoiding thermal decomposition a low temperature growth technique which does not need an auxiliary phase would generally be a good choice: sublimation growth. However experience with sublimation growth is scarce in the literature. Remarks may be found that usually only thin flakes are obtained. This is obviously not true for high vacuum sublimation in a small temperature gradient, where the author has not found a single case where otherwise rather anisotropically growing material did not form bulky crystals (examples are 1,4-dibromonaphthalene, 9,10-dichloroanthracene and phenazine for crystals with tendency towards needle-like growth and naphthalene, anthracene, pyrene, and p-terphenyl for crystals with tendency towards growth in thin platelets).

It has been reported elsewhere[181, 203] that triplet exciton lifetime, (a rather general criterion for purity and crystal perfection, see paragraph 4.1.6) was generally lower for sublimed crystals than for Bridgman crystals. This was not confirmed; vacuum-sublimed crystals (grown by plate sublimation) of anthracene reached more than 20 ms, mixed crystals with deuterated anthracene up to 36 ms, pyrene 34 ms, (cf. table 5, p. 79), i. e. values similar to those reported for Bridgman crystals.

Sublimed crystals often exhibit a much higher crystallographic and optical perfection than Bridgman crystals (cf. Fig. 39 p. 62). Low temperature spectroscopy gave narrower lines for sublimed crystals than for Bridgman crystals[204]. Dislocation density was found to be lower as well[205, 206].

However, it is much easier to grow large Bridgman crystals (usually 5–20 cm^3). Sublimed crystals of a size greater than 0.1 cm^3 need very well controlled growth conditions and a growth period of more than one week.

Stress and strain problems arising from different thermal expansion/contraction of an ampoule or support with respect to the crystal are reduced with growth techniques from which result more or less freely grown crystals as compared to those techniques where the shape of the ampoule is necessarily adopted. Slow, near room temperature solution growth of freely suspended seed crystals seems to be optimal under this aspect. Pulling techniques and sublimation methods yielding crystals surrounded by natural faces may be less satisfactory, but they should usually be superior to Bridgman growth.

Oriented seeding is most easily done with the pulling techniques, whereas it causes difficulties with the Bridgman technique. Repeating unsuccessful trials is very conveniently done with sublimation. With Bridgman crystals repeated remelting may cause decomposition problems.

Doping profiles are most even under sublimation; addition of dopants during growth does not cause problems; however the doping range is somewhat restricted by vapour pressure limitations.

For annealing of growth defects at temperatures very close to the melting point the Bridgman set-up provides advantages.

Chemical impurity levels tend to be highest in solution-grown crystals.

4 Characterization

4.1 Chemical Defects, General

Those defects which are produced by chemically different molecules shall be classified as "chemical defects"; if these molecules have not been added on purpose one commonly speaks of "impurities", otherwise they are called dopants. Detection of impurities and dopants is often possible by the same methods. However under the aspect of the tremendous variety of organic molecules under question one is often uncertain how universal a given method is. Therefore it may be very difficult to make a statement on the overall purity of a sample considered. It is not surprising therefore that much is known about those impurities which betray their presence by distinct distortions while others may remain undetected.

The action of foreign molecules may be either a direct or an indirect one in that they are active themselves, or cause the surrounding "host" molecules to alter their properties (e. g. to lower their energy below a crystals energy band, thus producing so called X-traps).

Chemical impurities may also be formed in situ by photochemical reactions such as dimerization, a case that has been extensively investigated for anthracene, cf.[197,198].

4.1.1 Analysis of Chemical Purity by Elemental Analysis

Elemental analysis might be considered at a first glance for the detection of trace amounts of impurities. However it is generally not applicable if no hetero-atoms are present. The C/H ratio of hydrocarbons is barely sensitive to small additives of foreign hydrocarbons, and totally insensitive to isomers. However, elemental trace analysis may be used for the determination of nitrogen, phosphorous, sulfur and halogens, cf.[19]. Flame photometry and atomic absorption are suitable methods for sensitive detection of many metals[207, 208]. There exist dye reactions which are specific for certain functional groups. This is a broad field which has barely been investigated by the crystal grower to date.

4.1.2 Gas Chromatography

Gas chromatography is a very versatile and extremely sensitive analytical method (cf.[18]), which is based on a sequence of adsorption and desorption processes on a suitable adsorbent evenly distributed on the internal wall of (or at a powder filled into) a thin metal or glass tube (so called "column"). Depending on adsorption free energies for different components of a small vaporized volume of the sample to be analysed, fractionation occurs more or less quickly during carriage to the other end by a stream of inert gas. A small hydrogen flame at the column outlet will burn hotter whenever organic material arrives. An increased ion-current through the flame is indicative of rates as small as 10^{-10} g s^{-1} and less ("flame ionization detector").

Gas chromatography is a retention time analysis. Under given apparatus parameters, (column length, filling, temperature, gas flow etc.) every chemically uniform stable compound has its characteristic retention time. Gas chromatographic peaks are a quantitative measure of the amount of material arriving at the detector. Dynamic range can be as high as $1:10^6$. The ion yield must be determined individually for each substance. Detector nonlinearity and partial sample decomposition must be accounted for by calibration[118]. Except for retention times, gas chromatographic peak detection is essentially compound-unspecific. Therefore it is comparatively easy to determine whether there are impurities present and in which concentration range. But their chemical composition remains unknown. (Detectors of slightly increased selectivity are available e. g. for the detection of compounds containing halogens, phosphorous or nitrogen). The need for a compound-specific detector has lead to coupling of fast mass spectrometers to gas chromatographs ("GC-MS" coupling). This technique is used in some places for trace analysis in organic crystals.

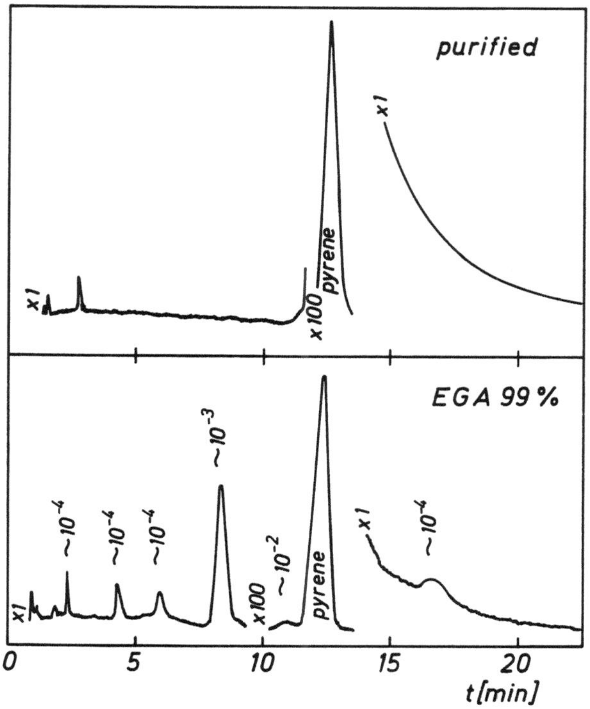

Fig. 46. Comparison of commercial and zone-refined pyrene by gas chromatographic analysis. The pyrene peak is plotted with x 100 scale reduction[134])

Gas chromatography can detect impurities down to $\sim 10^{-6}$ mol/mol. Fast mass spectrometry can yield a complete mass spectrum, i. e. the masses of the parent ion and of specific fractionation products of the special chemical compound contained in each gaschromatographic peak. However, MS detection limit is around 10^{-5} mol/mol. Therefore concentration of impurities may be necessary for the determination of the chemical nature of trace impurities.

An example for gas chromatographic analysis is given in Fig. 46, where commercial pyrene, specified 99% (obtained from EGA, Steinheim, Germany) is compared to zone-refined pyrene. In the purified material there is a first peak marking the pressure change at injection and a second one due to air. – A weak shoulder in front of the main peak is indicative of the impurity, mass 208, mentioned on page 40, which is not efficiently removed during zone refining and which requires potassium fusion and liquid chromatography[128]). Other typical examples of gas chromatograms of impure starting material may be found in the literature, e. g.[8, 118, 119, 125, 152]).

Figure 47 displays a plot of the distribution of the impurity β-methyl-naphthalene in naphthalene, analyzed by gas chromatography after 50 zone passes, as indicated.

Figure 48 presents gas chromatograms for three samples taken from a fluorene zone refining tube (after 50 zone passages) at the zone entrance, at the center of the

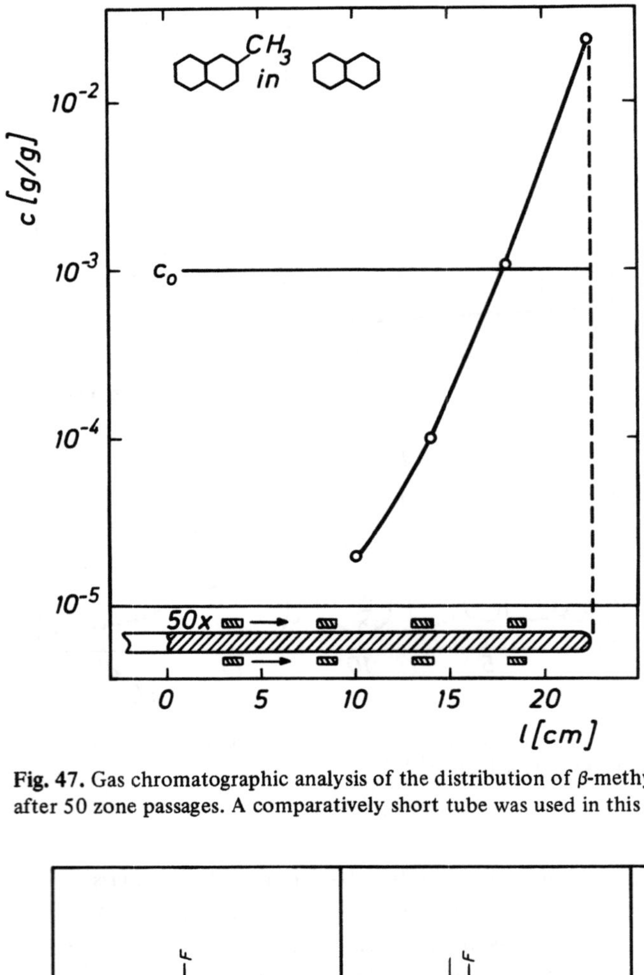

Fig. 47. Gas chromatographic analysis of the distribution of β-methylnaphthalene in naphthalene after 50 zone passages. A comparatively short tube was used in this experiment[128])

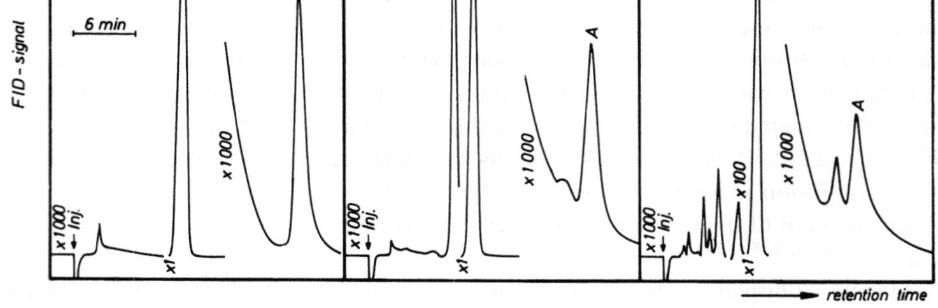

Fig. 48. Gas chromatographic analysis of impurity distributions in a zone refining tube containing "natural" fluorene, after 50 zone passages; from[107])

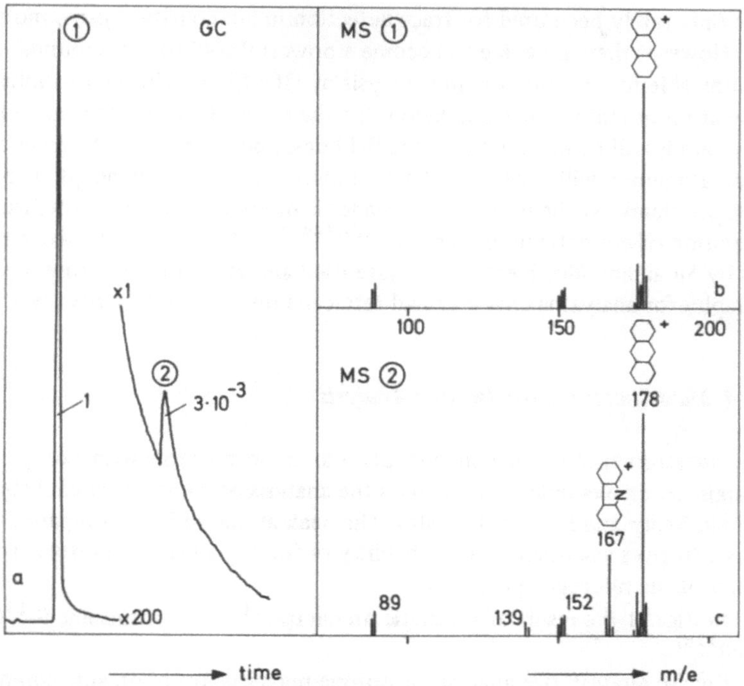

Fig. 49 a–c. (c) Mass spectrum of the GC impurity peak 2 (carbazole) compared to that of the host (anthracene) peak 1 (**b**) of the gas chromatogramm (**a**) of selected commercial anthracene[134]. Notice that the GC separation of anthracene/carbozole is incomplete

ingot and at the end. Notice that all impurities have a distribution coefficient $k < 1$ and hence travel with the molten zones, except for anthracene ($k > 1$). This system has been discussed in more detail in point 2.1.4e, p. 38.

In order to reach highest GC separation power, solid sample injection should be preferred to solution injection. This is especially important for substances which are poorly soluble and in which case the solvent would excessively load the system. Solid sample injection has been described, among other things, in[118].

Use of a mass spectrometer as a selective detector for gaschromatographic peak identification is demonstrated in Fig. 49. A gas chromatogram of selected commercial anthracene (Schuchardt, München, "for scintillation") is shown in Fig. 49a. Mass spectra obtained for the GC peaks labeled 1 and 2 are plotted schematically in Fig. 49b and in Fig. 49c. It is found that the impurity has the mass and fractionation pattern of carbazole. High resolution mass spectrometry has been used to determine the empirical sum formula of a number of tetracene impurities[125].

4.1.3 Analytical Liquid Chromatography

Two versions of analytical liquid chromatography should be mentioned here: Thin film liquid chromatography and high pressure liquid chromatography. Both methods have only rarely been used for trace detection in high purity organic molecular crystals. However they are going to become a powerful tool for compounds which are too unstable for evaporation and analysis by GC: Liquid chromatography can be done at room temperature and below. To the extent to which thermally less stable compounds will be included into the field described in this article, analytical liquid chromatography will necessarily be used more frequently and adapted to the special trace problems. At the moment the reader is referred to the basic literature on the separation of aromatic hydrocarbons[140–144; 209–213] and to the work by McGhie[147] and by Sloan and McGhie[125] who have used analytical high pressure liquid chromatography for analyzing chrysene and tetracene on a low detectivity level.

4.1.4 Mass Spectrometric Isotope Analysis

Mass spectrometric isotope analysis plays an important role with isotopically mixed crystals. As an example, Fig. 50 gives the analysis of commercial naphthalene-d_8 (Merck, Sharp and Dohme, Canada). The peak at mass 135 as compared to that at mass 136 gives essentially the probability of finding an H atom instead of a D atom at any of the hydrogen positions.

In Fig. 51 the result of synthetic preparation[120] of anthracene-9-^{13}C is analyzed[128].

Precise quantitative analysis of isotope mass spectra of organic compounds needs thorough consideration of all contributions to parent-ion peaks by fractionation of isotopically different parent-ions (e. g. naphthalene-d_8-^{13}C minus d leads also to mass 135 in the example Fig. 50). The book by Budzikiewicz et al.[214] may be consulted for basic reading.

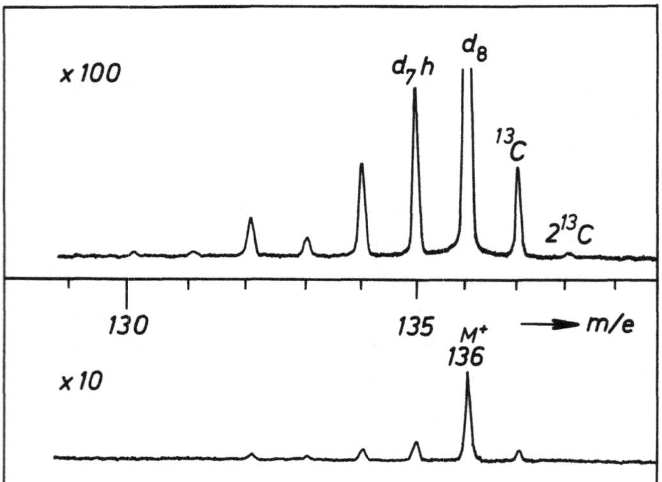

Fig. 50. Mass spectrometric analysis of isotope content of commercial perdeuterated naphthalene[134]

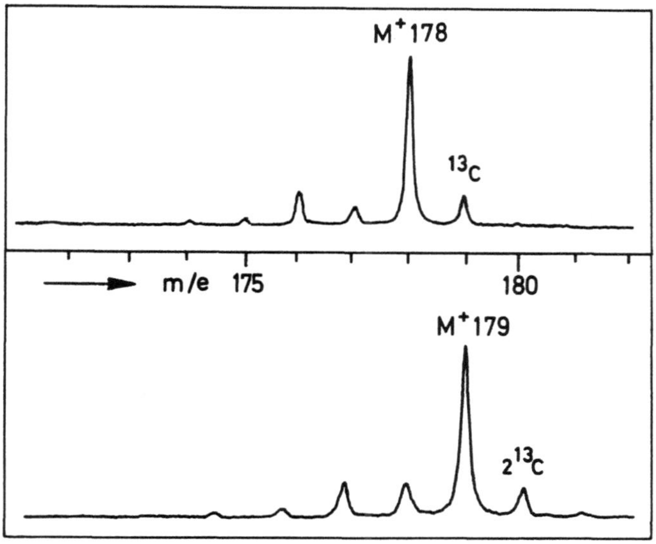

Fig. 51. Mass spectrometric analysis of isotope content of synthetized anthracene-9-^{13}C [128], *lower Fig.*, compared to normal anthracene, *upper Fig.*

4.1.5 Trace Detection by Optical Spectroscopy [k]

The sensitivity of optical spectroscopic methods generally increases from electronic absorption spectroscopy ($\sim 10^{-5}$ mol/mol, [41,45]) via electronic singlet and triplet[221] excitation spectroscopy, singlet (fluorescence) (10^{-7})[41, 42, 45, 222] and triplet (phosphorescence) emission spectroscopy[223] to delayed fluorescence spectroscopy (10^{-9} mol/mol)[47, 224, 225].

For many organic crystals lines narrow with decreasing temperature, allowing high spectral resolution at low temperature (e. g. at 4.2 K). Therefore analysis of low-temperature spectra can lead to detailed information on chemical defects, cf.[225].

The high sensitivity of emission spectroscopy has its origin in "energy migration". If excitonic crystal states are initially excited then these states can diffuse around in the crystal lattice, scanning appreciable areas during their lifetimes. (Typical mean free paths for singlet excitons in well purified crystals are ~ 0.1 μm; those for triplet excitons range around 1 μm). Whenever traps are encountered during diffusion of excitons these will be occupied by "energy transfer" with comparatively large interaction cross section, leading to so called "sensitized emission"[41, 42, 224]. Sensitized emission exhibits the characteristic spectrum of the trap molecule. Most efficient scanning is done by long lived triplet excitons occuring in high purity crystals.

---

[k] For an introduction to basic aspects of spectroscopy of organic crystals the reader is referred to Refs. 215–220

This fact is used in "delayed fluorescence"[47, 224] which is the most sensitive optical method for trace detection. Delayed fluorescence is based on the fact that two triplet excited states, one excitonic in nature and free to move in the lattice, and one trap-localized, can fuse very efficiently upon encounter, leading to an excited singlet state. This excited singlet state is often (but not necessarily) also localized at the impurity molecule. Hence it can radiate fluorescence emission, carrying information on the chemical idendity of the trap. The increased sensitivity of delayed fluorescence as compared to phosphorescence is due to the fact that phosphorescence quantum yield is generally very low whereas fluorescence quantum yields often approach unity.

Fluorescence emission of naphthalene at 4.2 K has been used[134] as a routine test for checking the purity of naphthalene after potassium fusion and zone refining. An example is given in Fig. 52. In the purest naphthalene samples no 2-methyl-naphthalene line can be detected any more (detection limit is $\sim 10^{-8}$ mol/mol, cf.[226]).

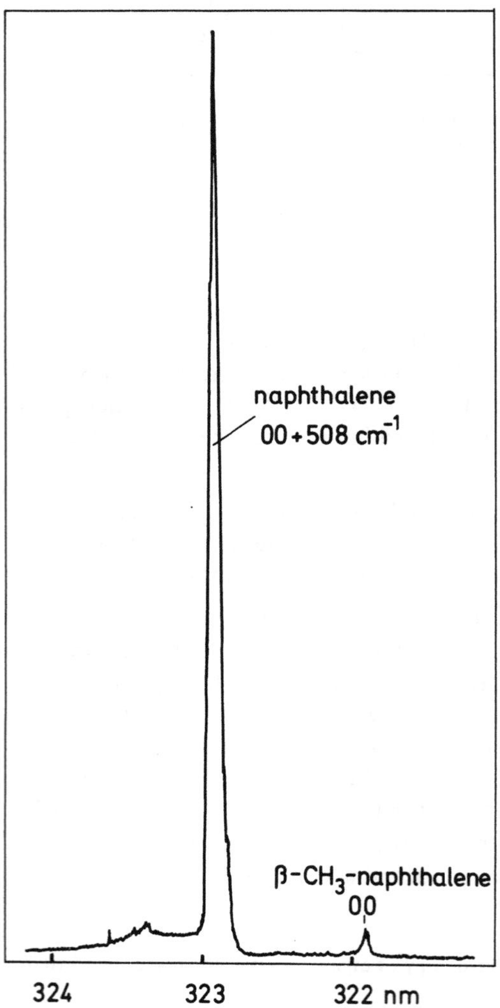

Fig. 52. Fluorescence spectrum of naphthalene at 4.2 K. From the ratio of the intensity of naphthalene vibronic line (labeled 00+508) cm^{-1} (*left*) to that of the pure electronic line (00) of 2-methyl-naphthalene (*right*) a concentration of the latter impurity molecule of $\sim 3 \cdot 10^{-7}$ mol/mol can be derived

4.1.6 Lifetime of Triplet Excitons and Free Charge Carriers

All foreign ("guest") molecules (and all disturbed host molecules) which form energy levels below the exciton or charge carrier transport bands lead to trapping of the excitation energy and hence to localization of the excited quasiparticle. A more or less monoexponential decay is often observed, allowing one to specify a "trapping lifetime".

Triplet exciton lifetimes $\tau_{Tr} = \frac{1}{\beta}$ can be measured by delayed fluorescence under *low* triplet excitation density, where, after a pulsed excitation, the decay rate

$$-\frac{dT_1}{dt} = \beta T_1 + \gamma T_1^2 \tag{19}$$

is dominated by the linear term, whereas the concentration is monitored by the second term leading to delayed fluorescence via

$$T_1 + T_1 \rightarrow S_1^* \tag{20}$$

The lifetime τ is a sensitive function of crystal purity. Triplet exciton lifetimes for selected purified compounds are collected in Table 5.

An easy way of measuring lifetimes[161)] ($\tau > 1$ ms) is to populate the triplet states under question by a photoflash equipped with appropriate filters to cut off short wavelength light of a photon energy which would be sufficient to directly popu-

Table 5. Lifetime of triplet excitons, measured by delayed fluorescence (s. paragraph 4.1.5) for selected well purified compounds[a] [134)]. z: measured in the zone refining tube before opening; b: Bridgman crystal; p: crystals obtained by plate sublimation; n: naphthalene; a: anthracene and py: pyrene filter combinations, see text

Material	Form	τ_{Tr} [ms]	Filter combination
naphthalene	z	420	n
fluorene, synth.	z	120	n
2,3-dimethylnaphthalene	z	40	n
anthracene-h_{10}	b	25	a
anthracene-d_{10}	b	45	a
anthracene-d_{10} + 0.5% h_{10}	p	36	a
anthracene-d_{10} + 1.5% h_{10}	p	32	a
anthracene-d_{10} + 6% h_{10}	p	30[b]	a
pyrene-h_{10}	b, p	60	py
pyrene-d_{10}	p	22	py
p-terphenyl	z	90[c]	n
triphenylbenzene	z	6	n

[a] Triphenylbenzene was not purified very extensively
[b] cf. Fig. 53
[c] after pretreatment with maleic-anhydride 200 ms have been reached[314)]

late the singlet levels. Weak delayed fluorescence is detected by a photomultiplier through appropriate filters which must be highly blocking for the exciting light. Suitable filter combinations in connection with a S5-photocathode are: n) (for naphthalene etc.): excitation through Schott filters GG 420/3 mm + KG 1/2 mm + BG 18/2 mm ($\lambda > 420$ nm), detection: UG 11/4 mm ($\lambda < 370$ nm). a) (for anthracene etc.): excitation through RG 610/2 mm + KG 1/2 mm ($\lambda > 610$ nm); detection: BG 28/2 mm + BG 3/2 mm ($\lambda < 450$ nm). p) (for pyrene), same as a), except that RG 610 is replaced by OG 590/3 mm.

Instead of calculating lifetimes from logarithmic plots of signal height minus baseline as a function of time (t) it has turned out to be more practical for routine lab-

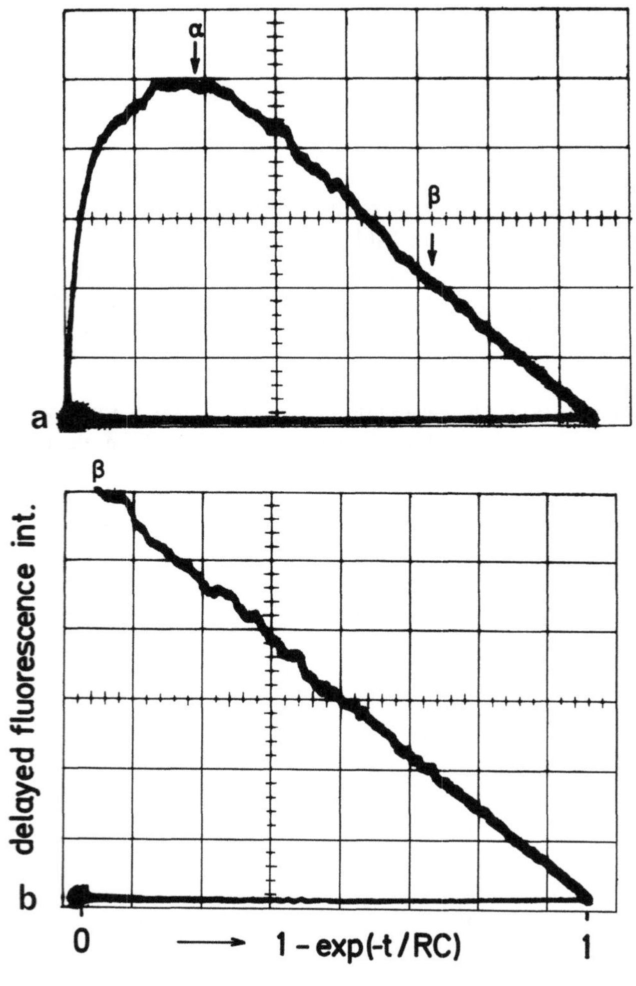

Fig. 53. **a** The complete delayed fluorescence function, obtained as indicated in the text. At time α the exciting flash has completely decayed. Sufficiently low triplet density for the bimolecular triplet decay channel to be negligable is reached at time β, where the finally considered decay curve **b**) is triggered,[161]. τ_{Tr} = 30 ms (anthracene-d_{10} + 6% anthracene-h_{10})

oratory control to use a coordinate transformation in the time axis. We therefore display the decay directly on an X-Y-oscilloscope against the function $\Theta = 1 - e^{-t/RC}$ on the abscissa[227]. The function Θ is generated (in principle) by charging a capacitor from the gate pulse of the otherwise unused time base of the oscilloscope. By button selection of an appropriate RC circuit (of $\tau_{RC} = RC$, with τ_{RC} being digitally displayed[228]) conditions are quickly found, where the apparent decay is linear, indicating $\tau_{RC} = (2\,\beta)^{-1} = \tau_{Tr}/2$. (The factor 2 results from monitoring via the quadratic channel). An example is given in Fig. 53. (Notice that the point labeled 1 on the abscissa corresponds to time $t = \infty$).

The distribution of the quantity "triplet lifetime" along pyrene zone refining tubes purified in two successive steps as indicated in the figure is plotted in Fig. 54. Note that there is present a persistent impurity with $k > 1$ which is only very slowly removed. Lifetimes for an anthracene tube are given in the caption, Fig. 21a.

Lifetime of free charge carriers can be measured by observing charge carriers moving across a crystal slab from one side to the other after having been excited optically by a short flash or laser pulse, or by a pulsed electron beam in a time short compared with the transit time (so called Kepler technique[43], cf.[31]). The driving electrical field must be kept sufficiently low in order to keep charge carriers drifting for a period of time comparable to the trapping time.

Trapping times for electrons (or by field reversal for holes) range up to about 1 ms in the best anthracene crystals. These times are sensitive to sub-ppm impurity levels.

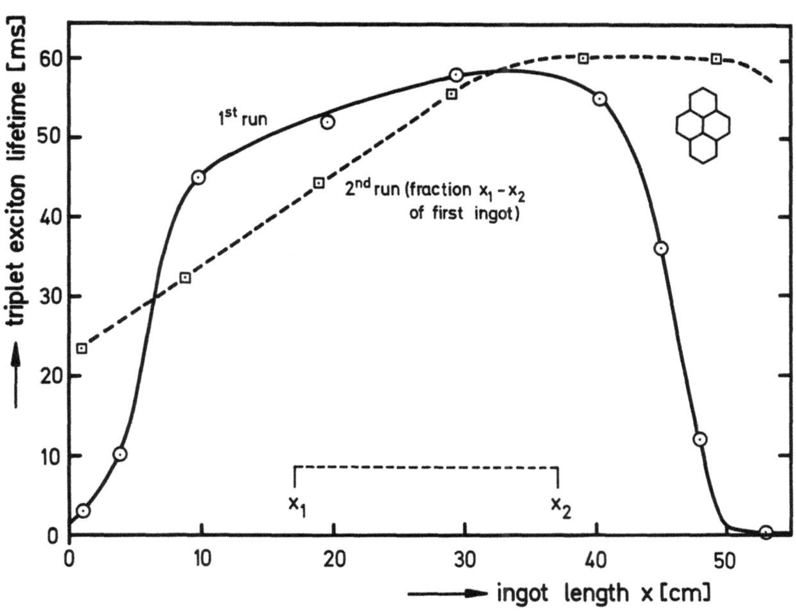

Fig. 54. Distribution of triplet exciton lifetimes in unopened pyrene zone refining tubes. First run: 170 "normal" zone passes. The second tube has been prepared from the fraction $x_1 - x_2$ of the first tube. It has obtained 80 reverse passes. Zone entrance is designated $x = 0$ for both cases[134]

Exciton[229] and charge carrier[205] trapping times have often been used in the literature as an integral characterization of purity and quality. In spite of the simplicity and usefulness of these methods, it should be remembered that impurity states may be present with levels lying above the bands which therefore remain hidden.

4.1.7 Selective Charge Carrier Detrapping

Charge carrier detrapping can be obtained either optically or thermally. Thermally stimulated currents ("TSC") obtained by increasing the temperature linearly after trap filling at low temperatures can detect a number of chemical impurities and/or structural defects even in the purest available organic crystals. Under appropriate conditions, and with a suitable evaluation of TSC curves, individual trap energies in the range 0.2–0.6 eV can be obtained, cf.[31].

For trap depths and temperature intervals where a dynamic equilibrium between trapping and thermal detrapping with sufficiently short relaxation times is established (regime of so called "shallow traps"), charge carrier transport may be sensitively influenced in that reduced "effective" mobilities are obtained. This result is due to a reduction of the average drift velocity by successive intervals of trap localization and drifting:

Electron and hole mobilities in c*-direction for a well purified (top) and a tetracene-doped ($\sim 10^{-7}$ mol/mol) anthracene crystal are compared in Fig. 55 (from Probst and Karl[46]). The activation energies indicated are characteristic for tetracene.

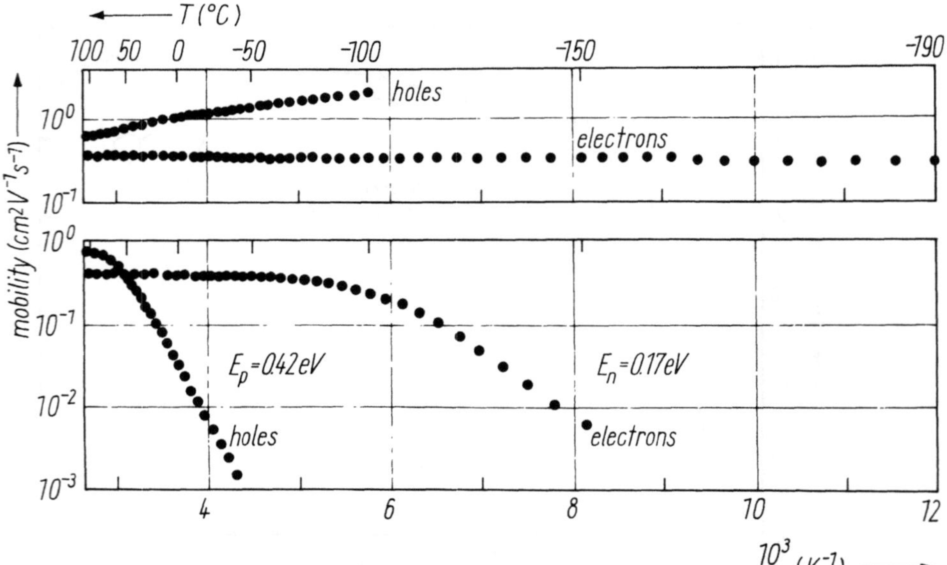

Fig. 55. Charge carrier mobilities in anthracene as a function of temperature. Transport of charge carriers is very sensitive to small amounts of impurities. The upper curves were obtained for well purified anthracene crystals. The lower set of curves was obtained for a crystal made from the same material, but deliberately doped with $\sim 10^{-7}$ mol/mol tetracene. The activation energies are characteristic for tetracene in anthracene. From Probst and Karl[46]

Optical detrapping of charge carriers can take place via intermediate excitation of the trap radical ion doublet states, leading to well structured characteristic photocurrent excitation spectra of the ionic transitions[230–232]. Direct trap to band transitions have also been observed[232]. Estimates give a lower detection limit of charged traps of the order 10^{-12} mol/mol[231].

4.1.8 Orientation of Guest Molecules by ESR and TTR

Electron-spin-resonance (ESR) techniques have been used to determine the fine structure tensor of triplet excited states localized at traps, e. g. at anthracene molecules in naphthalene and tetracene molecules in anthracene crystals[233]. Since the fine structure tensor is tied to the molecular symmetry, the orientation of the above-mentioned symmetric molecules can be established. Calculations by the use of atom-atom potentials[234] are in agreement with the experiments.

Triplet-triplet resonances (TTR) as a function of the magnitude and orientation of an applied magnetic field have been observed in the photocurrent in anthracene: PMDA, similarly allowing the determination of the orientation of guest or distorted host molecules. Microwaves are not needed with this technique[235].

4.1.9 X-Traps

An impurity molecule which does not appear as a trapping center itself, may distort neighbouring host molecules to act as traps. Since the chemical nature of the impurity has often remained unknown, those states were called "X-traps"[47], cf.[225]. In spectroscopy it is characteristic for X-traps that host spectra are reproduced on a slightly shifted energy scale. The microscopic environment of X-traps (in their triplet state) can be explored by ENDOR (electron nuclear double resonance) measurements[258]. X-traps may also be due to structural defects. Structural or "physical" defects will be discussed in the next section.

4.2 Physical Defects

Even a chemically pure crystal may contain defects such as empty lattice sites (Schottky defects), dislocation lines and mosaic structure. Small angle grain boundaries and twinning are another type of defects of more macroscopic nature. Methods are therefore required for detection and quantitative measurement of microscopic defects and for checking macroscopic single crystallinity.

4.2.1 Lattice Vacancies

From densest packing arguments it may be safely assumed that point defects in organic molecular crystals will occur predominantly as lattice vacancies and not as interstitial molecules. Lattice self-diffusion proceeds via the point defects in the lattice and the experimental self-diffusion coefficients are a direct measure of the concentration and mobility of these defects. Radiotracer self-diffusion studies by Sherwood[236, 237] have revealed two contributions, a faster one along dislocations and sub-boundaries and a slower one through the lattice. In the class of plastic crystals, composed of almost spherical molecules, a high temperature rotator phase occurs in which diffusion rates are $\sim 10^4$ times greater than in crystals of linear aromatic hydrocarbons. Typical lattice self-diffusion coefficients for the latter class of substances are on the order of 10^{-11} cm^2 sec^{-1} (at the melting point) with total activation energies which are about twice the sublimation energy. This fact indicates that the jump activation energy is roughly equal to the formation energy.

Plastic deformation at higher temperatures takes place via dislocation climb mechanisms which are controlled by the self-diffusion of point defects. Therefore creep studies can be applied to investigate point defects. The activation energy of creep reflects the diffusion activation energy[237].

Nuclear magnetic resonance can detect diffusion by motional narrowing of lines and via altered spin lattice relaxation times. Other contributions like rotations must be taken into account; cf.[237].

Positronium annihilation has been used to obtain information on the size of the free volume in the neighbourhood of foreign molecules, influencing the lifetime of orthopositronium and the width of the momentum distribution of para-positronium[238].

4.2.2 Dislocations

Dislocation lines penetrating a surface can be made visible by sublimation, or by solvent or chemical etching[239], developing stress fields around the dislocation line. The density of etch-pits so obtained ranges from 10^2 cm^{-2} for vapor-grown anthracene crystals, over 10^3 for sulution-grown samples, to 10^5-10^6 for melt-grown crystals[239, 152].

Low temperature plastic deformation occurs predominantly by dislocation glide. There are preferred slip systems such as (001) [010] (slip in the (001) plane in [010] direction), (001) [110], Ref.[237] and (010) [001], Ref.[239] in anthracene. A great deal of work has been devoted to the problem of characterizing slip systems and obtaining a more profound understanding on a microscopic basis of the physical defects created,[237,239–243]. Certain slip systems may be inhibited[237] but also enhanced[241] in impure crystals. For highest purity and quality anthracene and naphthalene crystals an especially pronounced tendency towards gliding prohibits (room temperature) cleavage,[134] cf. Figs. 56 and 57; impurities tend to generate dislocations and to suppress gliding.

In general "as grown" crystals contain excess point defects and dislocations. In part these defects are frozen in at higher temperatures, in part they are formed by stress-induced strain during cooling. Therefore annealing of freely suspended crystals often leads to a considerable improvement of crystal quality.

Low density dislocation distributions can be made visible directly by X-ray topography, cf.[21]. Klapper[20] has extensively applied this technique to organic crystals such as benzil; cf.[49].

Transmission electron microscopy[244–246] is developing into a powerful tool for the observation of physical defects in organic crystals.

4.2.3 Mosaic Structure and Grain Boundaries

Mosaic structure and grain boundaries are very common defects in many apparently single crystalline samples. With optically anisotropic crystals these defects are best seen with polished plane parallel slices placed between crossed polarizers under a polarizing microscope. Misorientations appear as different brightness (or colour) in the main extinction directions and may range over a few degrees or less. An inert gas environment during growth can be one possible source[157].

Homogeneity of the orientation can also be checked by X-rays, monitoring the range of Bragg reflections from a selected lattice plane while scanning the crystal (so called "rocking-curve" technique). For Bridgeman anthracence and naphtalene crystals halfwidths of 0.5° and angular variations of the peak position of 0.5° could be reached[1].

4.2.4 Physical Defects and the Lifetimes of Excited States

Even at complete absence of chemical traps there would be trapping of mobile excited crystal states such as excitons and charge carriers: Structural defects may lead to local lowering of the corresponding energy levels. Triplet exciton trapping has been studied by Arnold et al. at dislocations produced in anthracene crystals by plastic bending[247, cf. 248]. The interactions of structural traps with charge carriers have been investigated by Silinsh[34] and Sworakowski[249].

4.2.5 Thermal Detrapping

In the presence of traps the concentrations of excitons or free charge carriers are strongly temperature dependent. From maxima in heating curves ("glow peaks", "thermally stimulated current" (TSC)-peaks) and from thermal activation energies obtained from initial rise of such peaks, trap energies can be derived in principle[152, 250–254]. Crystals can be characterized by the number, energies, relative and absolute height of glow peaks, however the microscopic interpretation is not always straight forward[255].

Since trapping of unipolar charge leads to space charge, the evaluation of non-Ohmic, space charge limited currents ("SCLC") can also lead to information about trapping centers[34, 256, 257].

---

1 Rocking curves were kindly measured for crystals grown in the Stuttgarter Kristallabor, by Prof. J. Kalus, University of Bayreuth

5 Preparation of Oriented Samples

For physical measurements with generally highly anisotropic organic molecular crystals well oriented samples are usually needed. If one symmetry-unique cleavage plane is present (a situation which is often encountered), then one crystallographic direction is already fixed. Otherwise and for the other two independent directions, polarization-optical or X-ray methods must be applied. Cutting and polishing of generally rather soft organic molecular crystals need special techniques. Storage of high quality samples is a problem which often requires special precautions. In what follows a few basic aspects are collected.

5.1 Cleaving

In many high quality crystals, e. g. of naphthalene, anthracene and p-terphenyl, gliding is so pronounced that cleavage cannot be performed without serious deterioration of the physical perfection. An example is given in Fig. 56, showing a slightly bent naphthalene crystal after cleaving, from left to right, by a razor blade. Traces of the glide plane are indicated by step edges, which point along [100]. Moving the edge of the cleaving tool in this direction should therefore be prefered in order to minimize distortions by gliding. Cleavage at reduced temperature may also be helpful.

Handling of cut samples also needs precautions if gliding is possible. An illustrative example is depicted in Fig. 57.

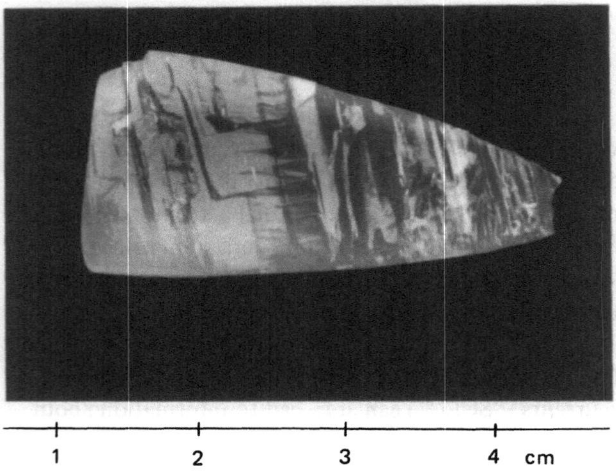

Fig. 56. Defects due to gliding introduced by the action of cleaving a high quality naphthalene crystal. The lines appearing on the (001) cleavage plane are traces of the (010) plane, indicating the direction [100]; slip system is (010) [001][161]

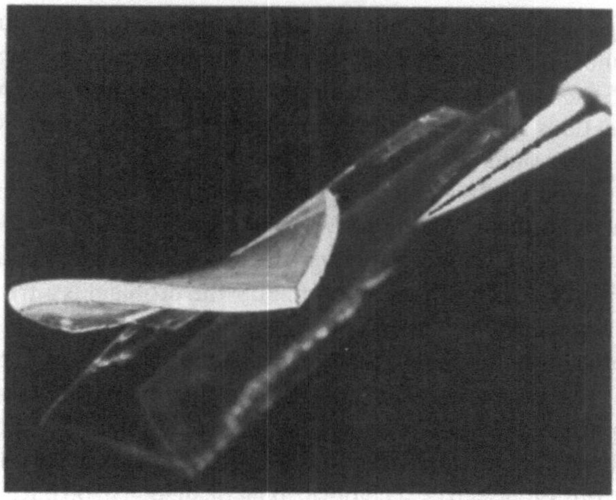

Fig. 57. Deformation by gliding of a plane parallel cut anthracene slice (arbitarily – not (001) – oriented) by removing it from adhesive tape[161]

In contrast to the examples mentioned, there exist crystals such as 2,3-dimethyl-naphthalene and 2,3-dimethyl-anthracene which combine elastic stiffness with well pronounced cleavage, recalling the exceptional properties of mica.

5.2 Orienting

Given a plane parallel cleaved or cut and polished slice of an optically anisotropic (i. e. non-cubic) crystal, polarization-optical observation in parallel light detects immediately the two main vibrational directions (extinction directions), which can be determined to better than 1 deg. using a polarizing microscope[m]. For (001) cleavage planes of (monoclinic) naphthalene or anthracene these directions are [100] and [010] from symmetry. However to distinguish which one is which, requires additional information, which can be gained from conoscopic viewing, preferentially with monochromatic light: Isogyres and isochromatic interference lines reveal symmetry planes, C_2, and C_3, C_4 or C_6 rotation axes present in the crystal. (In the naphthalene/anthracene example a vertical symmetry plane is observed which is parallel to [100].) In addition, optical axes may be seen if the crystal orientation and the aperture used are appropriate. In a monoclinic crystal, a symmetry which is very common, the optic axial plane may coincide with the symmetry plane, (class 2/m and m) or the plane perpendicular to the twofold rotation axis, (010), (by convention), (class 2). This situation holds for naphthalene, anthracene, α- and β-perylene. Or it may be perpendicular to (010), a case which is realized with pyrene and 1,4-dibromonaphthalene. In these monoclinic cases the orientation of the acute bisectrix allows one to dis-

m Except for the unlikely situation where an optical axis is accidentally parallel to the plate normal tinguish between +a and –a. An arbitrarily cut plane-parallel slice of a cleavable monoclinic crystal is easily oriented by measuring the angle between the face normal and the cleavage planes(s) and by determining in the cleavage planes(s) the extinction directions and the approximate direction of the acute bisectrix, cf.[31]. These angles are best noted in a Wulff net.

For a triclinic crystal optical orientation is still possible, if the full orientation matrix of the optical tensor, represented by the indicatrix, is known relative to the crystallographic axes. However the procedure is more complicated because of refraction. – For orthorhombic symmetry the problem is a trivial one. However, for trigonal, tetragonal and hexagonal symmetry only the direction of the c-axis can be obtained. Optical orientation of cubic crystals is not possible with the method described.

The conoscopic determination of the orientation of a sublimed, 1,4-dibromonaphthalene crystal is given as an example in Fig. 58. Notice the additional faint Fabry-Perot interferences in the center indicating the coincidence of [010] and the obtuse bisectrix. The optic axial plane is ~(100), Ref.[259]–. The conoscopic interference figure is very sensitive to distortions of the single-crystallinity[260]. For questions of fundamental crystal optics, the books by Rinne-Berek[261], Wahlstrom[262] and Nye[263] may be consulted. A compilation of crystal optical data was published by Winchell[23].

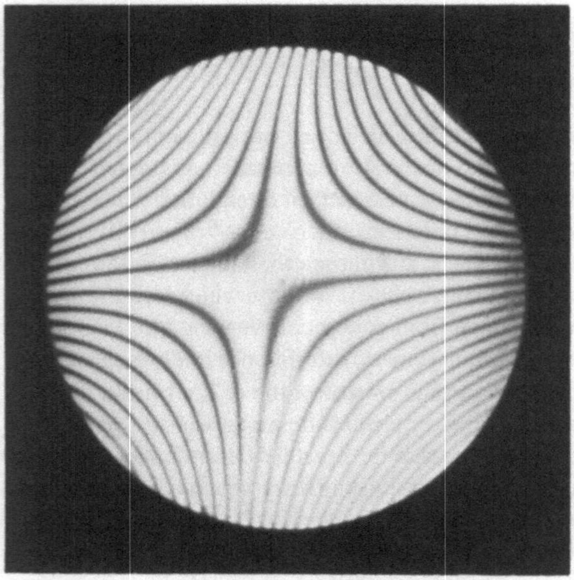

Fig. 58. Determination of the orientation and single crystallinity of a sublimed 1,4-dibromonaphthalene crystal by conoscopy on a polarization microscope in Na_D-light; picture taken in diagonal position. The crystal is viewed along [010], which by symmetry is one principal axis direction of the optical tensor (indicatrix). The trace of the optic axial plane, bc, is indicated by the convergence of the "isochromatic" interference lines to the optical axes (falling outside the used aperture of 0.85). Notice the additional faint Fabry-Perot interferences in the center.

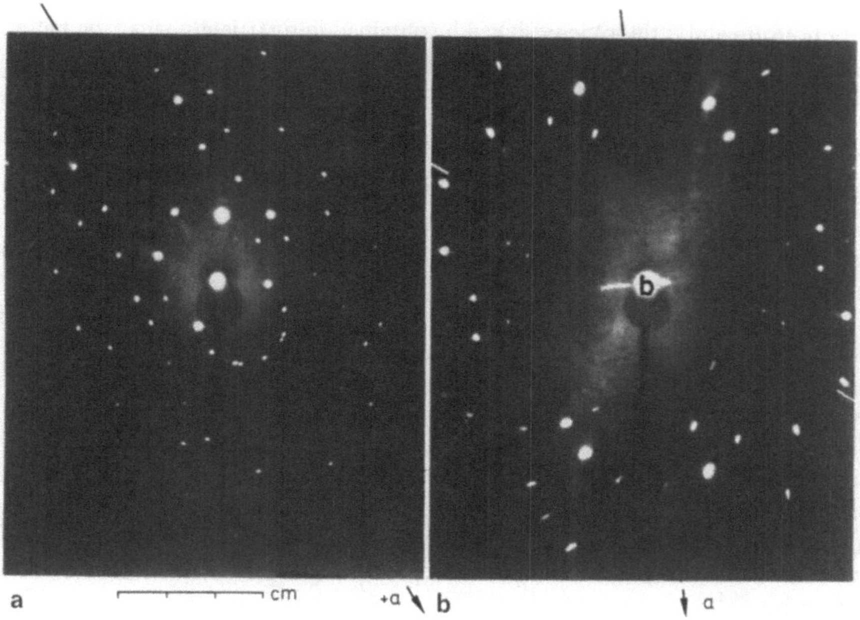

Fig. 59 a, b. Laue photographs of anthracene taken with the x-ray beam in + c*-direction (**a**), revealing the ac-symmetry plane (11 o'clock ⟷ 5 o'clock), and for the b-direction (**b**) which is the 2 fold symmetry axis. Crystal distances for the original photos were 3.5 cm (**a**) and 5 cm (**b**). In **a** the reader looks in the direction of the X-ray. The two strong (h0l) reflexes are $(20\bar{1})$, $2\,\theta_{theor.} = 25.47°$, and $(20\bar{2})$, $2\,\theta_{theor.} = 27.03°$

For X-ray orientation the Laue method can be used. However with most organic molecular crystals the conventional set-up for back reflection leads to diffuse reflections only, because of thermal motion and great penetration depths. Laue forward scattering with thin (1 mm) samples leads to useful results if symmetry elements are found. This is tedious and cumbersome with photographic registration because of the low symmetry displayed by most organic crystals. The photographic method is only useful if one can rely on morphological characteristics such as a known face or a cleavage plane. With the aid of a HeNe-laser beam parallel (and opposite) to the X-ray beam the plane can be positioned. A very useful alternative technique is "Laue-life" by the aid of an X-ray image converter and intensifier[134].

Examples of Laue photographs of a monoclinic crystal are reproduced in Fig. 59,[161].

5.3 Cutting

It is generally advisable to cut soft organic crystals by a thread saw, using a solvent-soaked thread (e. g. a cotton thread). Under a sufficiently small pressure crude mechanical cutting is more or less replaced by gentle dissolving. Surface roughness amounts to 0.1–0.25 mm. Xylene as a solvent has some advantages: it has a low vapor pressure and little toxicity; it can be obtained in sufficient purity; the molecule is comparatively bulky so that penetration into most crystal lattices is rather unlikely.

5.4 Polishing

Again a dissolution technique is preferable to using abrasives. Polishing can be done on a solvent-soaked lens tissue on a glass optical flat. A polishing tool equipped with a micrometer screw is indispensable for obtaining good surfaces. The appropriate choice of the solvent used is essential. It should have a medium solubility for the compound under consideration. Screw rapport, optimal polishing intervals and polishing speed must be found by experiment. Too high rapport and/or speed give rough and distorted surfaces. If the speed is too low the edges begin to become rounded. Using xylene as a solvent a flatness of a few μm per cm^2 can be obtained for anthracene[260]. The flatness can be checked by interference methods in reflected light, using a precision optical flat as support; interference pictures can even be obtained with a normal polarization microscope, equipped for reflection, and a sodium vapor lamp[260].

Cut and polished anthracene crystals are depicted in Fig. 60. (The ray double refraction vanishes for the orientation (010) of the sample Fig. 60b. The size of the crystal Fig. 60a is ~1 cm^3.)

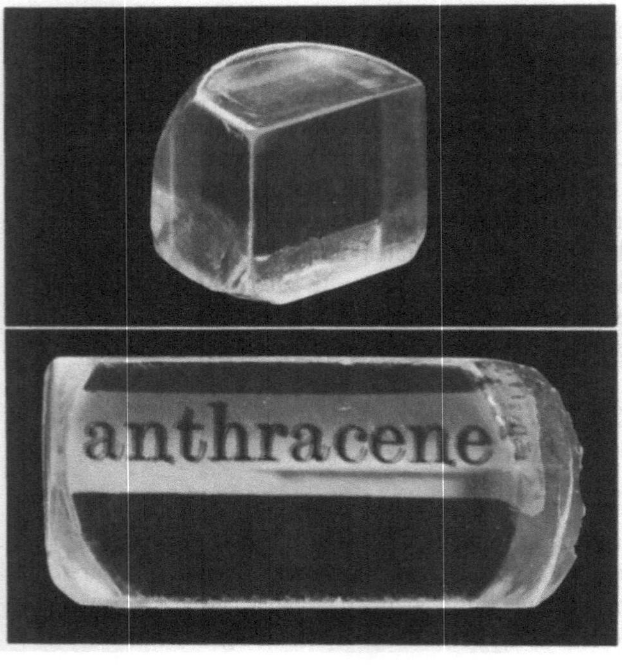

a

b

Fig. 60

Defect formation during polishing has been monitored in anthracene by triplet exciton lifetime measurements. Subsequent annealing can heal most of the structural defects introduced[264].

5.5 Storage

In order to avoid rapid temperature changes, high quality crystals should be thermally shielded e. g. by Styrofoam. So enclosed, high vapor pressure material can be stored in a refrigerator. It has been conjectured by several authors that oxygen affects triplet exciton and charge carrier lifetimes. However for good anthracene crystals stored in dark the triplet exciton lifetime has not seriously diminished in 10 years[161]; (a rapid small initial drop cannot be excluded). But many crystals suffer serious deterioration from simultaneous action of oxygen and light. Photochemical dimerization, another source of deterioration, has been mentioned earlier.

6 Properties and Applications

The physics of organic molecular crystals is characterized by weak van-der-Waals interaction between large molecules preserving their molecular properties to a great extent in the solid state. Each molecule contributes many intramolecular vibrational degrees of freedom. Intermolecular (lattice) vibrations are partially coupled to internal molecular motions. Most organic molecular crystals studied so far are distinguished by comparatively small binding energy "conjugated" π electrons which are strongly coupled in each molecule. These low-lying states are responsible to a great extent for all low energy properties of the crystalline state, because these properties are strongly determined by overlap, coulomb and exchange interaction of the π-electrons. The electronic excitations are strongly coupled to intra- and inter-molecular vibrations. Electronic excitations can lead to neutral (Frenkel) excitonic states, being more or less localized at individual molecules. Excitations with charge transfer are precursory steps to electronic conduction. After final separation into "free" electrons and holes, fairly localized states are formed as well. Localization and efficient scattering by strong electron-phonon coupling restrict exciton and charge carrier motion to incoherent hopping with low diffusivities and mobilities respectively, except for the lowest temperatures, where coherent transport tends to take over (cf. Ref.[265] for triplet excitons and Ref.[266] for electrons as examples). A concise model of charge carrier transport is still an open problem, cf.[267] and references given therein.

High purity is of extreme importance for obtaining meaningful results for all questions which are connected with energy and charge transfer. In weakly bound van-der-Waals solids these properties are very subtle. Evaluation is strongly affected by trapping of movable excited states at chemical or physical imperfections.

The geometric and electronic structure of these imperfections, cf.[233, 258], which are always present to some extent, are interesting in themselves since they influence the crystal properties to a great extent. On the other hand, understanding of the basic physical mechanisms allows one to tailor-make systems with the desired properties.

So far activities in the field of organic molecular crystals have pertained mostly to fundamental research. *Technical* applications of organic crystals have been rare in the past. A number of medium purity organic crystals with short-lived luminescence emission and near unity quantum yield have found widespread use as scintillation crystals for many years, cf.[268, 269].

Radiation damage of organic crystals has been investigated by optical and ESR techniques, see for example Ref.[270], in order to understand basic aspects of material degradation and carcinogenesis by high energy radiation.

However, there were numerous *scientific* applications of undoped and doped, one- and two-component crystals whose purification, growth and characterization has been described in this review article. Exciting results were obtained in the last few years. In what follows a brief compilation of some important topics is given:

High resolution absorption, emission, and reflection spectroscopy (examples: Shpolskii matrix spectroscopy[271], exciton amalgamation in isotopically mixed crystals[272], Stark effect with zero field splitting[273], charge-transfer in one-component crystals, cf.[274] and Refs. therein) and ultrahigh resolution laser excitation spectroscopy, cf.[204, 275]; coherent spectroscopy (by photon echo methods etc.)[276]; room temperature laser emission from dopant levels[169] and from Frenkel exciton states[277, 278]; energy transfer[225]; photochemistry[279], including the possibility of energy and information storage; photoelectron emission[280–282]; photoconductivity[232, 283], (see also reviews[30, 31, 284, 285]), and charge carrier transport[46, 232, 266, 286]; quasimetallic conductivity[287–289] and the first example of an organic superconductor[315]; electron spin resonance, ESR, (Refs.[233, 258, 265, 290–297] give some examples of important results), including ENDOR[258] optically detected magnetic resonance, ODMR,[295, 296] and spin-echo[296] techniques, and phenomena of triplet excitons[291, 295], pair states[290], coherent, one-dimensional exciton motion[296], spin polarisation[297], X-traps[258], triplet excimers[294] and charge transfer[292, 293]; nuclear magnetic resonance, NMR,[298]; optical and electrical magnetic field effects, cf.[299–303, 235] and the review Ref.[304]; electrochemistry at crystal surfaces[305]; organic chemistry in the solid state[306, 307].

The results obtained with regularly ordered and well defined crystals are not confined to the solid state; they have lead to a better understanding of the whole world of organic molecules including biological processes such as photosynthesis and vision and may soon stimulate new ideas of technical applications. Electrophotography[308], using photoconducting organic polymers[309], has already become a powerful technique. This point may be mentioned as one of the developments of research done with single crystals whose preparation and characterization have been described in this article.

Acknowledgement. The author would like to gratefully acknowledge helpful discussions with A. R. McGhie, J. N. Sherwood, G. J. Sloan and H. C. Wolf. His own contributions to the field were based on thorough work done by D. Deylig, M. Gerdon, M. Herb, A. Huber, G. Pampel, and W. Tuffentsammer of the "Stuttgarter Kristallabor". The technical support of these colleagues in the scientific work of several of the author's students is also greatly appreciated.

Financial support by the Physikalisch-Chemische Institut der Universität Freiburg, (Prof. F. Matossi) and the Fraunhofer-Gesellschaft München (1965–1968), the Deutsche Forschungsgemeinschaft (1967–1969), including the Sonderforschungsbereich 67 (since 1969) and, last but not least, by the Physikalische Institut, Teil 3, under the direction of Prof. H. C. Wolf, Universität Stuttgart (since 1968) is greatly appreciated.

Thanks are also due to Mrs. E. Wanner who has patiently typed the manuscript, and to J. J. Stezowski who kindly helped to improve the grammatical and orthographical "purity".

7 References

1. Herington, E. F. G.: Zone Melting of Organic Compounds, Oxford: Blackwell Sci. Pub. 1963
2. Wilcox, W. R., Friedberg, R., Back, N.: Chem. Rev. *64*, 187 (1964)
3. Landolt-Börnstein, Vol. II/Part 3: Schmelzgleichgewichte und Grenzflächenerscheinungen. Berlin: Springer 1956
4. Fox, D., Labes, M. M., Weissberger, A. (eds): Physics and Chemistry of the Organic Solid State, Vol. I, II, III. New York: Interscience 1963, 1965, 1967
5. Westrum, E. F., McCullough, J. P.: Thermodynamics of Crystals. In: Physics and Chemistry of the Organic Solid State, Vol. I, 1. New York Interscience 1963
6. Bondi, A.: Physical Properties of Molecular Crystals, Liquids and Glasses. New York: J. Wiley 1968
7. Sloan, G. J. In: Physics and Chemistry of the Organic Solid State, ref.[4], Vol. I, 1963, p. 179
8. Sloan, G. J., Mol. Cryst. *1*, 161 (1966)
9. Aleksandrov, S. B., Freimanis, J. F., Grishin, E. P., Jurel, S. P., and Buchbinder, J. E., Latvija PSR Zinatnu Akad. Vestis *366*, 23 (1978) (Russian)
10. Reynolds, G. F. In: Physics and Chemistry of the Organic Solid State, Ref.[4], Vol. I, 1963, p. 223
11. Wilke, K.-Th.: Kristallzüchtung, Berlin: VEB Deutscher Verlag d. Wiss. 1973
12. Porter, M. W., Spiller, R. C.: The Barker Index of Crystals, Vol. 7, Cambridge: W. Heffer 1951
13. Robertson, J. M.: Organic Crystals and Molecules, Ithaca N. Y.: Cornell Univ. Press 1953
14. Kitaigorodskii, A. I.: Organic Chemical Crystallography, New York: Consultant Bureau 1961
15. Kitaigorodskii, A. I.: Molecular Crystals and Molecules, New York: Acad. Press 1973
16. Wyckoff, R. W. G.: Crystal Structures, Vol. 5, Vol. 6 I, II, New York: Interscience 1966, 1971
17. Landolt-Börnstein, New Series Group III, Vol. 5a, b: Structure Data of Organic Crystals, Berlin: Springer 1971
18. Kaiser, R.: Chromatographie in der Gasphase I–IV, B. I. Hochschultaschenbücher 22/22a, Mannheim 1960
19. Ehrenberger, F., Gorbach, S.: Methoden der organischen Elementar- und Spurenanalyse, Weinheim: Verlag Chemie 1973
20. Klapper, H.: Untersuchung der Geometrie eingewachsener Versetzungen in lösungsgezüchteten Kristallen, Habilitationsschrift, Aachen 1975; Defects in Nonmetal Crystals, Nato Advanced Study Institute: The Characterization of Crystal Growth Defects by X-Ray Methods; Durham, U. K. 1979
21. Tanner, B. K.: X-Ray Diffraction Topography, New York: Pergamon Press 1976
22. Amorós, J. L., Amorós, M.: Molecular Crystals: Their Transforms and Diffuse Scattering, New York: J. Wiley 1968
23. Winchell, A. N.: The Optical Properties of Organic Compounds, New York: Academic Press 1954
24. Landolt-Börnstein, New Series Vol. 3, Schmillen, A., Legler, R.: Luminescence of Organic Substances. Berlin: Springer 1967
25. Berlman, I. B.: Handbook of Fluorescence Spectra of Aromatic Molecules, 2. ed. New York: Academic Press 1971

26. Clar, E.: Polycyclic Hydrocarbons, Vol. I, II. New York: Academic Press 1964
27. Wolf, H. C.: L. Genzel (ed.), Die feste Materie. Frankfurt/M: Umschau-Verlag 1973, p. 157
28. Birks, J. B.: Photophysics of Aromatic Molecules. London: Wiley-Interscience 1970
29. Birks, J. B.: Organic Molecular Photophysics Vol. 1, 2. London: J. Wiley 1973/1975
30. Karl, N.: Organic Semiconductors. In: H. J. Queisser (ed.), Adv. Solid State Physics XIV, Oxford/Braunschweig: Pergamon, Vieweg 1974, p. 261
31. Karl, N.: Halbleitereigenschaften des organischen Molekülkristalls Anthracen. Habilitationsschrift, Univ. Stuttgart 1975
32. Kurik, M. B.: Organic Semiconductors. Kiew: Akad. Nauk USSR, Inst. Fisiki 1976 (Russian)
33. Kurik, M. B.: Photonics of Organic Semiconductors. Kiew: Akad. Nauk USSR, Inst. Fisiki, 1977 (Russian)
34. Silinsh, E.: Electronic States of Organic Molecular Crystals, Monograph. Riga, Zinatne 1977 (Russian), Springer 1980 (Engl.)
35. Gutman, F., Lyons, L. E.: Organic semiconductors. New York: J. Wiley 1967
36. Briegleb, G.: Elektronen-Donator – Akzeptor-Komplexe. Berlin: Springer 1961
37. Foster, R.: Organic Charge Transfer Complexes. New York: Academic Press 1969
Foster, R.: (ed.), Molecular Complexes, Vol. I, II London: Elek. Science 1973, 1974
Foster, R.: Molecular Association, Vol. 1. New York: Academic Press 1975
38. Herbstein, F. H.: Crystalline π-Molecular Compounds, Spectroscopy and Crystallography. In: J. D. Dunitz, J. A. Ibers (eds.), Perspectives in Structural Chemistry, Vol. IV, New York: J. Wiley 1971
39. Lipsett, F. R.: Bibliography, 1956–1968. Mol. Cryst. a. Liqu. Cryst. *6*, 175 (1969) and references given in the introduction
40. Sheka, E. F.: Spectroscopy of. Mol. Cryst., Bibliography, 1967–1975. Mol. Cryst. a. Liqu. Cryst. *46*, 25 (1978), and references given in the introduction
41. Benz, K. W., Wolf, H. C.: Z. Naturforsch. *19a*, 177 (1964)
42. Wolf, H. C.: Bates and Estermann (eds.), Atomic and Molecular Physics, Vol. III, New York: 1967, p. 119
43. Kepler, R. G.: Phys. Rev. *119*, 1226 (1960)
44. Oyama, K., Nakada, I.: J. Phys. Soc. Jap. *24*, 792 (1968)
45. Schmillen, A., Falter, W. W.: Z. Phys. *218*, 401 (1969)
46. Probst, K.-H., Karl, N.: Phys. stat. sol. *a27*, 499 (1975)
47. Wolf, H. C., Benz, K. W.: Pure Appl. Chem. *27*, 439 (1971)
48. Decamps, E. A., Durand, M., Granger, M. M.: Bull. Soc. Sci. Bretagne (France), Experimental crystallization colloquium, Conf. Paper, Vol. 48, 103 (1973) (French)
49. Bleay, J., Hooper, R. M., Narang, R. S. and Sherwood, J. N., J. Cryst. Growth *43*, 589 (1978)
50. Aasden, D. R., Simmons, R. O.: Bull. Am. Phys. Soc. *21*, 422 (1976)
51. Fraas, B. A., Heald, S. M., Simmons, R. O.: J. Cryst. Growth *42*, 370 (1977)
52. Lamotte, M., Joussot-Dubien, J.: J. Chem. Phys. *61*, 1892 (1974)
53. Pfister, C., Kahane-Paillous, J.: J. Chim. Phys. *65*, 876 (1968)
54. R. Nitsche, A. Räuber (eds.), Information on Crystal Growth (in Germany), Freiburg 1976: XII: Hellner, E., Treutmann, W., p. 284; XIII/XIV: Haussühl, S., p. 217 k/h; XV: Ulmer, K., p. 260a; XVI: Engel, W., p. 222
55. Wegner, G.: Z. Naturforsch. *24b*, 824 (1969); idem in: W. Hatfield (ed.): NATO-ARI Molecular Metals, Les Ascs 1978, Plenum Press, (1979)
56. Wunderlich, B.: J. Crystal Growth *42*, 241 (1977)
57. Anthonj, R., Karl, N., Robertson, B. E. and Stezowski, J. J., J. Chem. Phys., *72*, 1244 (1980)
58. Kistenmacher, T. L., Phillips, T. E., Cowan, D. O.: Acta Cryst. *30B*, 763 (1974)
59. Berlinsky, A. J.: Contemp. Phys. *17*, 331 (1976)
60. Anzai, H.: J. Cryst. Growth *33*, 185 (1976)
61. Kaplan, M. L.: J. Cryst. Growth *33*, 161 (1976)
62. Kitaigorodskii, A. I., Mirskaya, K. V.: Mat. Res. Bull. 7, 1271 (1972); Mason, R.: Mol. Cryst. *9*, 3 (1969)
63. Williams, D. E., Starr, T. L.: Comput. and Chem. *1*, 173 (1977)

64. Hartmann, P.: Physics and Chemistry of the Organic Solid State, Vol. I. New York: Interscience 1963, p. 369
65. Sinclair, V. C., Robertson, J. M., Mathieson, A.McL.: Acta Cryst. *3,* 251 (1950)
66. Camerman, A., Trotter, J.: Acta Cryst. *18,* 636 (1965)
67. Colson, S. D., Wheeler, G. I.: J. Chem. Phys. *60,* 4634 (1974)
68. Fourme, R., Clec'h, G., Figuiere, P., Ghelfenstein, M., Szwarc, H.: Mol. Cryst. a. Liqu. Cryst. *27,* 315 (1974)
69. Mirsky, K., Cohen, M. D.: Acta Cryst. *A 34,* 346 (1978)
70. Bertinelli, F., Stremmenos, C.: J. Chem. Soc., Faraday Transact II, *69,* 889 (1973)
71. Robertson, B. E., Stezowski, J. J.: Acta Cryst. *B34,* 3005 (1978)
72. Trotter, J.: Acta Cryst. *12,* 54 (1959)
73. Mnyukh, Yu. V., Panfilova, N. A., Petropavlov, N. N., Uchvatova, N. S.: J. phys. chem. sol. *36,* 127 (1975)
74. Herbstein, F. H.: Acta Cryst. *18,* 997 (1965)
75. Zallen, R., Griffiths, C. H., Slade, M. L., Hayek, M., Brafman, O.: Chem. Phys. Letters *39,* 85 (1976)
76. Jones, W., Ramdas, S., Thomas, J. M.: Chem. Phys. Letters *54,* 490 (1978)
77. Soltzberg, L. J., Piliero, P. A., Shea, M. R.: Mol. Cryst. a. Liqu. Cryst. *29,* 115 (1974)
78. Goworek, T., Gustaw, W., Rybka, C., Wawryszczuk, J.: phys. stat. sol. *b65,* K97 (1974)
79. Welberry, T. R.: Proc. Roy. Soc. London *A 334,* 19 (1973)
80. Messager, J. C., Cailleau, H., Yelon, W. B.: Acta Cryst. *A 34,* 304 (1978)
81. Trotter, J.: Acta Cryst. *12,* 232 (1959)
82. Burns, D. M., Iball, J.: Proc. Roy. Soc. *227A,* 200 (1955)
83. Dideberg, O., Dupont, L., Andre, J. M.; Acta Cryst. *B28,* 1002 (1972)
84. Banerjee, A.: Acta Cryst. *B29,* 2070 (1973)
85. Trotter, J.: Can. J. Chem. *39,* 1574 (1961)
86. Scott, R. A. M., Laridiani, S. R., Morantz, D. J.: J. Cryst. Growth *22,* 53 (1974)
87. Morantz, D. J., Mathur, K. K.: J. Cryst. Growth *16,* 147 (1972)
88. Kitaigorodskii, A. I.: Dokl. Akad. Nauk. SSSR *113,* 604 (1957)
89. Morawetz, E.: J. Chem. Thermodyn. *4,* 455 (1972)
90. Morawetz, E.: J. Chem. Thermodyn. *4,* 461 (1972)
91. Irving, R. J.: J. Chem. Thermodyn. *4,* 793 (1972)
92. Malaspina, L., Gigli, R., Bardi, G.: J. Chem. Phys. *59,* 387 (1973)
93. Pfann, W. G.: Trans. AIME *194,* J. Metals *4,* 747 (1952)
94. Pfann, W. G.: Zone melting. New York: J. Wiley 1958
95. Rokos, K., Robinson, P. M., Scott, H. G.: Mol. Cryst. a. Liqu. Cryst. *24,* 331 (1973)
96. Matz, G.: Chem.-Ing.-Techn. *36,* 381 (1964)
97. Robinson, P. M., Rossell, H. J., Scott, H. G., Legge, C.: Mol. Cryst. a. Liqu. Cryst. *11,* 105 (1970)
98. Robinson, P. M., Scott, H. G.: Mol. Cryst. a. Liqu. Cryst. *18,* 143 (1972)
99. Burton, I. A., Prim, R. C., Slichter, W. P.: J. Chem. Phys. *21,* 1987 (1953)
100. Steybe, H.: Ingenieurarbeit, Universität Stuttgart 1969
101. Burris, L. Jr., Stockman, C. H., Dillon, I. G.: Trans. AIME, J. Metals *7,* 1017 (1955)
102. Braun, I., Marshall, S.: Brit. J. Appl. Phys. *8,* 157 (1957)
103. Lord, N. W.: Trans. AIME *197,* J. Metals *5,* 1531 (1953)
104. Reiss, H.: Trans. AIME *200,* J. Metals *6,* 1053 (1954)
105. Braun, I.: Brit. J. Appl. Phys. *8,* 457 (1957)
106. Schildknecht, H.: Zonenschmelzen. Weinheim: Verlag Chemie, 1964
107. Karl, N., Feederle, H.: Unpubl.; see Feederle, H.: Diplomarbeit, Univ. Stuttgart 1964
108. Developed in the Stuttgarter Kristallabor by M. Gerdon and the author
109. Karl, N.: Diplomarbeit, Univ. Freiburg 1965
110. Fischer, D.: Mat. Res. Bull. *8,* 385 (1973)
111. Benz, K. W., Port, H., Wolf, H. C.: Z. Naturforsch. *26a,* 787 (1971)
112. Karl, N.: Presented at the annual meeting of the Deutsche Gesellschaft für Kristallwachstum und Kristallzüchtung, DGKK, Freiburg, Germany, Sept. 1972

113. Benyon, J. H., Saunders, R. A.: Brit. J. Appl. Phys. *11*, 128 (1960)
114. Ziebold, G.: Diplomarbeit, Univ. Stuttgart 1968
115. Probst, K.-H.: Diplomarbeit, Univ. Stuttgart 1970
116. McGhie, A. R., Sloan, G. J.: J. Crystal Growth *32*, 60 (1976)
117. Fischer, D.: J. Appl. Phys. *44*, 1977 (1973)
118. Karl, N., Probst, K.-H.: Mol. Cryst. a. Liqu. Cryst. *11*, 155 (1970)
119. McArdle, B. J., Sherwood, J. N., Damask, A. C.; J. Crystal Growth *22*, 193 (1974)
120. Tuffentsammer, W.: Stuttgarter Kristallabor, Unpubl.
121. Sherwood, J. N., Thomson, S. J.: Trans. Faraday Soc. *56*, 1443 (1960)
122. Lee, C. H., Kevorkian, H. K., Reucroft, P. J., Labes, M. M.: J. Chem. Phys. *42*, 1406 (1965)
123. Burns, G., Sherwood, J. N.: J. Chem. Soc. *68*, 1036 (1972)
124. Nickel, B.: private communication, and our gas chromatographic and mass-spectroscopic results
125. Sloan, G. J., McGhie, A. R.: Mol. Cryst. a. Liqu. Cryst. *18*, 17 (1972)
126. Boyens, J. C. A., Herbstein, F. H.: J. Phys. Chem. *69*, 2153 (1965)
127. Haarer, D., Karl, N.: Chem. Phys. Letters *21*, 49 (1973)
128. Pampel, G., Tuffentsammer, W., Karl, N.: unpubl.
129. Klemm, L. H., McCoy, D. R., Olson, D. R.: J. Heterocycl. Chem. *7*, 1347 (1970)
130. Zief, M., Wilcox, W. R. (eds.): Fractional Solidification. New York: Marcel Dekker Inc. 1967
131. Forsyth, J. S., Wood, J. T.: Trans. Inst. Chem. Engrs. (London) *33*, 122 (1955)
132. Wilcox, W. R.: Thesis, Univ. California, Berkeley 1960: UCRL-9213
133. Zief, M., Wilcox, W. R. (eds.): Fractional Solidification, Appendix III: New York: Marcel Dekker Inc. 1967, pp. 650
134. Concepts realized and results obtained in the Stuttgarter Kristallabor by Gerdon, M., Herb, C., Huber, A., Pampel, G., Tuffentsammer, W. and the author
135. Sherwood, J. N.: M. Zief (ed.). Purification of Inorganic and Organic Materials. New York: Marcel Dekker Inc. 1969 p. 157
136. Thomas, J. F., Sanborn, E. N., Mukai, M., Tebbens, B.: Anal. Chem. *30*, 1954 (1958)
137. Gosling, K., Bowen, R.: Anal. Chem. *45*, 1574 (1973)
138. McGhie, A. R., Garito, A. F., Heeger, A. J.: J. Crystal Growth *22*, 295 (1974); cf. McGhie, A. R., US-patent 3, 840, 349 (1974)
139. Sizmann, R.: Angew. Chem. *71*, 243 (1959)
140. Wilk, M., Rochlitz, J., Bende, H.: J. Chromatog. *24*, 414 (1966)
141. Mair, B. J., Hwang, P. T. R., Ruberto, R. G.: Anal. Chem. *40*, 838 (1968)
142. Oelert, H. H.: Z. Anal. Chem. *244*, 91 (1969)
143. Hsieh, B. C. B. et al.: Anal. Chem. *41*, 1066 (1969)
144. Woelm Information 32, M. Woelm, D-344 Eschwege
145. Sangster, R. C., Irvine, J. W.: J. Chem. Phys. *24*, 670 (1956)
146. Meier, R., Fletschinger, J.: Angew. Chem. *68*, 373 (1956)
147. McGhie, A. R.: Mol. Cryst. a. Liqu. Cryst. *27*, 431 (1974)
148. Deyl, Z., Rosmus, J., Juricová, M., Kopecký, J.: J. Chomatogr., Suppl. *3*, 1067 (1973)
149. Knox, J. H.: Ann. Rev. Phys. Chem. *24*, 29 (1973)
150. Hammer, A., Wolf, H. C.: Phys. stat. sol. *33*, K 25 (1969)
151. Williams, R. G., Lowry, B. A.: J. Chem. Phys. *56*, 5736 (1972)
152. Reucroft, P. J., McGhie, A. R., Hillman, E. E., and Damiano, V. V.: J. Cryst. Growth *11*, 355 (1971)
153. Johnson, E. A.: J. Chem. Sol. *57*, 994 (1962)
154. Warta, W.: Diplomarbeit, Univ. Stuttgart 1978; cf. [232)]
155. Bridgman, P. W.: Proc. Am. Acad. Arts Sci. *60*, 305 (1925)
156. Lipsett, F. R.: Can. J. Phys. *35*, 284 (1957)
157. Lupien, Y., Williams, J. O., Williams, D. F.: Mol. Cryst. a. Liqu. Cryst. *18*, 129 (1972)
158. Sherwood, J. N., Thomson, S. J.: J. Sci. Instr. *37*, 242 (1960)
159. Chang, E., Yip, V. F. S., Wilcox, W. R.: J. Cryst. Growth *22*, 247 (1974)

160. Gerdon, M.: Unpubl., Stuttgarter Kristallabor 1973, (cf. ref. 31)
161. Karl, N.: Unpubl.
162. Chadwick, G. A., Zief, M., Wilcox, W. R. (eds.): Fractional Solidification, Vol. I. New York: Marcel Dekker Inc. 1967, p. 113
163. Goldacker, W., Schweitzer, D., Zimmermann, H.: Chem. Phys. *36*, 15 (1979)
164. Radomska, M., Radomski, R., Pigon, K.: Mol. Cryst. a. Liqu. Cryst. *18*, 75 (1972)
165. Kitaigorodskii, A. I., Kozhin, V. M.: Soviet Phys. Crystallogr. *4*, 187 (1959)
166. Robinson, P. M., Scott, H. G.: Mol. Cryst. a. Liqu. Cryst. *5*, 387 (1969)
167. Robinson, P. M., Scott, H. G.: Mol. Cryst. a. Liqu. Cryst. *11*, 239 (1970)
168. Flack, H. D.: Phil. Trans. Roy. Soc. Lond. *A 266*, 559, 583 (1970)
169. Karl, N. In: Lasers in Physical Chemistry and Biophysics. Amsterdam: Elsevier 1975, p. 61
170. Honigmann, B.: Z. Elektrochem., Ber. Bunsenges. Phys. Chemie *58*, 322 (1954)
171. Honigmann, B., Heyer, H.: Z. Krist., Mineralog., Petrogr. *106*, 199 (1955)
172. Honigmann, B., Heyer, H.: Z. Elektrochem., Ber. Bunsenges. Phys. Chemie *61*, 74 (1957)
173. Florian, W., Heyer, H.: Ber. Bunsenges. Phys. Chemie *77*, 1083 (1973)
174. Sloan, G. J.: Mol. Cryst. *2*, 323 (1967)
175. Sears, G. W., Hopke, E. R.: J. Am. Chem. Soc. *71*, 1632 (1949)
176. Scholz, H.: Philips Techn. Rundschau *28*, 220 (1967); Acta Electronica *17*, 1 (1974)
177. Scholz, H., Kluckow, R.: Crystal Growth (Suppl. J. Phys. Chem. Sol. Proc. Int. Conf. Cryst. Growth Boston, June 1966) Pergamon Press 1967, p. 475
178. Scholz, H.: Private communication
179. Gerdon, M., Karl, N.: Unpubl.
180. Vitovskii, B. V., Dobrzhanskii, G. F., Netesov, G. B., Chernyshev, K. S.: Sov. Phys. Cryst. *13*, 481 (1968)
181. Jones, K. H.: Mol. Cryst. *3*, 393 (1968)
182. Meuser, W.: I. Phys. Inst. der Univ. Gießen, Diplomarbeit 1973 and private communication
183. Zwerdling, S., Halford, R. S.: J. Chem. Phys. *23*, 2215 (1955)
184. Ebisuzaki, Y.: J. Cryst. Growth *43*, 648 (1978)
185. Karl, N.: Presented at the 1st annual meeting of the Deutsche Gesellschaft für Kristallwachstum und Kristallzüchtung, Bonn 1971
186. Karl, N., Haarer, D.: Presented at the 2nd annual meeting of the Deutsche Gesellschaft für Kristallwachstum und Kristallzüchtung, Freiburg 1972
187. Karl, N.: 1st Europ. Conf. on Cryst. Growth Zürich (1976) abstract book p. 80; idem, 5th Internat. Conf. on Cryst. Growth, Boston 1977, abstract No. 207
188. Mark, P.: Z. Naturforsch. *16a*, 950 (1961)
189. Glockner, E.: Dissertation, Univ. Stuttgart 1974
190. Karl, N.: Presented at the annual spring meeting of the Deutsche Physikal. Ges., Freudenstadt 1978
191. Handbook of Chemistry and Physics
192. Klapper, H.: Private communication 1976
193. Herb, M.: Stuttgarter Kristallabor 1978
194. Kallmann, H., Pope, M.: Rev. Sci. Instr. *29*, 993 (1958)
195. Cobas, A., Chalverus, J., Szmant, H. H., Trester, S., Weisz, S. Z.: Rev. Sci. Inst. *37*, 232 (1966)
196. Pott, G. T., Kommandeur, J.: Mol. Phys. *13*, 373 (1967)
197. Julian, M. M.: Acta Cryst. *29*, 116 (1973)
198. Aptekar', I. L., Galashin, A. E.: Pis'ma Zh. Eksp. Teor. Fiz. (USSR) *28*, 421 (1978)
199. Khanna, S. K., Ehrenfreund, E., Garito, A. F., and Heeger, A. J.: Phys. Rev. *B10*, 2205 (1974)
200. Ferraris, J., Cowan, D. O., Walatka, V. Jr., Perlstein, J. H.: J. Am. Chem. Soc. *95*, 948 (1973)
201. Coleman, L. B., Cohen, M. J., Sandman, D. J., Yamagishi, F. G., Garito, A. F., Heeger, A. J.: Solid State Commun. *12*, 1125 (1973)
202. Kruse, K. M. M.: J. Phys. E: Sci. Instr. *8*, 592 (1975)
203. Lipsett, F. R., Goode, D. H.: The Triplet State. Cambridge 1967, p. 425

204. Port, H., Rund, D., Small, G. J., Yakhot, V.: Chem. Phys. *39*, 175 (1979)
205. McGhie, A. R., Voshchenkov, A. M., Reucroft, P. J., Labes, M. M.: J. Chem. Phys. *48*, 186 (1968)
206. Thomas, J. M., Williams, J. O.: Mol. Cryst. a. Liqu. Cryst. *9*, 59 (1969)
207. Robinson, J. W.: M. Zief (ed.) Purification of inorganic and organic materials. New York: Marcel Dekker 1969
208. Welz, B.: Atom-Absorptions-Spektroskopie, 2. Aufl. Verlag Chemie 1975
209. Waters Associates: Separation of Polynuclear Aromatics by Liquid Chromatography, AN 131 (1973)
210. Giger, W., Blumer, M.: Anal. Chem. *46*, 1663 (1974)
211. Waters Associates: Light Distillates and their Product Fuels. Benzene and Hydrocarbon Group Quantitation, J 03 (1977)
212. Waters Associates: Polynuclear Aromatic Hydrocarbons. Trace Level Determination in Workplace Air, H 90 (1977)
213. Johnson, E. L., Liquid Chromatography Bibliography, Varian (1977)
214. Budzikiewicz, H., Djerassi, C., Williams, D. H.: Interpretation of Mass Spectra of Organic Compounds. San Francisco: Holden-Day 1964
215. McClure, D. S.: Sol. State Phys. *8*, 1 (1959)
216. Wolf, H. C.: Sol. State Phys. *9*, 1 (1960)
217. Craig, D. P., Walmsley, S. H.: Excitons in molecular crystals. In ref. [4], Vol. I, p. 586
218. Avakian, P., Merrifield, R. E.: Mol. Cryst. *5*, 37 (1968)
219. Rice, S. A., Jortner, J.: Comments on the theory of the exciton states of molecular crystals. In Ref. [4], Vol. III, p. 199
220. Davydov, A. S.: Theory of Molecular Excitons, New York: Plenum Press 1971
221. Avakian, P., Abramson, E., Kepler, R. G., Caris, J. C.: J. Chem. Phys. *39*, 1127 (1963)
222. Parker, C. A.: Proc. Soc. Anal. Chem. Conf. Nottingham, Engl. 1965 p. 208
223. Brilliante, A., Craig, D. P., Mau, A. W.-H., Rajikan, J.: Chem. Phys. Letters *30*, 5 (1975)
224. Wolff, H., Schmillen, A.: Z. Naturforsch. *33a*, 1115 (1978)
225. Wolf, H. C., Port, H.: J. Luminesc. *12/13*, 33 (1976)
226. Braun, A., Pfisterer, H., Schmid, D.: J. Luminesc. *17*, 15 (1978)
227. Häcker, W.: Diplomarbeit, Universität Stuttgart 1969
228. Ziegler, F. W., Karl, N. Unpublished
229. Lupien, Y., Williams, D. F.: Mol. Cryst. *5*, 1 (1968)
230. Rohrbacher, H., Karl, N.: Phys. stat. sol. *a29*, 517 (1975)
231. Karl, N., Feederle, H.: Phys. stat. sol. *a34*, 497 (1976)
232. Karl, N.: Ionic States in Nonionic Molecular Crystals. In Sci. Papers Inst. Org. Phys. Chem. No. 16, Wroclaw Tech. Univ. 1978, p. 44
233. Dörner, H., Hundhausen, R., Schmid, D.: Chem. Phys. Lett. *53*, 101 (1978)
234. Ramdas, S.: Chem. Phys. Letters *60*, 320 (1979)
235. Ziegler, J., Karl, N.: Chem. Phys. *40*, 207 (1979)
236. Sherwood, J. N., Thomson, S. J.: Trans. Faraday Soc. *56*, 1443 (1960)
237. Sherwood, J. N.: Mol. Cryst. a. Liqu. Cryst. *9*, 37 (1969); idem: Lattice Defects, Self Diffusion, and the Plasticity of Plastic Crystals. In J. N. Sherwood (ed.) The Plastically Crystalline State. New York: J. Wiley 1979
238. Goworek, T., Rybka, C., Wawryszczuk, J.: Phys. Stat. sol. *b89*, 253 (1978)
239. Williams, J. O., Thomas, J. M.: Trans. Faraday Soc. *63*, 1720 (1967)
240. Robinson, P. M., Scott, H. G.: Mol. Cryst. *11*, 13 (1970)
241. Narang, R. S., Shah, B. S., Sherwood, J. N.: J. Cryst. Growth *22*, 201 (1974)
242. Sloan, G. J., Thomas, J. M., Williams, J. O.: Mol. Cryst. a. Liqu. Cryst. *30*, 167 (1975)
243. Kojima, K.: Phys. stat. sol. *a51*, 71 (1979)
244. Williams, J. O., Thomas, J. M.: Mol. Cryst. a. Liqu. Cryst. *16*, 223 (1972)
245. Jones, W., Williams, J. O.: J. Mat. Sci. *10*, 379 (1975)
246. Jones, W., Thomas, J. M., Williams, J. O., Hobbs, L. W.: J. Chem. Soc., Faraday Transactions II *71*, 138 (1975)
247. Arnold, S., Whitten, W. B., Damask, A. C.: J. Chem. Phys. *53*, 2878 (1970)

248. Sasaki, A., Hayakawa, S.: Jap. J. Appl. Phys. *12*, 1806 (1973)
249. Sworakowski, J.: Mol. Cryst. a. Liqu. Cryst. *33*, 83 (1976)
250. Thomas, J. M., Williams, J. O., Cox, G. A.: Trans. Faraday Soc. *64*, 2496 (1968)
251. Thomas, J. M., Williams, J. O.: Mol. Cryst. a. Liqu. Cryst. *9*, 59 (1969)
252. Samoc, M., Zboinski, Z.: Phys. stat. sol. *a46*, 251 (1978)
253. Williams, J. O., Zboinski, Z.: J. Chem. Soc. Faraday Trans. II *74*, 618 (1978)
254. Samoć, A., Samoć, M., Sworakowski, J.: Phys. stat. sol. *a36*, 735 (1976) Samoć et al.: Phys. stat. sol. *a37*, 271 (1976) Samoć, A., Samoć, M., Sworakowski, J.: Phys. stat. sol. *a39*, 337 (1977)
255. Samoć, M., Samoć, A.: Phys. stat. sol. *a51*, K195 (1979)
256. Nespurek, S., Silinsh, E. A.: Phys. stat. sol. *a34*, 774 (1976)
257. Nespurek, S., Sworakowski, J.: J. Appl. Phys., in press
258. Dörner, H., Schmid, D.: Chem. Phys. Letters *54*, 154 (1978); *61*, 171 (1979)
259. Castro, G., Hochstrasser, R. M.: J. Chem. Phys. *47*, 2241 (1967)
260. Karl, N., Schmid, E., Seeger, M.: Z. Naturforsch. *25a*, 382 (1970)
261. Rinne, F., Berek, M.: Anleitung zu optischen Untersuchungen mit dem Polarisationsmikroskop, Stuttgart: E. Schweizerbart'sche Verlagsbuchhandlung 1953
262. Wahlstrom, E. E.: Optical Crystallography. New York: J. Wiley 1969
263. Nye, J. F.: Physical Properties of Crystals, their representation by tensors and matrices Oxford: The Clarendon Press 1957
264. Weston, W., Mey, W., Pope, M.: J. Appl. Phys. *44*, 5615 (1973)
265. Francis, A. H., Harris, C. B.: Chem. Phys. Let. *9*, 181 (1971)
266. Schein, L. B., Duke, C. B., McGhie, A. R.: Phys. Rev. Letters *40*, 197 (1978)
267. Schein, L. B., McGhie, A. R.: Chem. Phys. Letters *62*, 356 (1979)
268. Birks, J. B.: Scintillations in Organic Solids. In Fox, Labes, Weissberger, Vol. II [4)]
269. Schmillen, A.: Physikalische Grundlagen und Anwendungen der Lumineszenz organischer Stoffe. In N. Riehl (ed.) Einführung in die Lumineszenz. München: 1971 Verlag Karl Thiemig
270. Böhme, U. R., Wolf, H. C.: Chem. Phys. Letters *17*, 582 (1972)
271. Merle, A. M., Pitts, W. M., Sayed, M. A.: J. Luminesc. *18*, 111 (1979)
272. Wolf, H. C., Port, H.: Molecular Spectroscopy of Dense Phases. Proc. 12. Europ. Congr. Mol. Spectr. Strasbourg 1975, Amsterdam: Elsevier 1976 p. 31
273. Haarer, D., Philpott, M. R., Morawitz, H.: J. Chem. Phys. *63*, 5238 (1975)
274. Syassen, K., Philpott, M. R.: Chem. Phys. Letters *50*, 14 (1977)
275. Port, H., Mistelberger, K., Rund, D.: Mol. Cryst. Liqu. Cryst. *50*, 11 (1979)
276. Aartsma, Th. J., Wiersma, D. A.: Chem. Phys. Letters *36*, 1360 (1976); Hesselink, H., Wiersma, D. A.: Chem. Phys. Letters *56*, 227 (1978)
277. Valera, A., Karl, N.: In press
278. For low temperature coherent emission see also references given in ref.[169)], especially the work by Broude
279. Schmidt, G. M. J.: D. Ginsburg (ed.) Solid State Photochemistry, Vol. III. Weinheim: Verlag Chemie 1976
280. Caywood, J. M.: Mol. Cryst. a. Liqu. Cryst. *12*, 1 (1970)
281. Seki, K., Hirooka, T., Kamura, Y., Inokuchi, H.: Bull. Chem. Soc. Japan *49*, 904 (1976)
282. Grobman, W. D., Koch, E. E.: Photoemission from Organic Molecular Crystals. Topics Appl. Phys. Vol. *27*. Berlin: Springer Verlag 1979, p. 261
283. Chance, R. R., Braun, C. L.: J. Chem. Phys. *64*, 3573 (1976)
284. Hanson, D. M.: Crit. Rev. Sol. State Sci. *3*, 243 (1973)
285. Kepler, R. G.: Organic Molecular Crystals: Anthracene. In N. B. Hannay (ed.) Treatise on Solid State Chemistry, Vol. 3. New York: Plenum Publ. Corp. 1976, p. 615
286. Burshtein, Z., Williams, D. F.: Phys. Rev. *B15*, 5769 (1977)
287. Miller, J. S., Epstein, J. (eds.) Synthesis and Properties of Low Dimensional Materials. New York: Academic Press 1978
288. Devreese, J. T., Evard, R. P., van Doren, V. E. (eds.), Highly Conducting One-Dimensional Solids. New York: Plenum Press (1979)

289. Hatfield, W. E. (ed.), Molecular Metals. New York: Plenum Press 1979
290. Schwoerer, M., Wolf, H. C.: Mol. Cryst. *3*, 177 (1967)
291. Haarer, D., Wolf, H. C.: Mol. Cryst. a. Liqu. Cryst. *10*, 359 (1970)
292. Möhwald, H., Böhm, A.: Chem. Phys. Letters *43*, 49 (1976)
293. Keijzers, C. P., Haarer, D.: J. Chem. Phys. *67*, 925 (1977); Chem. Phys. Letters *49*, 24 (1977)
294. Furrer, R., Heinrich, M., Stehlik, D., Zimmermann, H.: Chem. Phys. *36*, 27 (1979)
295. Steudle, W., von Schütz, J. U.: J. Luminesc. *18/19*, 191 (1979)
296. Botter, B. J., Dicker, A. I. M., Schmidt, J.: Mol. Phys. *36*, 129 (1978)
297. Hausser, K. H., Wolf, H. C.: Adv. Magnetic Resonance *8*, 85 (1976)
298. Haeberlen, U.: High Resolution NMR in Solids. In: (J. S. Waugh ed.) Adv. Magn. Resonance, Suppl. 1. New York: Academic Press 1976; Aravamudhan, S., Haeberlen, U.: In press
299. Johnson, R. C. Merrifield, R. E., Avakian, P., Flippen, R. B.: Phys. Rev. Letters *19*, 285 (1967)
300. Bouchriha, H., Delacote, G., Delannoy, P., Schott, M.: J. Phys. (Paris) *35*, 577 (1974)
301. Chabr, M., Zschokke-Gränacher, I.: J. Chem. Phys. *64*, 3903 (1976)
302. Klein, G.: Chem. Phys. Letters *57*, 202 (1978)
303. Müller, N., Papier, G., Charlé, K.-P., Willig, F.: Ber. Bunsenges. Phys. Chem. *83*, 130 (1979)
304. Swenberg, C. E., Geacintov, N. E.: In Birks (ed.)[29], Vol. I, ch. 10; Vol. II, ch. 8
305. Willig, F.: Electrochemistry at Organic Insulator Crystals. In H. Gerischer, C. Tobias (eds.), Advances in Electrochemistry and Electrochemical Engineering, in press
306. Cohen, M. D., Green, B. S.; Chemistry in Britain *9*, 490 (1973)
307. Schmidt, G. M. J.: Solid State Photochemistry, Vol. 8, D. Ginsburg (ed.). Weinheim: Verlag Chemie (1976)
308. Schmidlin, F. W.: Electrophotography. In J. Mort, D. M. Pai (eds.), Photoconductivity and related phenomena, Elsevier Sci. Publ. Comp., New York 1976, p. 421
309. Gill, W. D.: Polymeric Photoconductors. In op. cit.[307], p. 303
310. Stezowski, J. J.: J. Chem. Phys. *73*, 538 (1980)
311. Heym, H., Karl, N.: Unpubl. results. Heym, H., Diplomarbeit, Univ. Stuttgart 1980
312. McGhie, A. R., Sloan, G. J.: 5th Mol. Cryst. Sympos. Philadelphia 1970 Abstr. Book, p. 109
313. Nakada, I.: Private communication (1975)
314. Kontani, M.: Private communication (1980)
315. Jerome, D., Mazaud, A., Ribault, M., Bechtgaard, K.: J. Physique Letters *41*, L-95 (1980); see also *41*, L-397 (1980)

Received September 27, 1979

Rare-Earth Germanates

Ludmila N. Demianets, Anatoly N. Lobachev, and Gennadi A. Emelchenko

Institute of Crystallography, Academy of Science of the USSR, Leninsky Prospect 59, USSR – Moscow B, 117333

Table of Contents

Introduction

The intense rise of contemporary science and technology has engendered the development and investigation of different materials possessing valuable physical properties. Of special interest are the rare-earth compounds, chiefly because of the possibility of utilizing them in diverse areas of modern engineering (for example: quantum generators, optical converters, scintillators, luminophores, etc.)[1–7]. However, only the oxides and fluorides have been thoroughly studied in practice. To date, most acute is the problem of producing and investigating other rare-earth compounds, particularly those of silicates and germanates. In virtue of their specific composition, RE germanates can find broad application in radioelectronics (as laser materials), in luminophore production (luminophores covering the whole optical radiation range, X-ray luminescent screens, thermoluminiscent dosimeters), in glass-ceramic manufacture (special kinds of glass and ceramics), in the nuclear instrumentation industry (nuclear radiation detectors), and in numerous other capacities.

A more complete review of the data on the preparation, structure, and certain physico-chemical, as well as optical, properties of single crystals of rare-earth Germanates is presented in[7].

1 Structure of Rare-Earth Germanates

In the IV b subgroup of the periodic system, germanium, due to its properties, is the nearest analogue to silicon. In most of its compounds, germanium follows the behavior of that element. Many germanates are isostructural with the corresponding silicates.

At the same time, the large size of the quadruple-charged germanium, in comparison with Si (ionic radius of Ge^{4+} = 0.44 Å, and that of Si^{4+} = 0.39 Å), allows the Ge^{4+} ion to play a "double" role in the compounds. Like its analogue, silicon, germanium can form as a coordination polyhedron a four-ligand tetrahedron; in this case germanium behaves like a typical anion. In other cases the coordination polyhedron of germanium is an octahedron, and, therefore, the element performs a more "cation" part, constructing the structural skeleton similar to that of other cations. Such a duality of germanium was firstly noted in[8]. The double role of germanium is more markedly seen in the presence of low-charged cations (mono-bivalent cations). When germanium forms compounds together with 3-, 4-valent cations, the "cation behavior" is considerably inhibited, at which point Ge behaves exactly like a typical anion. In combination with rare-earth elements, germanium usually displays anion behavior.

The classification of RE germanium compounds assumed in this paper results from the tetrahedral coordination of germanium, and is based on the germanium-oxygen radical construction similar to the classification of silicates[9,10]. Table 1 presents the basic X-ray characteristics of representatives of the known structural types of RE germanates: non-alkali (14 structural types, alkali-RE (13) germanates,

Table 1. X-Ray characteristics of rare-earth germanates

Composition	Symbol, Ln	Compound	Cell parameters a, Å	b, Å	c, Å	α, β, γ	Z	Space group	Ge-radical	Ref.
1	2	3	4	5	6	7	8	9	10	11
$Ln_2GeO_5 = Ln_2[GeO_4]O$	Q$_1$ Ln = La–Tb (Gd)	Sm_2GeO_5	9.4	7.2	6.9	β = 108°	4	$P2_1/c$	GeO_4	7
	Q$_2$ Ln = Y, Dy–Lu	Y_2GeO_5 Lu_2GeO_5	14.94 14.76	10.58 10.41	6.87 6.77	γ = 122° 02′ γ = 122° 23′	8	B 2/b	GeO_4	11
	Q$_3$ Ln = Sc	Sc_2GeO_5	10.927	10.656	10.486	γ = 93° 59′	12	B 2/b	GeO_4	7, 12
$7Ln_2O_3 \cdot 9GeO_2 =$ $= Ln_{4.67}[GeO_4]_3O$	D Ln = La–Dy	$Nd_{4.67}[GeO_4]_3O$	9.75		7.15		2	$P6_3/m$	GeO_4	7
$Ln_8[GeO_4]_6$	D Ln = La	$La_8[GeO_4]_6$	9.89		7.26		2	$P6_3/m$	GeO_4	13
$Ln_{13}[GeO_4]_6(OH)_3 \cdot$ $\cdot [O, (OH, F)_2]_6 =$ $= Ln(OH)_3 \cdot 6Ln_2[GeO_4] \cdot$ $\cdot [O, (OH, F)_2]$	H Ln = Nd–Eu	$Sm(OH)_3 \cdot 6Sm_2GeO_4 \cdot$ $\cdot [O, (OH, F)_2]$	15.90		6.98		2	$P6_3/m$	GeO_4	14
$Ln_4GeO_8 = Ln_4[GeO_4]O_4$	Ln = La, Gd	$La_4[GeO_4]O_4$	10.54		7.63				$[GeO_4]$	15
	M Ln = Nd	$Nd_4[GeO_4]O_4$	7.475	5.727	17.927		4	$Pmc2_1$	$[GeO_4]$	16

Table 1 (continued)

1	2	3	4	5	6	7	8	9	10	11
$Ln_2Ge_2O_7$	L Ln = La, Pr	$La_2Ge_2O_7$ = $La_4[Ge_3O_{10}]$ $[GeO_4]$	12.76	7.070	7.006	α = 90° 57′ β = 90° 21′ γ = 94° 06′	4	$P\bar{1}$	$[Ge_3O_{10}]$ + +$[GeO_4]$	17
	L_1 Ln = Nd–Gd	$Gd_2Ge_2O_7$ = $Gd_4[Ge_3O_{10}]$ $[GeO_4]$	18.50	6.800	6.858	α = 87° 53′ β = 91° 31′ γ = 94° 33′	6	P1	$[Ge_3O_{10}]$ + + $[GeO_4]$	18
	L_2 Ln = (Gd), Tb–Lu	$Gd_2Ge_2O_7$ $Er_2Ge_2O_7$	6.90 6.778		12.25 12.34		4 4	$P4_12_12$ $P4_12_12$	$[Ge_2O_7]$ $[Ge_2O_7]$	7 19
	L_3 Ln = Sc	$Sc_2Ge_2O_7$	6.550	8.690	4.900	β = 102° 1′	2	C2/m	$[Ge_2O_7]$	20
	F Ln = Yb, Sc	$Yb_2Ge_2O_7$	9.81						$[GeO_6]$	7
$Ln_4Ge_3O_9(OH)_6$	Ln = Sm	$Sm_4Ge_3O_9(OH)_6$	15.195	10.909	7.161		8	Cmcm	$[Ge^{VI}Ge_2^{IV}O_9]_\infty$	21
$NaLnGeO_4$	A Ln = Sm–Lu, Y	$NaSmGeO_4$	5.27	11.70	6.50		4	$Pbn2_1$	$[GeO_4]$	22
	B Ln = La, Ce–Nd	$NaEuGeO_4$	11.75		5.45		8	I4/m	$[GeO_4]$	7
$NaLn_3[GeO_4]_2(OH)_2$	E Ln = Nd–Gd	$NaNd_3[GeO_4]_2(OH)_2$ $NaSm_3[GeO_4]_2(OH)_2$	5.26 13.78	12.249 12.15	13.931 5.28	α = 90° 15′	4 4	I2/c, Ic	$[GeO_4]$ $[GeO_4]$	23 7
$NaLn[GeO_4]OH$	R_1 Ln = Pr, Gd	Na_2PrGeO_4OH	9.52	7.48	6.98		4	Pnma	$[GeO_4]$	25
	R_2 Ln = Sc	Na_2ScGeO_4OH	12.34	13.78	5.53		8	Pmab, $P2_1ab$	$[GeO_4]$	26

$NaLn_4[GeO_4]_2O_2OH$	Ln = Ho	$NaHo_4[GeO_4]_2O_2OH$	6.75	21.67	6.91		4	Pnma	$[GeO_4]$	27
$K_2Ln_2[Ge_2O_7](OH)_2$	K Ln = Sc	$K_2Sc_2Ge_2O_7(OH)_2$	7.06	11.35	10.67		4	Pbcn	$[Ge_2O_7]$	28
$Na_4Ln_2Ge_4O_{13}$	T Ln = Ho, Yb, Lu, Sc	$Na_4Yb_2Ge_4O_{13}$ $Na_4Sc_2Ge_4O_{13}$	14.45 14.30	5.60 5.46	15.35 14.65		4 4	$Pna2_1$ $Pna2_1$	$[Ge_4O_{13}]$ $[Ge_4O_{13}]$	7 29
$NaLn_3Ge_4O_{13}$ = $= Na_2Ln_6[Ge_2O_7]_2$ $[Ge_4O_{12}]$	S Ln = Pr, Nd	$Na_2Pr_6[Ge_2O_7]_2 \cdot$ $\cdot [Ge_4O_{12}]$	7.18	11.61	7.04	$\alpha = 98° 55'$ $\beta = 91° 37'$ $\gamma = 102° 30'$	2	$P\bar{1}$	$[Ge_2O_7] +$ $+ [Ge_4O_{12}]$	30
$K_2Ln_4[Ge_4O_{13}](OH, F)_4$	G Ln = Pr–Gd	$K_2Nd_4Ge_4O_{14}(OH, F)_4$	6.821	7.438	17.693	$\gamma = 121°$	2	$P2_1$	$[Ge_4O_{13}]$	31 7
$K_2Ln[Ge_4O_{10}](OH,F)$ = $= K_4Ln_2[Ge_8O_{20}](OH,F)_2$	P Ln = Yb	$K_2YbGe_4O_{10}(OH, F)$	12.105	11.896	8.576	$\gamma = 112° 04'$	2	$P2_1/m$	$[Ge_8O_{20}]$	32
$NaLnGe_2O_6$; $KLnGe_2O_6$	N Ln = Sc	$NaScGe_2O_6$	10.12	9.14	4.73	$\gamma = 108°$	2		$[Ge_2O_6]_\infty$	7
$Na_5LnGe_4O_{12}$ $Na_{15}Ln_3[Ge_{12}O_{36}]$	Ln = Gd–Lu, Y	$Na_5GdGe_4O_{12}$	22.781		12.994		16	$R\bar{3}c$	$[Ge_{12}O_{36}]$	33
$Me_3Ln_2Ge_4O_{12}$	garnet Ln = Ho–Lu, Y	$Sr_3Er_2Ge_3O_{12}$ $Ca_3Y_2Ge_3O_{12}$	13.065 12.805				8	Ia3d	$[GeO_4]$	34
$M_2Ln_3[GeO_4]_3Cl$	apatite Ln = La, Ce	$Ca_2Ce_3Cl[GeO_4]_3$ = $= Ca_4Ge_6[GeO_4]_6Cl_2$ $Ba_2La_7[GeO_4]_6O_{0.5}$	9.467 9.99		6.999 7.37		2 2	$P6_3/m$ $P6_3/m$	$[GeO_4]$ $[GeO_4]$	13 13
$MeLn_2GeO_6$	perovscite Ln = La	$MgLa_2GeO_6$	3.90				1	Pm3m	$[GeO_6]$	35

and germanates containing bivalent cations (3). The RE germanates will be considered below in the order of the complexity of the composition and the construction of the Ge-O radical (from finite through island to infinite radicals).

1.1 Non-Alkali Rare-Earth Germanates

The oxyorthogermanates $Ln_2GeO_5 = Ln_2[GeO_4]O$ may be broken down into three structural groups: La-Gd; (Tb); Y; Dy-Lu; Sc. In analyzing the x-ray and infrared spectra of $Ln_2[GeO_4]O$ and the silicates of the same composition, some authors[11, 36)] have concluded an isostructurality of the compounds of the corresponding first two subgroups. The structure of Sc_2GeO_5 has no analogue among the silicates.

A monoclinic crystal structure (space group $P2_1/c$[37)]) is characteristic for $Ln_2[SiO_4]O$ silicates, where Ln are rare-earth elements from La to Tb. Rare-earth silicates from Tb to Lu are also assumed to be monoclinic, but the structure is of another kind[38)]. (The same B 2/b space group, but the unit cell is approximately twice as large in size). In the case of germanates this morphotropic jump in the RE-series occurs earlier, at Gd; both structural modifications[15)] were disclosed for Gd_2GeO_5.

A typical form of Ge-O radical for all the strucures indicated is the isolated tetrahedron $[GeO_4]$. The distinction lies in the coordination, form, and order of the Ln-O polyhedron joints. In $Gd_2[SiO_4]O$, there are alternating layers consisting of 7- and 9-cornered polyhedra, in the center of which reside Gd atoms[37)]; in $Yb_2[SiO_4]O$, two crystallographically independent cations, Yb^{3+}, have a coordination of 7 and 6, with respect to oxygen. A further decrease in the coordination of the trivalent cation is specific to the third indicated structural type ($Sc_2[GeO_4]O$); Sc is situated in octahedra (coordination number 6).

Schematically, the structure of RE orthogermanates can be described as a mutual packing of $[GeO_4]$ and $[OLn_4]$ tetrahedra, the latter are constructed from Ln^{3+} cations surrounding free oxygen atoms.

Ln_4GeO_8 orthogermanates have no structurally analogue silicates. Spectroscopic data (infrared spectra of this series of germanates) and x-ray patterns of their powders indicate that there exist at least two structural types (Ln = La–Gd and Ln = Tb–Lu, Sc[1,15,36)]) characteristic of the isolated distribution of Ge–O tetrahedra. The structural formula of these compounds can be written in the form of an orthogermanate with additional oxygen anions, $Ln_4[GeO_4]O_4$.

According to[15)], the germanates of the first structural subgroup are characterized by a hexagonal symmetry, although a variation in the symmetry within this series is quite possible. Indeed, in a recent investigation of Nd_4GeO_8, obtained by hydrothermal techniques, it was observed that the symmetry is not hexagonal, but orthorhombic[16)]. Six kinds of Nd polyhedra (coordination numbers 8, 7), in contact along the edges, unite in a rather rigid three-dimensional skeleton containing $[GeO_4]$ tetrahedra.

Orthogermanates of the $7\,Ln_2O_3 \cdot 9\,GeO_2$ *composition* are characteristic of the left side of the rare-earth series (Ln = La–Dy[7,36,39,40)]). These compounds, as well as analogous silicates have an apatite structure (space group $P6_3/m$). Isolated germanium-

Fig. 1. Kinds of cation polyhedra in the apatite structure[40)]

oxygen tetrahedra create (together with the empty polyhedra) a column along the twofold screw axis. The basic types of cationic polyhedra are illustrated in Fig. 1.

Compounds $Ln(OH)_3 \cdot 6Ln_2[GeO_4][O, (OH, F)_2]$ having an H-type structure are characteristic of large rare-earth cations ranging from Nd to Eu[7)]. Analyses of samarium germanates[14)] revealed their close relationship to the apatite-like structures, i. e., they are built of blocks (Fig. 2) similar to those seen in apatite (columns of triple Sm polyhedra around the 6_3 axis and columns of Sm_4-9-cornered polyhedra

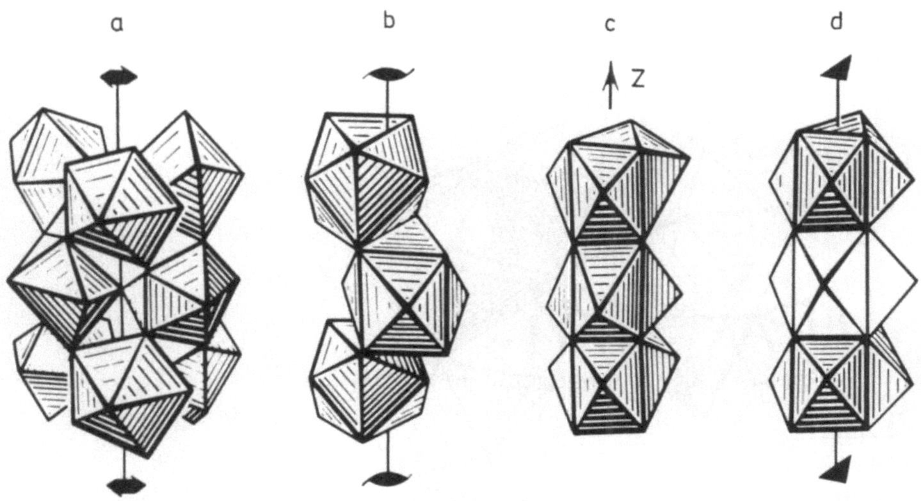

Fig. 2a–d. Fragments of the structure of H-phase $Sm(OH)_3 \cdot 6Sm_2[GeO_4] \cdot \{O, (OH, F)_2\}$[14)]. **a** infinite columns of Sm_1 polyhedra along 6_3 axis; **b** infinite column of Sm_3 polyhedra along 2_1 axis; **c** infinite column of Sm_2 polyhedra along C axis; polyhedra connected by m planes; **d** alternation of empty and occupied polyhedra along $\bar{3}$ axis

on triple axes). A formula developed for the H-type compounds (for a unit cell), and based on x-ray structural analysis, DTA, and NMR in combination with a valency balance can be written as: $2Sm(OH)_3 \cdot 3Sm_2[GeO_4](OH)_2 \cdot 4Sm_2GeO_5 \cdot 5Sm_2[GeO_4]F_2$.

If we take into account the possibility of mutual substitution of OH^- and F^-, we can write down the formula at Z = 2 in the following form:
$Sm(OH)_3 \cdot 6Sm_2[GeO_4][O, (OH, F)_2]$.

Diorthogermanates $Ln_2Ge_2O_7$ break down structurally into four subgroups: La–Pr; Nd–Gd; Tb–Lu; Sc.

Primarily, the isolation of the structural subgroups was effected on the basis of spectroscopic data[1]. Afterwards, the structures of the extreme members of the subgroups $La_2Ge_2O_7$[17], $Gd_2Ge_2O_7$[18], and $Er_2Ge_2O_7$[19] were resolved. High-temperature modifications were found for Yb and Sc germanates, making it possible to isolate a fifth structural type characteristic of compounds of gross composition $Ln_2Ge_2O_7$.

Of the five isolated subgroups, three (low-temperature form of germanates with Ln = La and Pr, germanates of the series Nd–Gd, and high-temperature modification, $Ln_2Ge_2O_7$) can be categorized as diorthogermanates only in terms of their gross composition. The structures of such compounds do not contain any diorthoradicals. In the first two subgroups, germanium creates a triorthogroup, $[Ge_3O_{10}]$ (Fig. 3), and isolated tetrahedra. Both structures ($La_2Ge_2O_7 = La_4[G_3O_{10}][GeO_4]$ and $Gd_2Ge_2O_7 = Gd_4[Ge_3O_{10}][GeO_4]$ are triclinic and characterized by a low degree of symmetry, but the unit cell $Gd_2Ge_2O_7$ exceeds the $La_2Ge_2O_7$ cell four times in volume. The octahedron is the typical form of a Ge radical for the high-temperature modification, $Ln_2Ge_2O_7$[41].

The other two subgroups (Tb–Ln; Sc) include typical diorthogermanates with a radical $[Ge_2O_7]$ and are characterized by a tetragonal unit cell ($Er_2Ge_2O_7$) and a monoclinic ($Sc_2Ge_2O_7$, thortveitite structures) cell.

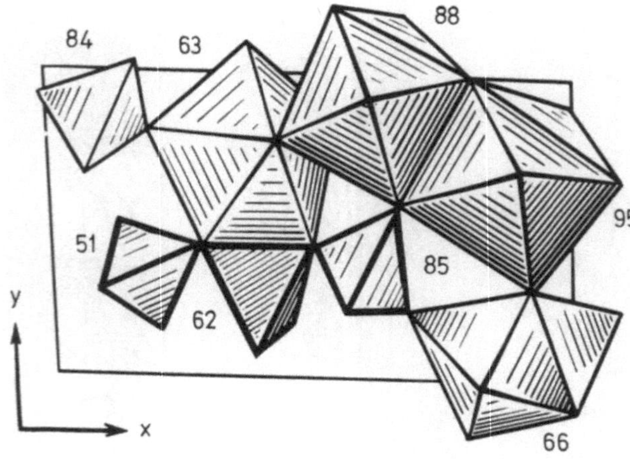

Fig. 3. Structure of $La_2Ge_2O_7 = La_4[Ge_3O_{10}]GeO_4$; x-y projection. The numbers correspond to the z coordinate in hundredth fraction of the period c[17]. The triorthogroup $[Ge_3O_{10}]$ is distinguished by thick lines

In a recently investigated germanate, *$Sm_4Ge_3O_9(OH)_6$*[21], the germanium atoms occupy two types of coordination polyhedra: tetrahedra and octahedra. Ge octahedra link up along the [001] axis in an infinite chain of $[GeO_5]$ "trimmed" with Ge tetrahedra; each octahedron joins 2 tetrahedra bound by a mirror plane of symmetry. The radical formula can be written in the following way: $[Ge^{VI}Ge_2^{IV}O_9]_\infty$. In the cation skeleton one may single out in the OYZ projection olivine-like bands characteristic of numerous alkali-rare-earth germanates. The stems of the band consist of Sm_1-7-pointed polyhedra; Sm_2-8-fold polyhedra are located at the dents. In the a-axis direction, such bands are double-decked (Fig. 4).

The germanate $La_2Ge_3O_9$ was first discussed in [42]; both infrared spectra and x-ray data supported the individuality of this compound[43]. Thus far, there is no indication of germanates of such a composition in combination with other rare-earth elements. Moreover, attempts made by other authors to obtain a lanthanum metagermanate ended in failure[1, 15, 44]. It is possible that the lanthanum metagermanate is metastable or that it is formed in the presence of impurities. Pointing to a considerable drop in frequency in infrared spectra upon transition from a quadruple coordination for germanium to a sextuple one[44], and to similar dependences for aluminium and titanium[45,46], the authors in[43] came to the conclusion that in the $La_2O_3 \cdot 3GeO_2$ compound, part of the germanium atoms is located in octahedra, part in tetrahedra. Since no exact x-ray data are available for La_2Ge_3O, this compound is not presented in the table.

Fig. 4. Structure of $Sm_4Ge_3O_9(OH)_6$, y-z projection. Radical formula is $[Ge_3O_9] = [Ge^{VI}Ge_2^{IV}O_9]$. Germanium octahedra lie in the mirror plane at x/a = 0. Bands of Sm polyhedra and Ge tetrahedra are found on both sides of the mirror plane at a hight x/a 0.14 and −0.14[21]

1.2 Alkali-Rare-Earth Germanates

The most commonly found composition of alkaline-rare-earth germanates corresponds to an equi-atomic cation ratio $ALnGeO_4$, where A is Li, Na[22,47–49]; the concrete structural type of the compound depends on the type of the A and Ln cations. $NaLnGeO_4$ germanates of the A-type are found for rare-earth elements from Sm to Lu and Y[47]. According to the structural type, they can be attributed to olivine $(Mg, Fe)_2SiO_4$.

A band pattern consisting of Na and Sm octahedra (Fig. 5) and characteristic of olivine is present in the $NaSmGeO_4$ structure[22]; the triangular holes of these bands accommodate germanium atoms surrounded by 4 oxygen atoms; Ge-O tetrahedra are discrete. In germanates containing heavy rare-earth elements, the band stem is made up of Na octahedra, with the Sm polyhedra located on the band dents. The authors in[49] presume that the substitution of sodium for lithium does not cause any change in the structural type.

As will be shown below, the "olivine" bands made up of alkali and rare-earth polyhedra are typical of a whole series of structural alkali-rare-earth germanate types.

In the second modification, $NaLnGeO_4$ (B-type), the rare-earth polyhedra are located at the dents and appear as a ring surrounded by sodium cations (Fig. 6); the structure becomes more symmetrical, i. e., it becomes tetragonal. Germanates of the B-type are known for Ln = La, Ce, Pr, Nd, Eu[22, 47, 48, 50]. As in the A-type structure compounds, germanium occupies discrete tetrahedra[48].

The basic of the E-type structure, $NaLn_3[GeO_4]_2(OH)_2$, where Ln = Nd–Gd[47,23], is an infinite olivine-like band parallel to (100), the stem of which is made up to alternating (Na + Sm) polyhedra; Sm-7-cornered polyhedra are located at the band dents (Fig. 7). The germanium atoms assemble in tetrahedra consisting of oxygen atoms. One of the hydroxyl ions belongs to two Sm atoms. The other enters the coordination sphere of 3 cations: Sm_1, Sm_3, and Na[24].

The cationic band of the olivine-montichellite type also lies at the base of the architectural pattern of the germanate $Na_2Pr[GeO_4]OH$. However, unlike the classical variant of montichellite $CaMgSiO_4$, where the band stem is made up of polyhedra of light (Mg) and dents of heavy atoms (Ca), a reverse variant is seen in Na_2PrGeO_4OH where in the dents lighter Na-five-cornered polyhedra are found[25]. The identical (along the [001]) bands bind discrete tetrahedra $[GeO_4]$.

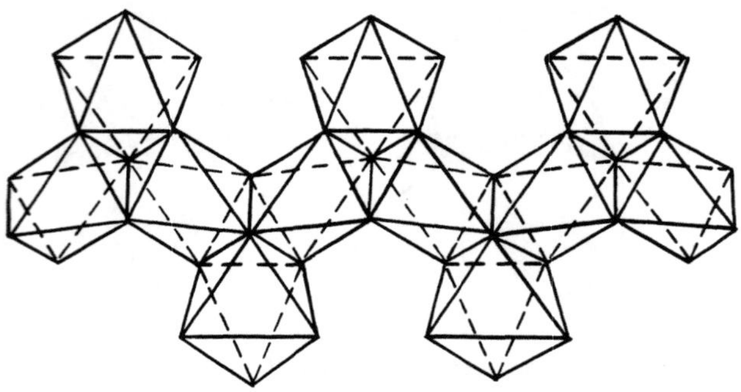

Fig. 5. Olivine-like band in A-type germanate[22]

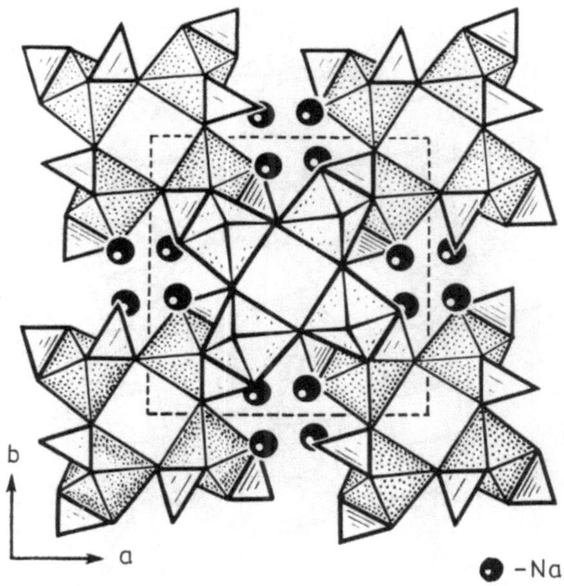

Fig. 6. Structure of B-type sodium silicate (germanate)[48].

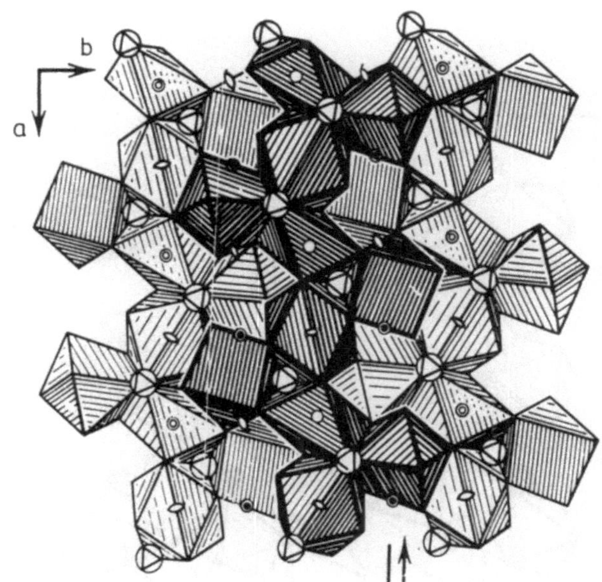

Fig. 7. Structure of $NaSm_3[GeO_4]_2(OH)_2$, E-type; z-y projection[24]. Light polyhedra bands correspond to the cation position at z = 0, dark bands – at z = 0.75

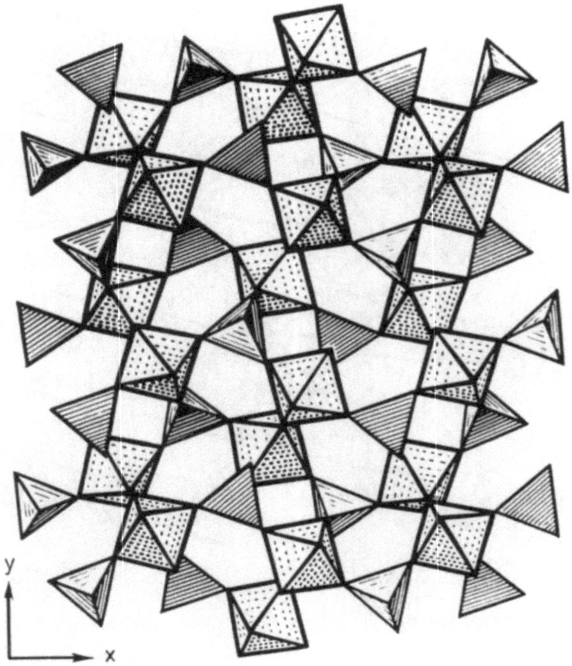

Fig. 8. Infinite bands of octahedra [ScO_6] closed by [GeO_4] tetrahedra in the Na_2ScGeO_4OH structure[26)]

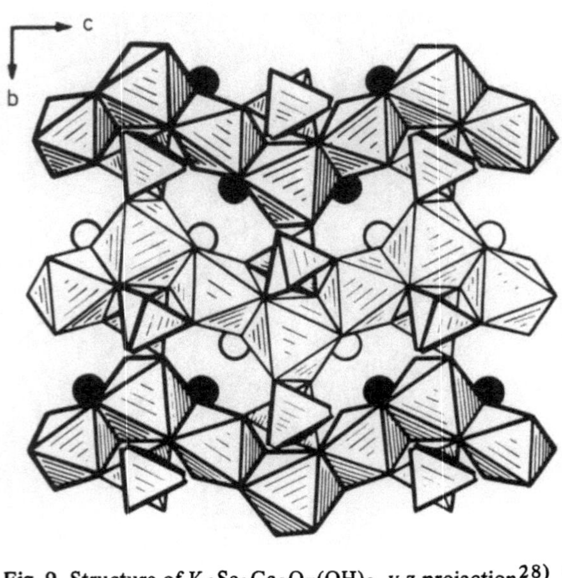

Fig. 9. Structure of $K_2Sc_2Ge_2O_7(OH)_2$, y-z projection[28)]. Dark bands of Sc octahedra and light circles of K cations are located at x = 0; lighter bands and full circles of K are at x = 1/2

The cationic band pattern is seen in another representative of alkali and rare-earth orthogermanates, Na_2ScGeO_4OH, but in this compound the bands are made up of Sc octahedra with common cis-vertexes (Fig. 8); the rest of the vertexes are closed by mutually isolated Ge-tetrahedra[26].

$NaHo_4[GeO_4]_2O_2(OH)$ is the terminating alkali-rare-earth orthogermanate known today. The study[27] describes the similarities between $[HoO_7]$ polyhedra of this structure and those observed in the structure of monoclinic holmium hydroxide.

Among the alkali-rare-earth germanates under investigation, we may attribute only one structure, $K_2Sc_2Ge_2O_7(OH)_2$, to diorthocompounds. The peculiarity of this structure is the rarely encountered diorthogroup $[Ge_2O_7]$ configuration: germanium tetrahedra being deployed one relative to another at an angle of approximately 120° (60°); the Ge-O-Ge angle is 120°. In this compound we again observe an olivine-like pattern, i. e., bands consisting of Sc-octahedra stretching out along the "C" axis (Fig. 9[28]).

The linear radical $[Ge_4O_{13}]$, unknown earlier in silicate crystal chemistry, was found in germanates $Na_4Ln_2Ge_4O_{13}$[29]. Compounds having such a structure are characteristic of "small" rare-earth cations Yb–Lu, Sc.

The new radical continues the island radical series in arithmetic progression: XO_4-X_2O_7-X_3O_{10}-X_4O_{13}- . . . In the $Na_4Sc_2Ge_4O_{13}$ structure such linear tetragroups are linked by pairs of the Sc-octahedra having common edges (Fig. 10[29]). The tetragroup $[Ge_4O_{13}]$ was also fixed for large cations. A bent chain of four Ge-O tetrahedra[31] was revealed in the G-type compound structure[50,51]. According to

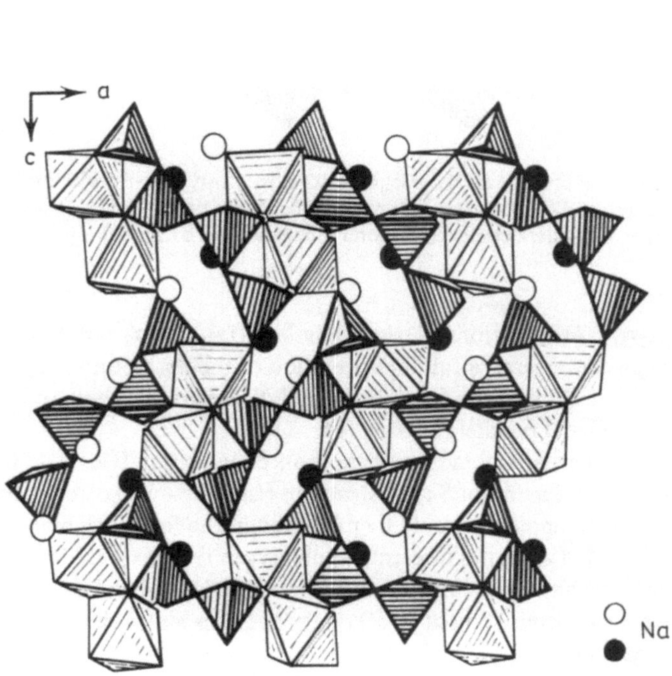

Fig. 10. Structure of $Na_4Sc_2Ge_4O_{13}$, x-z projection[29]. Tetragroups $[Ge_4O_{13}]$ are shown by thick lines. Tetragroups are connected by double on edges Sc octahedra

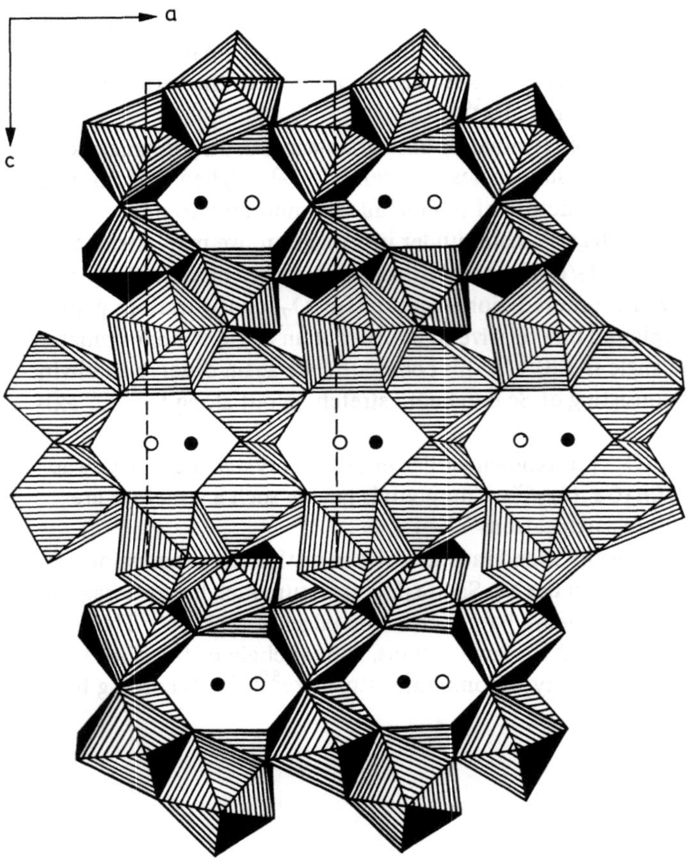

Fig. 11. Structure of $K_2Nd_4[Ge_4O_{13}](OH, F)_4$, G-type; x-z projection[31]. Joint stacking of Nd bands and Ge_4O_{13} tetragroups. Dark bands are on the level b · sin ~ 3/8, light ones – on the level ~ –3/8; dark circles correspond to K cations on b · sin = 0, open circles – on the level ~ 1/2

tentative investigations[51], the formula $K_2Ln_2Ge_2O_7(OH)_2$ was ascribed to the G-type germanates; a full interpretation of their structure has lead to the doubling of the subscript in the formula. Ln polyhedra group together in a ring band pattern, with the band parallel to [010] (Fig. 11).

The germanium radical of the gross composition $[Ge_4O_{13}]$ is characteristic of compounds having an *S-type structure* (Ln = Pr–Nd; however, the structural formula of these compounds should be presented in the following manner: $Na_2Ln_6[Ge_2O_7]_2 \cdot [Ge_4O_{12}]$. Two types of germanium-oxygen radicals were found in the Pr germanate possessing an S-structure[30]: diorthogroup $[Ge_2O_7]$ and the four-fold ring $[Ge_4O_{12}]$. Such a ring, consisting of $[GeO_4]$ tetrahedra, is a new structural element in the germanate class.

The island radical in the form of a ring with 12 $[Ge_{12}O_{36}]$ was discovered in the *$Na_5LnGe_4O_{12}$ germanate* (Fig. 12). In this structure (determined for $Na_5YSi_4O_{12}$[52]), double-sextuple rings are united in a common framework by single Y and Na poly-

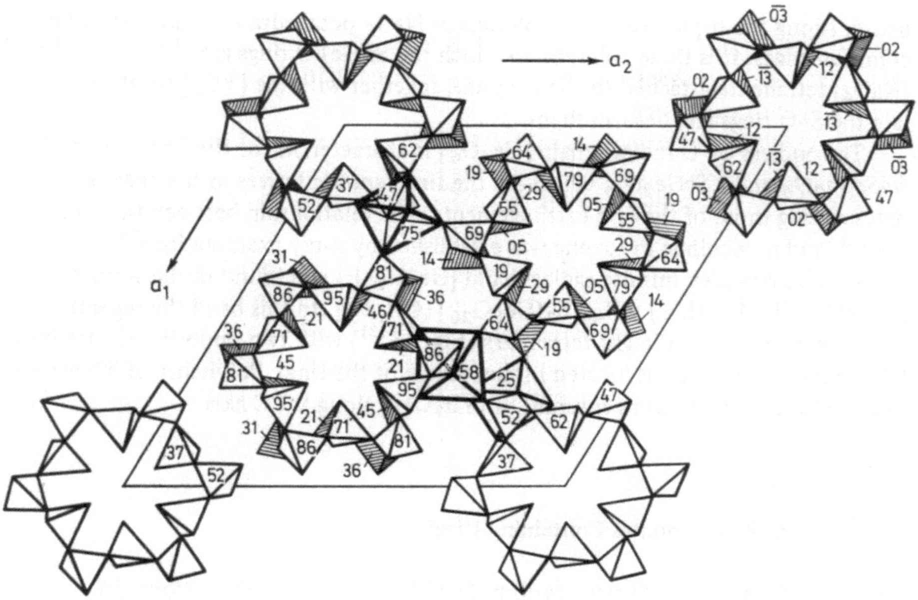

Fig. 12. Structure of $Na_5YSi_4O_{12}$ [52)]

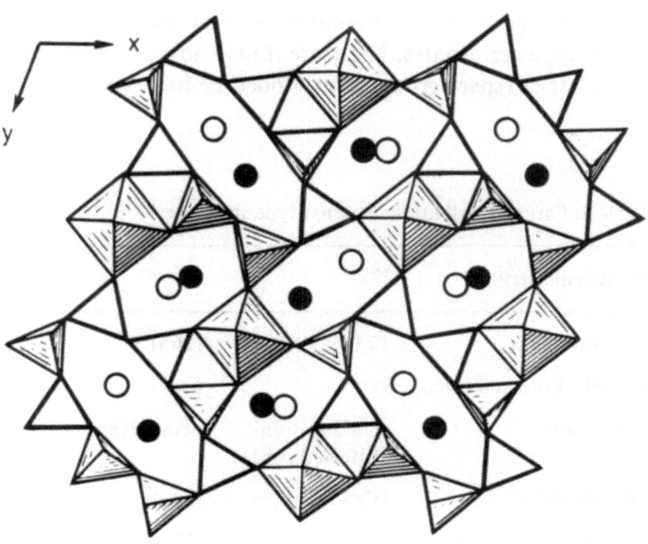

Fig. 13. Structure of $K_4Yb_2[Ge_8O_{20}]$; x-y projection[32)]. Cross section of octagonal tubular $[Ge_8O_{20}]$ columns and of Yb-columns $[Yb_2O_{10}]$ along C axis. Full and open circles are K atoms at z = 3/4 and z = 1/4 c, respectively

hedra. Along the triple axes these columns of NaO_6 octahedra are found linked by common edges. It is these columns on which the Si(Ge)-O rings are "beaded". The $[NaO_4]$ tetrahedra attached to the columns, together with the $[YO_6]$ octahedra, isolate the Si-O rings and deform them.

The endless Ge-O radical-chain $[Ge_2O_6]$ is characteristic of the Sc compounds, $NaScGe_2O_6$ and $KScGe_2O_6$, which are the first representatives of the chain germanates among those of the rare-earth elements. The relationship between these compounds and monoclinic pyroxene was established by x-ray examinations[53].

A more complex infinite radical-band $[Ge_8O_{20}]_\infty$ was found in the germanate $K_2Yb[Ge_4O_{10}]$ (OH, F) = $K_4Yb_2[Ge_8O_{20}]$ $(OH, F)_2$. In this band the repeating link represents a ring of eight tetrahedra (Fig. 13[32]) with two opposite orientations (the center of inversion is located in the middle of the ring). Polyhedra of Yb atoms united at the vertexes form columns of $[Yb_2O_{10}]$ along the C axis.

1.3 Rare-Earth Germanates Containing Divalent Cations

Germanates containing divalent cations, in addition to rare-earth cations, are, as a rule, characterized by an intricate composition and do not form new structural types inherent only to germanates. The structural similarities of these germanates with one or another already known structure were determined with the help of x-ray analyses, but no complete structural determination of these compounds was made.

Germanates containing Ln^{3+} and Me^{2+} ions have structures corresponding to those of garnets, apatites, and perovskites, the latter occur rarer than the first two.

A crystallo-chemical review of compounds of different classes with the structure of garnets is given in[54]; therefore, we shall discuss only certain peculiarities of the garnet-type germanates. There are three known types of cationic polyhedra in the cubic lattice (space group Ia3d): dodecahedron (coordination number = 8), octa-

Table 2. Cation distribution in garnet-type structures[54,59]

Pcint symmetry	222	$\bar{3}$	$\bar{4}$	$\bar{1}$
Position	{24c}	[16a]	(24d)	96
Coordination on oxygen	8	6	4	
Polyhedron	dodecahedron (distorted cube)	octahedron	tetrahedron	
Ideal formula	$\{Ca_3\}$	$[Al_2]$	(Si_3)	O_{12}
Germanate composition	$\{Gd_3\}$ $\{MgGd_2\}$ $\{Gd_3\}$ $\{CaY_2\}$	$[Mn_2]$ $[Mg_2]$ $[Mg_2]$ $[Mn_2]$	$(GaGe_2)$ (Ge_3) $(GaGe_2)$ (Ge_3)	O_{12} O_{12} O_{12} O_{12}

hedron (coordination number = 6), and tetrahedron (coordination number = 4). Trivalent Ln^{3+} cations usually occupy a C position in the garnet structure. Cations of smaller size (Dy–Ln, Y, Sc) can have a coordination number = 8 or 6 and occupy an a-position (Table 2).

Germanium is found usually in a tetrahedral environment of oxygen atoms located in the d-position of the structure. In rarer circumstances, Ge can replace ions in the octahedral a-position[55)].

The structure of apatite has already been considered in Sect. 1.1. This structure is supposed to be very capacious with regard to isomorphic substitutions and preserves itself in various substitutions in the cation, as well as in the anion part of the structure. In silicate (germanate) complex apatites (e. g. $Ca_4La_6(SiO_4)_6(OH)_2$), 4 Ca^{2+} ions, and 6 La^{3+} ions statistically occupy 4f and 6h positions in the unit cell.

La_2MgGeO_6 is one of the few representatives of the germanate class in which germanium occupies an octahedral position[35)]. There are two crystallographically non-equivalent positions for cations in this structure: large sized cations (La) surrounded by 12 oxygen atoms and small-sized ones (Mg, Ca) coordinated with 6 atoms. In a simplified cubic cell the large cations are located at the corners of the elementary cube, the small ones in its center, and the oxygen atoms in the center of the faces.

After an analysis of x-ray data on rare-earth germanates, it was possible to distinguish thirty structural types characteristic of such compounds; in only three of them (the $Ln_2Ge_2O_7$ structures, pyrochlorine type; $Sm_4Ge_3O_9(OH)_6$ and Ln_2MeGeO_6 structures, perovskite type) germanium occurs in a six-fold coordination with respect to oxygen. In the garnet and metagermanate structural type, $La_2Ge_3O_9$, germanium is likely to enter the octahedral position of the structure. In all other cases germanium is accommodated in the tetrahedral voids, thereby creating, as silicon does, different types of finite or infinite germanium-oxygen radicals. Among such compounds, orthocompounds, with their discrete accommodation of Ge-O tetrahedra, are most characteristic of germanates. In the non-alkali Ln-germanate groups, 8 structural types out of 14 are attributed to orthogermanates, whereas in the alkali-rare-earth group we find 7 out of 13, respectively. Other finite radicals are represented by di-, tri- and tetraorthogroups. Ring radicals are present in two germanates: ring $[Ge_4O_{12}]$ in $Na_2Pr_6 [Ge_2O_7]_2 [Ge_4O_{12}]$, S-type and ring $[Ge_{12}O_{36}]$ in $Na_5LnGe_4O_{12}$.

Infinite radicals $[GeO_4]$ are characterized by the adaptability of the Ge-O pattern to the structural constituents made up of larger cations. They can be the conventional pyroxenic bands $[Ge_2O_6]_\infty$, as in $NaScGe_2O_6$, $KScGe_2O_6$, or bands composed of germanium-oxygen tetrahedra and octahedra ($[Ge^{VI}Ge_2^{IV}O_9]_\infty$ in $Sm_4Ge_3O_9(OH)_6$), or a complex tubular radical $[Ge_8O_{20}]_\infty$ detected in the P-type germanate, $K_2Yb[Ge_4O_{10}](OH, F)$.

It is frequently possible to distinguish olivine-like bands in the cation pattern of the compounds under investigation. Such bands were observed both in non-alkali germanates (e.g., $Sm_4Ge_3O_9(OH)_6$) and in alkali-rare-earth germanates (e.g., $NaLnGeO_4$, A-type; $NaLn_3[GeO_4]_2(OH)_2$, E-type; $Na_2Pr[GeO_4]OH$, $K_2Sc_2[Ge_2O_7](OH)_2$).

2 Preparation of Single Crystals of Rare-Earth Germanates

The majority of the literature[1, 15, 20, 41, 56–60] is devoted to a solid-phase synthesis of RE germanates, and considerably less to the production of single crystals of these compounds.

There exist numerous ways of growing RE germanate single crystals: from the melt, from solutions at low temperatures, from flux and by hydrothermal techniques, the latter two being the most widespread methods. The method of growing crystals from the melt appears to be unsuitable in many cases because of the rather high melting temperatures but chiefly because of the volatility of GeO_2 and the incongruency of melting of most germanates. Thus far, no attempts have been made to produce RE germanates from melt under pressure. The method of growing crystals from solution at temperatures of less than 100 °C also proved to be unworkable because of the low solubility of RE germanates in water and aqueous solutions at low temperatures.

With the aid of flux it is possible to lower the temperature in synthesizing rare-earth germanate single crystals down to 800–1000 °C, thereby reducing the volatility of germanium oxide. The use of hydrothermal techniques (simultaneous effect of pressure and temperature in the active aqueous medium) considerably raises the solubility of the compounds under investigation, promotes their production in the form of single crystals, and makes it possible to conduct experiments at significantly lower temperatures (ca. 300–500 °C).

The present review pays a great deal of attention to this method, since it has been established that hydrothermal conditions for synthesizing and growing rare-earth germanate single crystals allow the production of a number of germanates, besides those compounds already known, which cannot be obtained by other methods. The hydrothermal technique appeared to be very effective, primarily in growing single crystals of mixed alkali-rare-earth germanates.

2.1 Crystallization of RE Germanates from Flux

Crystallization of RE germanates from a Bi_2O_3 flux is widely used. This method makes it possible[11] to obtain yttrium-germanate single crystals, Y_2GeO_5 and Y_4GeO_8. Whether one or the other germanate grows depends on the Y/Ge starting ratio in the nutrient. With $Y/Ge < 4/3$, we obtain Y_2GeO_5; with $Y/Ge > 2$, Y_4GeO_8 crystallizes ($T = 1250° – 1400$ °C). $Gd_2Ge_2O_7$ and $Er_2Ge_2O_7$ crystals were also grown from a Bi_2O_3 flux[15,61]. The maximum temperature for melting the polycrystalline powder mixture, $Ln_2Ge_2O_7 + Bi_2O_3$, was 1400 °C; the melt was cooled to 900 °C at the rate of 1–4 deg/h. The produced single crystals contained Bi_2O_3 as impurity phase (3.09% of Bi_2O_3 in $Eu_2Ge_2O_7$) and were as large as 2 x 2 x 3 mm (erbium germanate) and 1 x 2 x 2 mm (gadolinium germanate).

The utilization of Bi_2O_3 as the flux for producing lanthanum and scandium germanates is described in[61–63]. Syntheses were carried out in a programmable furnace for a gradual decrease in temperature. The rate of cooling from 1500 °C down to 900 °C was varied between 1 and 4 deg/h. During crystallization of gadolinium germanates, there appeared polycrystalline intergrowths, which in composition are gadolinium germanates containing 4% Bi_2O_3 impurities, and complex ternary compounds with a possible composition of $GdBiGeO_5$. The content of Bi_2O_3 in lanthanum germanates reaches 8%.

The authors[64,65] used crystallization from a PbF_2-flux to produce the rare-earth germanates $Ln_2Ge_2O_7$, Ln_2GeO_5, and Ln_4GeO_8; PbO_2 was introduced into the system in order to prevent a reduction of the lead. The crystals were obtained at temperatures between 1200–1270 °C by slowly evaporating the flux. At a ratio of $Ln_2O_3 : GeO_2 = 1:2$, the diorthogermanates, $Ln_2Ge_2O_7$, are the dominating solid phase; oxyorthogermanates do not form. Ln_4GeO_8 crystals are formed when there is insufficient GeO_2 because of the high volatility of germanium oxide. RE oxyorthogermanates and diorthogermanates have also been produced by using PbO + GeO_2 as the flux mixture; T = 1270–1290 °C, cooling to 800 °C at a rate of 5–6 deg/h. Paper[64] presents material indicating negative results obtained in growing rare-earth germanates from flux PbO, $Li_2Mo_2O_7$, PbO · $2Bi_2O_3$, $Pb_2V_2O_7$, and $Na_2B_4O_7$.

2.2 Hydrothermal Synthesis of RE Germanates in Ln_2O_3-GeO_2-R-H_2O Systems

The majority of "pure" rare-earth germanates were first produced under "dry" conditions, whereas all the alkali-rare-earth germanates known at present have been grown by hydrothermal techniques[22–32,47,48,50–53,66,67]. Besides the alkali-rare-earth germanates, a whole variety of non-alkali germanates have been grown now by this method. Therefore, the results obtained under hydrothermal conditions are discussed in greater detail in this paper.

Experiments dealing with the synthesis of rare-earth germanates under hydrothermal conditions have been mostly systematic in nature[46,66–76]. Besides producing new crystalline substances, the purpose of these studies has been to determine the behavior of RE oxides in combination with germanium under hydrothermal conditions. The medium contained a solvent-mineralizer, that is, compounds which ensured the alkaline reaction (NaOH, KOH, KF) under experimental conditions. Thus, it was possible to observe the effect of the A-size cation on the crystallization of the same type of compounds and to determine the role of fluorides, the strong complex-forming agents in hydrothermal crystallization.

Experimental conditions:

Nutrient	$GeO_2 + Ln_2O_3$
Molar ratio of oxides contained in the nutrient, $N = Ln_2O_3/GeO_2$	2:1 – 1:6
Solvent, R	NaOH 0–55 wt.% KOH 0–55 wt.% KF 0–60 wt.%
Ratio of solid to liquid phase (in the volume)	1:10(15)
Temperature, T	450–500 °C
Temperature drop, ΔT	30–50 °C
Pressure, P	1200–1500 atm.

The phase formation was investigated as a function of the atomic number of the rare-earth element, concentration, solvent, and the ratio of the initial oxides.

The formation of crystal phases in Ln_2O_3-GeO_2-R-H_2O systems was determined as a function of various parameters of crystallization (N, solvent concentration, C_R, T, P) and of the position occupied by the rare-earth element in the RE series.

2.2.1 Basic Regularities of Phase Formation in Ln_2O_3-GeO_2-NaOH-H_2O Systems

Depending on the atomic number of the rare-earth element in the lantanoid series for systems of this type, one can discern three subgroups differing in the type of crystallization phase diagrams, the character of the paragenesis, and composition of the crystallizing compounds (Table 3).

Examples of the phase diagrams for these subgroups are illustrated in Fig. 14.

The first subgroup includes the largest rare-earths (La–Pr) and is characterized by the formation of non-alkali germanates of the D-type, which occupy most of the field in the crystallization diagram.

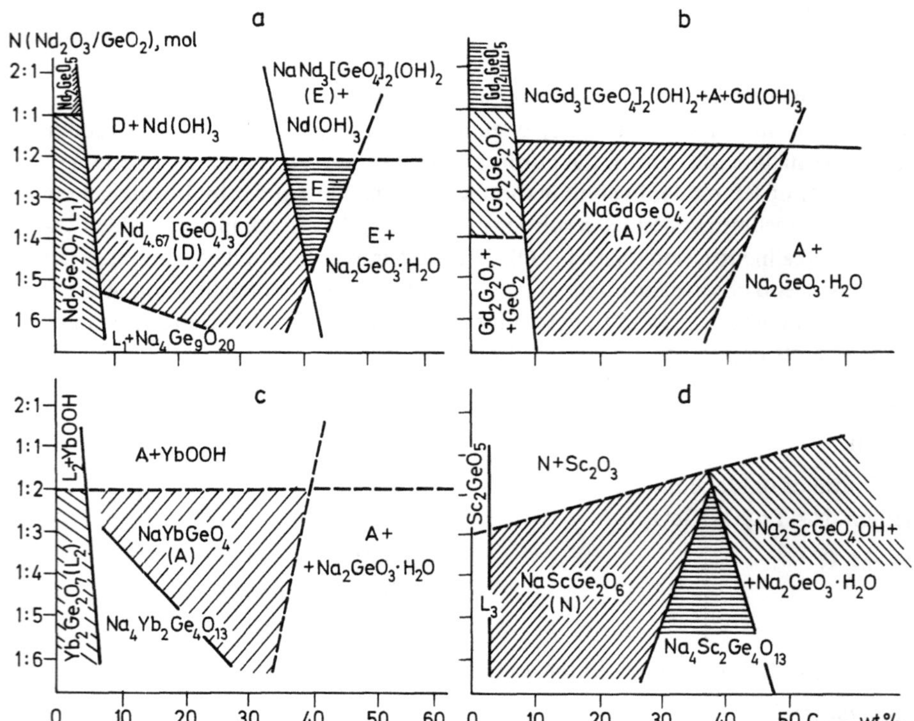

Fig. 14a–d. Examples of phase diagrams of crystallization in the systems Ln_2O_3-GeO_2-NaOH-H_2O at T_{sol} = 500 °C, ΔT = 30 °C. (a) $Ln_2O_3 = Nd_2O_3$ (b) Gd_2O_3 (c) Yb_2O_3 (d) Sc_2O_3

Table 3. The composition of crystalline products in the hydrothermal crystallization in the systems Ln_2O_3-GeO_2-NaOH-H_2O

Composition	Structure type
La [76]	
$La_2Ge_2O_7$	L
$La_{4.67}[GeO_4]_3O$	D
$Na_4Ge_9O_{20}$	
$La(OH)_3$	
Pr [74]	
$NaPrGeO_4$	A
Na_2PrGeO_4OH	R_1
$Na_2Pr_6Ge_4O_{12}[Ge_2O_7]_2$	S
$NaPr_3[GeO_4]_2(OH)_2$	E
$Na_4Ge_9O_{20}$	
$Pr(OH)_3$	
Pr [76]	
$Pr_{4.67}[GeO_4]_3O$	D
$Na_5Pr_2Ge_{10}O_{25}$	
$Na_2Pr_6Ge_4O_{12}[Ge_2O_7]_2$	S
$Na_4Ge_9O_{20}$	
$Pr(OH)_3$	
Nd	
$Nd_{4.67}[GeO_4]_3O$	D
Nd_2GeO_5	Q_1
$Nd_2Ge_2O_7$	L_1
Nd_4GeO_8	M
$NaNd_3Ge_4O_{13}$	T
$NaNd_3[GeO_4]_2(OH)_2$	E
$Na_2GeO_3 \cdot H_2O$	
$Na_4Ge_9O_{20}$	
$Nd(OH)_3$	
Sm	
Sm_2GeO_5	Q_1
$Sm_2Ge_2O_7$	L_1
$NaSmGeO_4$	A
$NaSm_3[GeO_4]_2(OH)_2$	E
$Na_2GeO_3 \cdot H_2O$	
$Sm(OH)_3$	
Eu	
$NaEuGeO_4$	A
$NaEuGeO_4$	B
$NaEu_3[GeO_4]_2(OH)_2$	E
$Na_4Ge_9O_{20}$	
$Na_2GeO_3 \cdot H_2O$	
$Eu(OH)_3$	
Gd	
Gd_2GeO_5	Q_1
$Gd_2Ge_2O_7$	L_2
$NaGdGeO_4$	A
$NaGd_3[GeO_4]_2(OH)_2$	E
$Na_2GeO_3 \cdot H_2O$	
$Gd(OH)_3$	
Tb, Dy	
$Tb_2Ge_2O_7$	L_2
$NaTbGeO_4$	A
Ho	
$Ho_2Ge_2O_7$	L_2
$NaHoGeO_4$	A
$Na_4Ho_2Ge_4O_{13}$	T
$Na_2GeO_3 \cdot H_2O$	
HoOOH	
Er	
$Er_2Ge_2O_7$	L_2
$NaErGeO_4$	A
$Na_4Er_2Ge_4O_{13}$	T
$Na_2GeO_3 \cdot H_2O$	
ErOOH	
Yb	
$Yb_2Ge_2O_7$	L_2
$NaYbGeO_4$	A
$Na_4Yb_2Ge_4O_{13}$	T
$Na_2GeO_3 \cdot H_2O$	
YbOOH	
Sc	
Sc_2GeO_5	Q_3
$Sc_2Ge_2O_7$	L_3
Na_2ScGeO_4OH	R_2
$Na_4Sc_2Ge_4O_{13}$	T
$NaScGe_2O_6$	
$Na_2GeO_3 \cdot H_2O$	

In the second subgroup (Sm–Gd), the D-type compounds with an apatite-like structure disappear and are replaced by two alkali-rare-earth germanates of the E- and A-types: $NaLn_3[GeO_4]_2(OH)_2$ and $NaLnGeO_4$. The system containing Nd can be regarded as a transition between these two subgroups, wherein the germanate $Nd_{4,67}[GeO_4]_3O$ is still in its crystallization stage (a feature of the first subgroup discussed); at the same time there appear E- and A-type germanates (features of the second subgroup).

In the transition to a rare-earth cation with a smaller radius (Tb–Lu), the E-type disappears and the basic field in the phase diagram of crystallization is occupied by the A-type ($NaLnGeO_4$); T-type germanates also appear with a more complex composition, $Na_4Ln_2Ge_4O_{13}$.

The formation of Ln_2GeO_5 germanates is characteristic of the first two subgroups: no Ln_2GeO_5 oxyorthogermanates were detected in systems containing small-sized rare-earth cations.

$Ln_2Ge_2O_7$ diorthogermanates crystallize in practically all the Ln_2O_3-GeO_2-$NaOH$-H_2O systems under study, with the exception of Ce, Pr, Eu. These compounds form in a morphotropic series with a morphotropic jump at gadolium: gadolinium diorthogermanate of tetragonal modification crystallizes under hydrothermal conditions; germanates of larger cations are attributed to triclinic and orthorhombic crystal systems. The Sc system resembles most those systems containing Ho_2O_3, Yb_2O_3, Lu_2O_3; their common features are the presence of $Ln_2Ge_2O_7$ crystallization regions and the formation of $Na_2Ln_2Ge_4O_{13}$ germanates. The remaining Na-Sc crystal phases detected in this system are not characteristic of typical rare-earth elements.

Thus, the basic structural transitions of the subgroups discussed (with the exception of Sc) include:

$$\begin{array}{lll} D + (E) \longrightarrow & E + A & \longrightarrow A + T \\ (La\text{–}Nd) & (Sm\text{–}Gd) & (Tb + Lu). \end{array}$$

The composition of the end products obtained by hydrothermal crystallization is determined to a considerable extent by the solvent concentrations, C_R. In all the systems under investigations one can abserve the transition from oxyortho- and diorthogermanates to a more complex composition of non-alkali or mixed germanates, when C_{NaOH} is raised and the solvent component enters the composition of the crystalline product of the reaction. Thus, for N = 1:2, the following change of stable crystallizing phases is observed with increasing C_R:

$$Nd_2Ge_2O_7 \rightarrow Nd_{4,67}[GeO_4]_3O \rightarrow NaNd_3[GeO_4]_2(OH)_2(+Na_2GeO_3 \cdot H_2O)$$
$$Gd_2Ge_2O_7 \rightarrow NaGdGeO_4(+NaGeO_3 \cdot H_2O)$$
$$Yb_2Ge_2O_7 \rightarrow NaYbGeO_4$$
$$Sc_2Ge_2O_7 \rightarrow NaScGe_2O_6 \rightarrow Na_4Sc_2(Ge_4O_{13}) \rightarrow NaScGeO_4OH(+Na_2GeO_3 \cdot H_2O).$$

For systems containing Ln^{3+} cations, the change in stable crystallizing phases is determined chiefly by the concentrations of alkali atoms in solution. As C_{NaOH}

increases, a depolymerization of the Ge radical takes place. Germanates with associated Ge radicals form, for example, non-alkaline diorthogermanate. These germanates have a linear tetragroup [Ge_4O_{13}] at small solvent concentrations. A small degree of polymerization of the germanium-oxygen radical is seen in the Sc system. Here one can observe a transition from an infinite [Ge_2O_6] chain radical in $NaScGe_2O_6$ via separate links of the Ge_4O_{13} chain, consisting of tetrahedrons bound by vertexes in $Na_4Sc_2Ge_4O_{13}$, to [GeO_4] orthoradicals in Na_2ScGeO_4OH when the alkali concentration increases.

The ratio between oxides in the initial composition plays an important role in the central subgroup (Ln = Sm, Eu, Gd), as well as in the "transitional" system with the participation of Nd_2O_3. The phase boundaries of the crystallization parallel to the C_{NaOH} axis are characteristic of these systems; the transition from A-type to E-type occurs in a broad range of C_{NaOH} variations, when the N value changes. At N = 1 : 1 and higher, the E-type germanates crystallize; at $N < 0.5$, the A-type germanates start to form. In this case the solid-phase composition adequately reflects the composition of the starting system: an excess of Ln_2O_3 is retained in the solid phases.

Germanates of complex composition and structure are formed, as a rule, when excessive Ge constituents exist in the starting compositions (for example, $NaLn_3Ge_4O_{13}$ phases in Nd-, Ho-, Yb-, Sc-containing systems and complex germanates in Pr systems).

2.2.2 Basic Regularities of Phase Formation in Ln_2O_3-GeO_2-KF-H_2O Systems

Depending on the atomic number of the rare-earth element in the series of the Ln_2O_3-GeO_2-KF-H_2O system, one can discern three subgroups: Ln = Pr-Eu; Gd; Tb-Lu distinguished by the type of crystallization phase diagrams, and by the composition of the crystal phase and their paragenesis (Fig. 15). As in Sc_2O_3-GeO_2-NaOH-H_2O, scandium is characterized by a distinctive type-C diagram, and also to a great extent by another crystal phase composition (Table 4).

The major distinguishing features of the various subgroups are: the first group is characterized by the crystallization of non-alkali germanates having an H-type structure ($Ln(OH)_3 \cdot 6Ln_2GeO_4[O, (OH, F)_2]$) and of alkali-rare-earth germanates having a G-type structure ($K_2Ln_4Ge_4O_{13}(OH, F)_4$); during the transition from Eu to Gd, the H-type germanate disappears. With further movement to the right in the rare-earth-element series, the main crystallizing phases become diorthogermanates and P-type germanates appear in a rather narrow N-C range. Thus, the decrease in size of the rare-earth cation leads to the following change in the basic stable crystallizing phases:

$$\underset{(Pr-Eu)}{H + G} \longrightarrow \underset{(Gd)}{G} \longrightarrow \underset{(Tb-Lu).}{L_2 + (F + P)}$$

It should be noted that such a subdivision is for the most part conditional. If we take into account the structure of the phase at a given composition, the Yb-system stands out as a separate subgroup, since $Yb_2Ge_2O_7$, a formula analog of Tb and Gd diorthogermanates, crystallizes in two polymorphic modifications. These modifica-

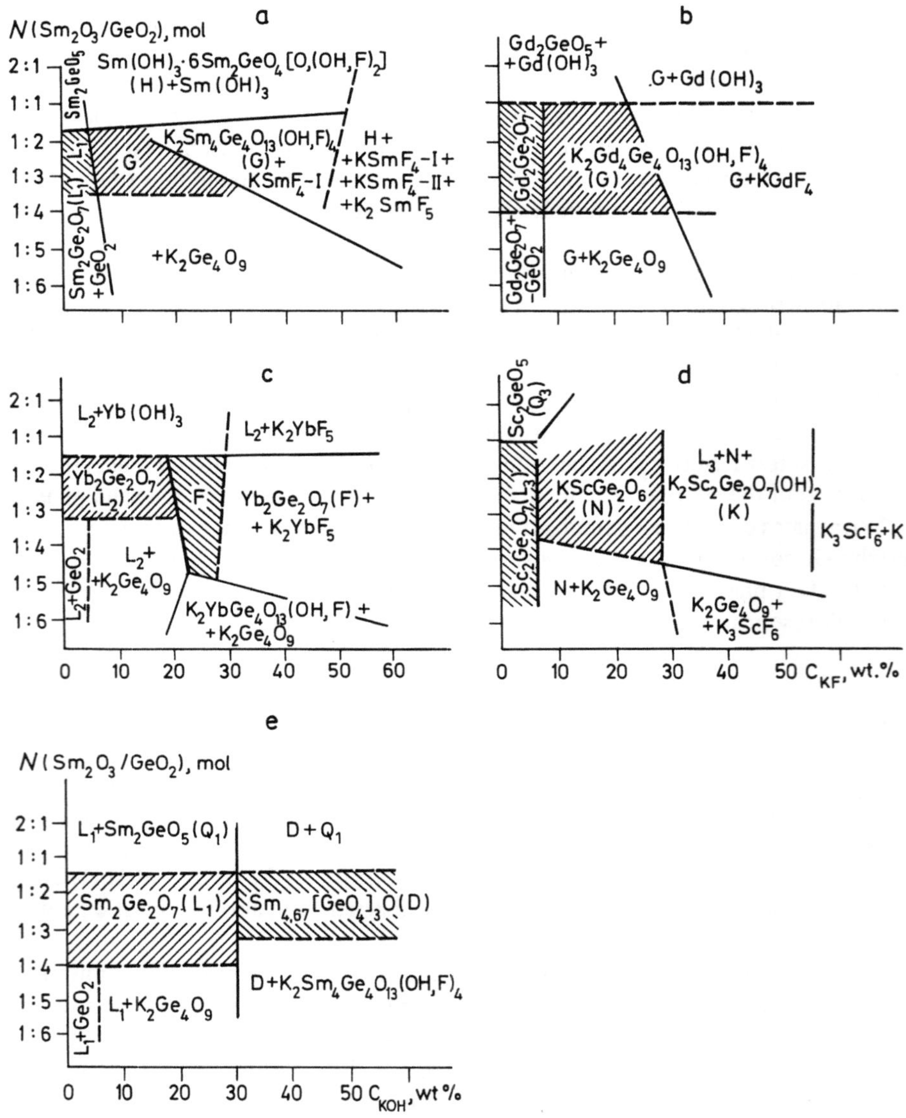

Fig. 15 a–e. Examples of phase diagrams of crystallization in the systems Ln_2O_3-GeO_2-KF-H_2O (a–d) and Sm_2O_3-GeO_2-KOH-H_2O (e). (a) $Ln_2O_3 = Sm_2O_3$ (b) Gd_2O_3 (c) Yb_2O_3 (d) Sc_2O_3

tions are tetragonal, as shown earlier for small rare-earth cations, and cubic, having a pyrochlorine structure. Moreover, in the Yb systems, a potassium-rare-earth germanate of the P-type ($K_2YbGe_4O_{10}(OH, F)$) was detected in a very narrow range of ratios of the starting components.

Diorthogermanates are characteristic of practically all the systems studied above (with the exception of the Eu-containing system, as in the case of R = NaOH); a varia-

Table 4. The composition of crystalline products in the hydrothermal crystallization in the systems Ln_2O_3-GeO_2-KF-H_2O and Sm_2O_3-GeO_2-KOH-H_2O

Composition	Structure type	Composition	Structure type
Pr		*Ho*	
$Pr_2Ge_2O_7$	L_1	$Ho_2Ge_2O_7$	L_2
$Pr(OH)_3 \cdot 6Pr_2GeO_4[O, (OH, F)_2]$	H	$K_2Ge_4O_9$	
$K_2Pr_4Ge_4O_{13}(OH, F)_4$	G	K_2HoF_5	
$Pr(OH)_3$		$Ho(OH)_3$	
Sm			
Sm_2GeO_5	Q_1	*Yb*	
$Sm_2Ge_2O_7$	L_1	$Yb_2Ge_2O_7$	L_2
$Sm(OH)_3 \cdot 6Sm_2GeO_4[O, (OH, F)_2]$	H	$Yb_2Ge_2O_7$	F
$K_2Sm_4Ge_4O_{13}(OH, F)_4$	G	$K_2YbGe_4O_{10}(OH, F)$	P
$K_2Ge_4O_9$		$K_2Ge_4O_9$	
$KSmF_4$-I			
$KSmF_4$-II		K_2YbF_5	
$Sm(OH)_3$		$Yb(OH)_3$	
Eu			
$Eu(OH)_3 \cdot 6Eu_2GeO_4[O, (OH, F)_2]$	H	*Sc*	
$K_2Eu_4Ge_4O_{13}(OH, F)_4$	G	Sc_2GeO_5	Q_3
$K_2Ge_4O_9$		$Sc_2Ge_2O_7$	L_3
$KEuF_4$-I		$KScGe_2O_6$	N
$KEuF_4$-II		$K_2Sc_2Ge_2O_7(OH)_2$	K
$Eu(OH)_3$		$K_2Ge_4O_9$	
Gd			
Gd_4GeO_8	M	*Sm (R = KOH)*	
Gd_2GeO_5	Q_1	Sm_2GeO_5	Q_1
$Gd_2Ge_2O_7$	L_1	$Sm_2Ge_2O_7$	L_1
$K_2Gd_4Ge_4O_{13}(OH, F)_4$	G	$Sm_{4.67}[GeO_4]_3O$	D
$K_2Ge_4O_9$		$Sm_4Ge_3O_9(OH)_6$	
$KGdF_4$-I		$K_2Sm_4Ge_4O_{13}(OH, F)_4$	G
$KGdF_4$-II		$K_2Ge_4O_9$	
$Gd(OH)_3$		$Sm(OH)_3$	

tion of the atomic number of Ln^{3+} changes the structural type of $Ln_2Ge_2O_7$ ($L_1 \longrightarrow L_2 \longrightarrow F$).

The scandium system does not fit into the general series of Ln-containing systems. Preserving the normal trend of interactions between the initial oxides, there appear new crystal phases characteristic only of scandium (pyroxen $KScGe_2O_6$, diorthogermanate $K_2Sc_2Ge_2O_7(OH)_2$).

In the systems under investigation, potassium-fluoride mineralizer acts not only as a transporting agent, but also as an active component in the system; in addition the concentration of KF substantially influences the composition and sequence of formation of the crystal phases in Ln_2O_3-GeO_2-KF-H_2O systems. The role of KF concentration is illustrated in the series of stable crystallizing compounds at T = 500 °C (N = 1 : 2, increasing C_{KF}):

$$Sm_2Ge_2O_7 \longrightarrow K_2Sm_4Ge_4O_{13}(OH, F)_4 \longrightarrow Sm(OH_3) \cdot 6\,Sm_2GeO_4[O(OH, F)_2] +$$
(L1) (G) (H)

$$+\ KSmF_4(K_2SmF_5)$$

$$Cd_2Ge_2O_7 \longrightarrow K_2Gd_4Ge_4O_{13}(OH, F)_4 + (KGdF_4)$$
(L2) (G)

$$Yb_2Ge_2O_7 \longrightarrow Yb_2Ge_2O_7 + (K_2YbF_5)$$
(L2) (F)

$$Sc_2Ge_2O_7 \longrightarrow KScGe_2O_6 \longrightarrow K_2Sc_2Ge_2O_7(OH)_2 + (K_3ScF_6).$$
(L3) (N) (K)

The phase boundaries of crystallization (with the exception of the boundaries outlining the formation of Ln_2GeO_5 and $Ln_2Ge_2O_7$) are usually significantly inclined relative to the C_{KF} axis. This indicates that, although the concentration of the mineralizer plays an important role, it is not always definitive in the formation of crystal phases. Most significant is the influence of C_{KF} in systems with smaller rare-earth cations (Yb, Sc).

The influence of the initial oxide ratios on the composition and successive changes in the crystal phases is very clearly seen in all the systems under investigation. The boundaries between the regions of oxyorthogermanate crystallization and of the diorthogermanate crystallization lie in all the systems at N = 1:1. The transition from G-type germanates to H-structure compounds also takes place with a variation of N (from 0.5 to 1) in a broad variation range of KF/H_2O.

2.3 General Regularities of Formation of Rare-Earth Germanates in Hydrothermal Systems Ln_2O_3-GeO_2-R-H_2O

Analyses of the end products of crystallization and of the N-C diagrams make it possible to break down the lantanoid series into three principle subgroups; the boundaries between them can be slightly shifted, depending on the type of solvent. Scandium stands out as a separate subgroup. Table 5 depicts the structural types of the major germanates obtained in the systems Ln_2O_3-GeO_2-R-H_2O, where R is NaOH and KF. The role of C and N is illustrated in Table 6.

The boundary of the "large" rare earths, with R = NaOH, lies between Nd to Sm; with R = KF it shifts to the right (Eu-Gd). The position of the boundaries is an indication of different "capacities" of individual structural types. From among the non-alkali Ln germanates, the D-type (apatite) structure possesses the least "capacity" for different RE elements. Germanates with such a structure were detected only for the largest rare-earth elements (La-Nd) and in one case (solvent: KOH) for Sm. The H-type was detected in five systems on the left side of the lantanoid series, with euro-

Table 5. The main structural types of Ln germanates synthesized in the systems Ln_2O_3-GeO_2-NaOH(KF)-H_2O

R \ Ln	La	Pr	Nd	Sm	Eu	Gd	Tb	Dy	Ho	Er	Tm	Yb	Lu	Sc
NaOH	L D	L D	L_1 D E	L_1 A E	A E	L_2 A E	L_2 A	L_2 A	L_2 A	L_2 A	L_2 A	L_2 A	L_2 A	L_3 N T, R_2
KF		L_1 H G	L_1 H G	L_1 H G	H G	L_2 H G	L_2 G	L_2	L_2	L_2	L_2 P	L_2 F P	L_2 P	L_3 N K

Table 6. Structure types of Ln germanates synthesized in 40 wt% NaOH and KF

R	N Ln_2O_3/GeO_2 \ Ln	La	Pr	Nd	Sm	Eu	Gd	Tb	Dy	Ho	Er	Tm	Yb	Lu	Sc
NaOH	2:1	D	D	E	E	E	E	A	A	A	A	A	A	A	N
	1:4	D	D	D + E	A	A	A	A	A	A	A	A	A	A	T
KF	2:1		H	H	H	H	G			L_2	L_2		L_2		L_3 + K
	1:4		G	G	G	G	G	L_2	L_2	L_2	L_2 + (P)		F + P	P	L_3 + K

pium as the smallest cation for this structural type. Since both structural types (D and H) possess fragments of the apatite structure; they may be regarded as being characteristic of large cations, regardless of the solvent type used in the hydrothermal reaction or the conditions for producing the crystals. The unusual characteristic associated with crystallization of most germanate as well as silicate systems occurring under hydrothermal conditions is the advantageous entry of the cation (anion) of the solvent into the composition of the solid phase of the reaction products. At small concentrations ($C_R \gtrsim 10$ wt.%), the stable crystal phases which are formed in pure water (Ln_2GeO_5, $Ln_2Ge_2O_7$) are preserved, regardless of the solvent type used in the system.

One may assume that in such concentrations the solvent serves only as a "kinetic" factor in accelerating the processes of synthesis and transfer. It is the rare-earth element that determines the compounds being formed: oxyortho- and diorthogermanates both in water and in weak concentrations of alkali-fluoride solvents. With an increase in C_R, the determining role is played by both the RE cation (formation of D-structure in alkali solvents and H-type in fluorides) and the $Me^{1+} + Ln^{3+}$ pair (formation of A-type compounds).

The change in the type of mineralizer and, consequently, the change in the basic form of transfer lead to an essential difference in the end product of the reaction.

Comparing the results of hydrothermal synthesis in Sm systems with different types of solvents (NaOH, KOH, KF) allows one to draw conclusions concerning the advantageous role of the steric factor in the process of formation of one or another structure. Thus, in the presence of KOH, a monotypical solvent with NaOH, no formation of the A-type germanates takes place; however, the compound characteristic of the fluoride system, a germanate with a G-structure, can be observed. Hence, the difference in the type of anion and in the complex-forming properties of the solvent (OH^--F^-) influences the types of N-C diagrams less than does the difference in the type of cations.

In the Na system, the stable phases (A, E) differ sharply from those in the K system. Probably no formation of either sodium rare-earth germanates, $NaLnGeO_4$, or $NaLn_3[GeO_4]_2(OH)_2$ can take place in a K-system through the simple substitution of sodium by potassium, because of their different sizes ($r_{Na}^{+} = 0.98$ A, $r_{K^+} = 1.33$ A). The larger K^+ cation forms one new compound (G-type), according to the structural pattern, but it does not enter into other phases (H and D). If one follows the influence of the type and composition of the solvent on the crystal phase formation in the rare-earth series, one may notice a decrease in the stability of potassium-rare-earth germanates for smaller Ln^{3+} ions. But in the Ho system (R = KF), no areas of stable crystallization of alkaline-rare-earth germanates have been established. In the Yb and Lu systems, there appears a new P-type germanate with a narrow region of crystallization. An exception is the Sc system, where in two new K-Sc germanates with a rather broad area of crystallization have been found.

The structural type of olivine (A-type), known for its capacity to perform a substitution, has been found in the Na systems in the whole rare-earth series from Sm to Lu.

Thus, when comparing the three types of mineralizers (NaOH, KOH, KF), one may conclude that the replacement of the solvent cation leads to a sharp change in the structural types of the RE stable germanates, while the change in the anion (OH^- for F^-) leads mainly to variations in the composition of accessory phases.

A great excess of one of the compounds in the system leads to the crystallization of compounds characterized by a stability in the subsystem which does not contain a second basic component (GeO_2 or Ln_2O_3). In this manner an excess of Ge in Ln_2O_3-GeO_2-KOH(KF)-H_2O systems always leads to the crystallization of tetragermanate $K_2Ge_4O_9$, characterized by a very high degree of stability in the subsystem GeO_2-KOH(KF)-H_2O. For Na-systems, the crystallization of the stable Na-germanates $Na_4Ge_9O_{20}$ and $Na_2GeO_3 \cdot H_2O$ is also evident. When the Na content in the system is raised, its content in the synthesis products also grows ($Na_4Ge_9O_{20}$ with a relation Na/Ge = 0.44 is replaced by $Na_2GeO_3 \cdot H_2O$ with a relation Na/Ge = 2).

2.4 Hydrothermal Synthesis of Mixed Germanates Containing Divalent and Rare-Earth Ions

Under hydrothermal conditions, germanates with a garnet structure, $Ca_3Sc_2Ge_3O_{12}$ and $Cd_3Sc_2Ge_3O_{12}$[77], have been synthesized. Germanium garnets were obtained at a temperature of ~ 550° and a pressure of 1300 atm. The Ca-Sc garnet was produced in solutions of 6% NH_4Cl from the nutrient $CaCO_3 + Sc_2O_3 + GeO_2$ (with molar ratios 3:1:3, 6:1:3, 6:1:2) and in a solution of $CaCl_2$ (10%) and ratios of the starting components in the nutrient of 3:1:3 and, 6:1:3. The yield of crystals is small.

Cd-Sc garnet was synthesized in NH_4Cl solutions (6%) from the nutrient CdO + $Sc_2O_3 + GeO_2$, (the molar ratio 6:1:2) and in $CdCl_2$ solutions (10%) from the nutrient $3\,CdO \cdot Sc_2O_3 \cdot 3GeO_2$ or $6CdO \cdot Sc_2O_3 \cdot 3GeO_2$.

An excessive content of oxides of large bivalent cations (Ca, Cd) in the starting mixture or solvent initiates the formation of garnets and increases their yield. The role of trivalent cations during synthesis, however, is not as great: if the garnet is not obtained from a nutrient with a molar ratio of $A_2O:Ln_2O_3:GeO_2$, equal to 3:1:3, the addition of mixtures 3:2:3 or 3:2:2 does not yield positive results. Under hydrothermal conditions, germanium garnets form more readily than silicate garnets. The substitution of Si^{4+} for a large isovalent Ge^{4+} leads to the loosening of the Ge-O bonds, as compared with the Si-O bonds, thereby making the necessary deformation of the coordination polyhedra easier.

The compound $Sr_3Sc_2Ge_3O_{12}$ was produced in the form of colorless transparent crystals of no distinct shape several hundredths of a fraction of a millimeter in size. The lack of a difference between the lattice parameters of the garnets produced ceramically and hydrothermally indicates the absence of any noticeable hydration of the garnets during their formation in aqueous media[34].

3 Some Physico-Chemical Characteristics of Rare-Earth Germanates

Since many studies conducted in the past have dealt with the production and investigation of the physical properties of rare-earth germanates, only a small portion of this paper will be devoted to that aspect of the subject.

Table 7. Physico-chemical properties of some Ln germanates[15,44,58,88]

Composition	Density g/cm^3	Melting temperature, °C	Refraction indices N_g	N_p
La_2GeO_5	5.97	1800	1.940	1.922
Nd_2GeO_5	6.55	1820 ± 25	1.948	1.939
Gd_2GeO_5	7.07	1880	1.941	1.920
Y_2GeO_5	4.83	1975 ± 15	1.848	1.832
Er_2GeO_5	7.28	2000 ± 50	1.904	1.894
La_4GeO_8	6.30	1720^a	1.973	1.953
Gd_4GeO_8	7.77	1650^a	1.975	1.955
Y_4GeO_8	5.10	2000	1.887	1.855
Er_4GeO_8	8.02	1600^a	N_m 1.93	
$La_{4.67}[GeO_4]_3O$	5.89	1770	1.930	1.909
$La_2Ge_2O_7$	5.67	1680^a	1.900	1.876
$Nd_2Ge_2O_7$	6.16	1820 ± 25^a		
$Gd_2Ge_2O_7$	6.70	1770	1.922	1.906
$Y_2Ge_2O_7$	4.69	1800 ± 25	1.840	1.830
α-$Er_2Ge_2O_7$	6.99	1875 ± 50	1.890	1.880
β-$Er_2Ge_2O_7$			1.938	1.927
$Sc_2Ge_2O_7$	4.46	1850	1.847	1.832
$La_2Ge_3O_9$	5.49	1300	1.995	1.980

^{a} melting with decomposition

From the general characteristics of rare-earth germanates the following properties should be mentioned: their high refraction indices and high melting points. Most refractory are the oxyorthogermanates. The alkali-rare-earth germanates are characterized by much lower melting points (Table 7).

Most fully studied and, perhaps, most interesting are the spectroscopic characteristics of Ln germanates, a further discussion of which will appear in greater detail below.

3.1 Infra-Red Spectra of Rare-Earth Germanates

Infra-red (IR) spectra of non-alkali rare-earth germanates (with the exception of germanates of the H-, F-types and Sc_2GeO_5) have been discussed at length in[1]. IR spectra of new non-alkali germanates (H- and F-types, Sc_2GeO_5) and of all the alkali-rare-earth compounds for the whole series of lantanoids have been considered in detail in[7,51,78] (factor-group analysis, position of bands in the spectrum, establishment of the coordiantion number, and Ge-O radical forms). Therefore, this review will mainly present typical spectra for compounds with different structural types which can be used for identifying the rare-earth crystal phases.

Two techniques have been used to measure the IR spectra in the middle (5000–400 cm^{-1}) and distant (400–33 cm^{-1}) ranges of the spectrum. IR spectra in the 5000–400 cm^{-1} range were taken from a UR-10 spectrophotometer; the samples were prepared by standard pressing techniques with KBr. The IR spectra in the range of 400–33 cm^{-1} were measured with a F1Z-3 spectrophotometer, the samples were prepared by dissolving them in molten paraffin. The samples were then studied by using a mirror reflection method. The IR spectra of non-alkali germanates are shown in Fig. 16, and the IR spectra of alkali-rare-earth germanates, in Fig. 17.

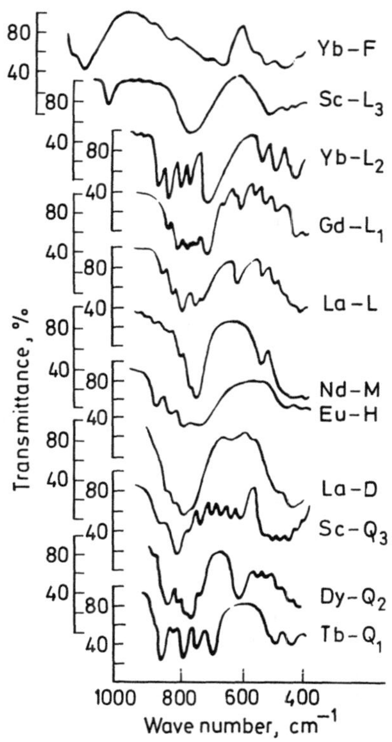

Fig. 16. Infra-red spectra of germanates, not containing alkali ions, of different structural types

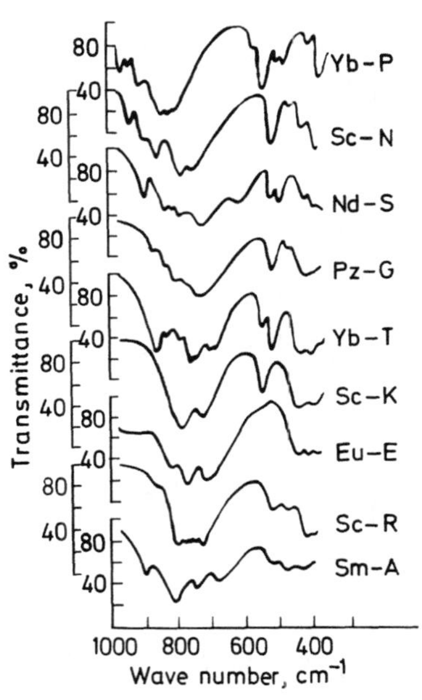

Fig. 17. Infra-red spectra of alkali-rare-earth germanates of different structural types

Table 8. Frequencies of the absorption bands in IR spectra of alkali-rare earth germanates

$NaSmGeO_4$, A-type		Na_2ScGeO_4OH, R_2-type		$NaSm_3[GeO_4]_2(OH)_2$, E-type		$K_2Sc_2Ge_2O_7(OH)_2$, K-type	
vibration type	ν, cm^{-1}	vibration type	ν, cm^{-1}	vibration type	ν, cm^{-1}	vibration type	ν, cm^{-1}
$\nu_s GeO_4$	840 m	$\nu_s GeO_4$	860	δ Ln-OH	930–850 m		790 sh
	785 vs		830 s		797	$\nu_s GeO_3$	775 vs
$\nu_{as} GeO_4$	734 s		813 s	$\nu_{as} GeO_4$	775		760 vs
	670 s	$\nu_{as} GeO_4$	798 s		754 s		720 s
	495 s		780 s		685 s	$\nu_s GeOGe$	568 s
$\delta_{as} GeO_4$	475 s		733 vs	γ Ln-OH	570 vw	δ GeO_3	475 vs
	450 sh		525 s		465 m	ν Sc-O	750 sh
		$\delta_{as} GeO_4$	496 s	$\delta_{as} GeO_4$	430 m		
		ν Sc-O	450 s		412 m		
			430 sh				

Table 8. (continued)

$Na_4Yb_2Ge_4O_{13}$, T-type		$K_2Sm_4Ge_4O_{13}(OH, F)_4$, G-type		$Na_2Nd_6[Ge_2O_7]_2Ge_4O_{12}$, S-type		$NaScGe_2O_6$, N-type		$K_2Yb[Ge_4O_{10}](OH, F)$, P-type	
vibration type	ν, cm^{-1}	vibration type	ν, cm^{-1}	vibration type	ν, cm^{-1}	vibration type	ν, cm^{-1}	vibration type	ν, cm^{-1}
ν_{as}GeOGe,	868 s	ν_{as}GeOGe,	871 w		893 m		920 m		931 m
$\nu_{as}GeO_2$,	842 w	$\nu_{as}GeO_3$	849 w	ν_{as}GeOGe	840 m	ν_{as}GeOGe,	885 sh		910 m
$\nu_s GeO_2$,	821 w	$\nu_s GeO_3$	840 w		827 m	$\nu_s GeO_2$,	868 s	ν_{as}GeOGe,	881 w
$\nu_{as}GeO_3$,	792 vs	$\nu_{as}GeO_2$,	805 w	$\nu_{as}GeO_2$,	800	$\nu_{as}GeO_2$	815 vs	$\nu_{as}GeO_2$,	854 sh
$\nu_s GeO_3$.	782 vs	$\nu_s GeO_2$	753 s	$\nu_s GeO_2$,	782 sh		785 vs	$\nu_s GeO_2$	836 vs
	748 s		730 sh	$\nu_{as}GeO_3$,	755 vs	ν_sGeOGe	512 s		823 vs
	725 s	ν_sGeOGe	547 s	$\nu_s GeO_3$	740 sh		488 m		808 vs
	580 m	δ Ge-O	498 w		714 sh	δ Ge-O,	463 s		606 sh
ν_sGeOGe	550 m	ν Ln-O	465 sh		690 m	ν Sc-O	425 vs		574 vs
	500 sh		437 m	ν_sGeOGe	565 sh			ν_sGeOGe	534 vs
ρGeO_3	465 sh		425 sh	ν''_sGeOGe	549 m				517 m
ρGeO_2	457 m		407 w	ρGeO_3	473 sh				461 w
δ GeOGe	438 m			δ GeOGe	456 sh			δ Ge-O	408 m
ν Ln-O	425 m			ν Ln-O	442 m				
	415 m				410				

s – strong, vs – very strong, m – mean, sh – shoulder, w – weak, vw – very weak

The positions of maxima of the absorption bands in the IR spectra of representative structural types of alkali-rare-earth germanates are presented in Table 8. The compounds are listed in the order of complexity of the Ge-O radical, from orthogermanates to compounds with differing degrees of Ge-O radical condensation (diorthogermanates, compounds with linear tetragroups, and compounds with infinite radicals).

3.2 Luminescence Properties of Rare-Earth Germanates

3.2.1 Luminescence Spectra

During photoexcitation, luminescence was observed both in unactivated rare-earth germanates of different structural types and in rare-earth germanate crystals doped with other rare-earth elements. We investigated luminescence of germanates of the A-, E-, G-, and H-types synthesized under hydrothermal conditions.

During photo-excitation of an unactivated $NaLnGeO_4$ crystal, luminescence was observed only for germanates of Eu and Tb ($NaGdGeO_4$ was not investigated)[79]. The first type of compound showed an orange luminescence, and the second, a green one (Fig. 18). Luminescence in both cases resulted from an electronic transition inside the 4f shell of trivalent rare-earth ions; for europium those transfers are $^5D_0 \longrightarrow {}^7F_4$. For terbium the lines 488, 491 nm correspond to transitions $^5D_4 \longrightarrow {}^7F_6$, and the lines 542, 547, 551 nm to transitions $^5D_4 \longrightarrow {}^7F_5$.

Europium germanates of the E-, G-, and H-type also possess photoexcited luminescence (Fig. 19a–c).

The luminescence and excitation spectra of Na(Li), Ln-germanates activated by Eu^{3+}, and Tb^{3+} ions have been reviewed in[49,80]. Matrices containing the rare-earth cations La, Gd, Y, and Lu (Table 9, 10) were selected for investigation.

All the phosphors are characterized by luminescence in the red region of the spectrum as a result of an electronic transition in Eu^{3+} from $^5D_0$ to $^7F_2$ levels.

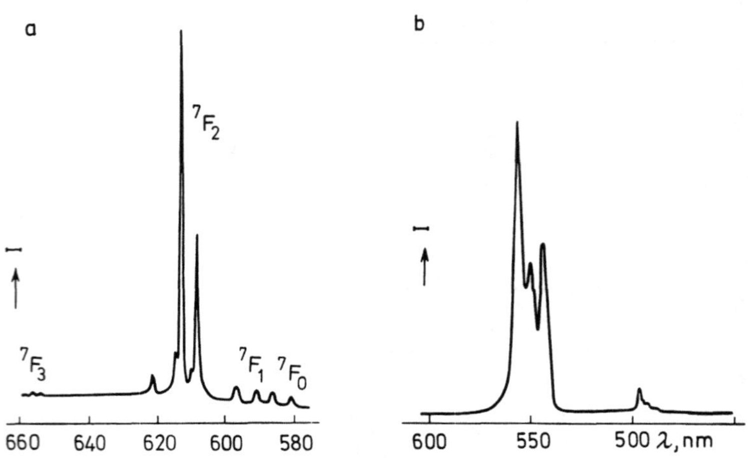

Fig. 18a, b. Luminescence spectra of $NaEuGeO_4$ (**a**) and $NaTbGeO_4$ (**b**)

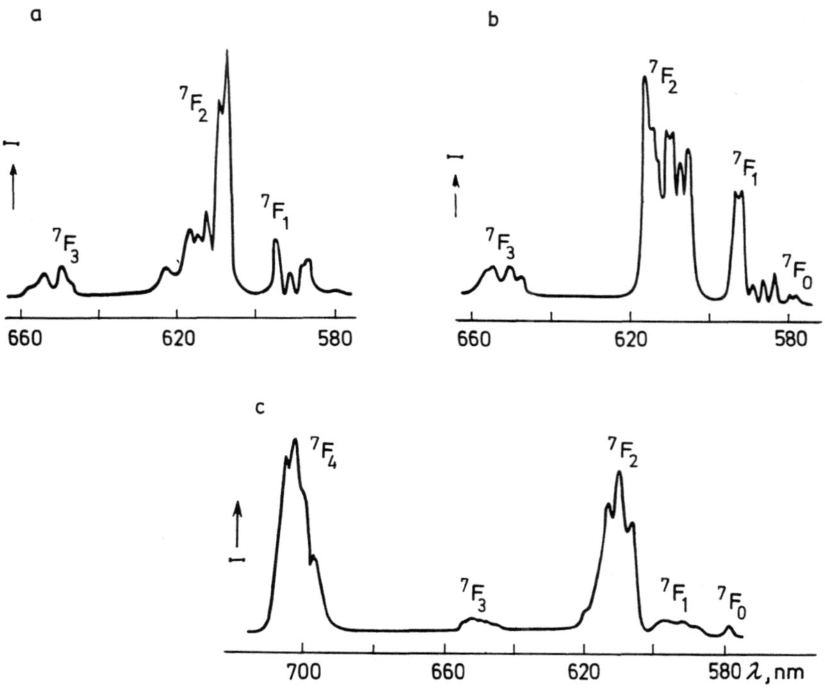

Fig. 19a–c. Luminescence spectra of europium germanates of E- (**a**), G- (**b**) and H-types (**c**)

Table 9. Luminescence spectra of $NaYGeO_4:Eu$ and $NaEuGeO_4$

$NaYGeO_4:Eu^{3+}$ (5%)		$NaEuGeO_4$			Transition	Polarization
λ, nm	I/I_{max}, % T = 300 °K	λ, nm	I/I_{max}, % T = 300 °K	I/I_{max}, % T = 120 °K		
579.5	9	577.5	3	−1	$^5D_0$-$^7F_0$ electric-dipole	π
586.0	5	584.2	4	9	$^5D_0$-$^7F_1$	π
593.0	4.5	591.4	5	8.5	magnetic-dipole	σ
598.3	4.5	596.2	5	8		π
610.0	36	608.5	40	75	$^5D_0$-$^7F_2$	π
610.8	6	609.2	10	10	electric-dipole	π
613.7	100	612.4	100	100		σ
614.6	8	613.0	15	22		π
623.5	5	621.7	5	6		σ

Table 10. Maximum quantum efficiency in the u. v. excitation region, radiant efficiency for cathode-ray excitation for some Tb^{3+} and Eu^{3+}-activated phosphors[49,80)]

Composition	Maximum quantum efficiency (% u.v. excitation)	Radiant efficiency (% c.r. excitation 20 kv)	Position of excitation band (hm)
$NaYGeO_4$:Tb	50	2	395
$LiYGeO_4$:Tb	30	3	402
$LiGdGeO_4$:Tb	40	0.3	436
$LiLaGeO_4$:Tb	30	0.2	410
$LiLaGeO_4$:Eu	10	0.1	282
$LiGdGeO_4$:Eu	55	0.5	242
$LiYGeO_4$:Eu	55	1.0	239
$LiLuGeO_4$:Eu	45	1.0	238
$LiLaGeO_4$:Eu	10	1.0	260
$NaGdGeO_4$:Eu	60	2.5	238 (232)[a]
$NaYGeO_4$:Eu	60	2.5	232 (228)[a]
$NaLuGeO_4$:Eu	70	2.5	230

[a] Data obtained by the authors for hydrothermal crystals

A detailed analysis of the luminescence spectra has been made for the olivine-like matrix with an Eu^{3+} activator[79)]. The splitting of levels and the intensity of the lines were considered on the basis of the crystal field model[81,82)].

In the first approximation the splitting of levels is determined only by the even part of the expansion of the crystal field potentials into a series with respect to the irreducible tensor operators. The Hamiltonian of the crystal field can be expressed by equivalent operators[83)] and has the following form:

$$\hat{H}_{cr} = \sum_{n,m} [A_{nm} \langle r^n \rangle \theta_n] O_n^m,$$

where $\theta_2 = dy$, $\theta_4 = \beta_y$, and $\theta_6 = \gamma_y$. These numerical coefficients depend, e. g., on the orbital quantum number of electrons in an unfilled shell and on the angular momentum J (or L and S).

An analysis of the luminescence spectrum of $NaEuGeO_4$ (Fig. 18a, b, Table 9) and the solution of the secular expression yielded the following crystal field potentials, thereby describing satisfactorily the experimental splitting of the $^7F_1$ and $^7F_2$ levels:

$A_{20} \langle r^2 \rangle = 100\ cm^{-1}$ $\qquad A_{42}^{+} \langle r^4 \rangle = 410\ cm^{-1}$

$A_{40} \langle r^4 \rangle = 190\ cm^{-1}$ $\qquad A_{44}^{+} \langle r^4 \rangle = -360\ cm^{-1}$

$A_{22}^{+} \langle r^2 \rangle = -790\ cm^{-1}$

The mean quadratic error of the calculated energy values is 15 cm^{-1}. The crystal field parameters were determined with the help of a special program designed to search for the minimum of a function of numerous variables[84)]. In the computations, we minimized the sum of squares of the energy-level values calculated from the experimental data. To calculate the proper energy levels as a function of varying parameters, we used the standard programs for determining the eigenvalues of a Hermitian matrix[85).

The analysis of the luminescence spectra of trivalent europium in the series of germanium compounds shows the absence of green luminescence of the Eu^{3+} ion (Transition from $^5D_1$ to a lower-term level of ^7Fy). This absence was detected even for concentrations of europium as small as 1 at.%[49) and may be associated with a high probability of the non-radiating transition $^3D_1-^5D_0$ (energy difference 1740 cm^{-1}).

The highest frequency of the lattice oscillation (GeO_4 tetrahedrons) for the RE germanates under consideration is in the range of 750 to 850 cm^{-1} [78)]; thus, the energy of 2–3 phonons becomes sufficient to scatter the energy in 1740 cm^{-1}. According to [86,87)] such a process of energy scattering takes place with a relatively high probability.

Among the non-alkali germanates the luminophore $Y_2GeO_5 \cdot Tb^{3+}$ is worth particular attention. It possesses a green glow, the brightness of which, relative to industrial luminophore ZnO · Zn, amounts to 26–30%, depending on the concentration of the activator and on heat treatment conditions[41, 88)]. The advantage of $Y_2GeO_5 \cdot Tb^{3+}$ is its narrow band spectrum.

Rare-earth germanate crystals also possess x-ray and thermoluminescences. Thermostimulated luminescence was detected for A and E-type-crystals which were both unactivated and activated by Eu^{3+} ions. Thermoluminescence is discussed in detail in[7,89)]. Additional investigations of $NaSmGeO_4$ crystals have demonstrated the advantage of employing them as working materials for a thermoluminescent dosimeter of mild x-ray radiation.

3.2.2 Excitation Spectra

Different investigators[90,91)] have noted a peculiarity associated with the excitation spectra of Eu^{3+}-activated oxide compounds: the existence of a broad band of high intensity in the UV band (less than 300 nm). A broad band was also detected in the germanates (Table 10, Fig. 20). If we attribute this band to the EuO_n group (it is

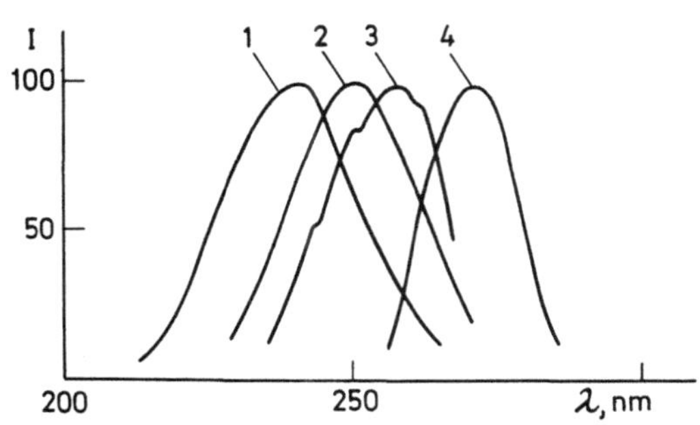

Fig. 20. Position of excitation bands due to the transfer of charge. 1-$NaYbGeO_4$: Eu^{3+}; 2-$NaYGeO_4$: Eu^{3+}; 3-$NaGdGeO_4$: Eu^{3+}; 4-$NaEuGeO_4$

absent in the EuF_3 spectrum), we can associate the probable mechanism of excitation with the transfer of charge from oxygen to europium[90–92].

$$| Eu^{III}O_n^{-2n} | \longrightarrow | Eu^{II}O_n^{-2n+1} | .$$

The shift of the broad band maximum in the direction of large wavelengths in the isostructural compound series $NaLnGeO_4$, $NaGdGeO_4$, and $NaEuGeO_4$ (Fig. 20) can be explained by a decrease in the potential at the localization points of the ligands (the distance of Ln-O increases) and by the decrease in the charge transfer energy[93].

Blasse [94] noted that the maximum effect in the UV range for Eu^{3+}-activated phosphors increases when the band maximum of the charge transfer of the Eu^{3+} center shifts in the direction of the short waves lengths. This correlation applies to those phosphors which become activated in the center of the glow. It is mainly to this class of phosphors that the rare-earth germanates belong. Table 10 illustrates the relationship indicated above; one may also expect that the samples prepared under hydrothermal conditions (presented in brackets) have a greater effect in comparison with those prepared by solid phase sintering.

In the Eu^{3+}-activated $NaGdGeO_4$ matrix, one observes a transfer of energy from Gd^{3+} to Eu^{3+} [95], i. e., an activator radiation sensitization takes place the matrices. The same effect occurs in hydrothermal Gd-germanate crystals[96].

The most effective excitation of terbium-activated phosphors occurs in the UV range in a broad band related to the interconfigurational transition[97]. In the excitation spectrum of the terbium-activated E-type germanate, $NaGd_3[GeO_4]_2(OH)_2 \cdot Tb^{3+}$, the most intensive lines in the spectrum are due to the transitions from the ground state of $^2F_6$ to the $^5F_4$ (274 nm), $^5H_6$ (310 nm), $^5G_6$ (367 nm), and $^5D_3$ (372, 377 nm) levels.

It was noticed that the effect of luminescence correlates with the position of the broad excitation band: when ν_{max} increases, the effect decreases[80].

3.3 Diffuse Reflection Spectra

Measurement of the diffuse reflection reveals the typical curve of powder dyes. The strong absorption in the UV range is also exhibited in diffuse reflection spectra of europium and terbium germanates[98].

Spectra of europium germanates of structure types A, E, H, and G are shown in Fig. 21 a. A broad absorption band of wave lenghts less than 270 nm present in the spectrum is typical of the above-mentioned compounds. As stated earlier, these bands are perhaps due to the absorption of the EuO_n group. It is not difficult to establish the quantitative dependence between the positions of the absorption band maximum and the interatomic distances of Eu-O due to the varied character of the compounds and the highly diffuse peaks of the maxima. Quantitatively, such a correlation is worth noting: when increasing the shortest distance of Ln-O (one may examine this in, for example, the nearest analogs to the europium compounds, the samarium germanates, for which there are structural data available), the wave length of the absorption band maximum will also increase. The shortest distance of Sm-O equals 2.23 Å in the

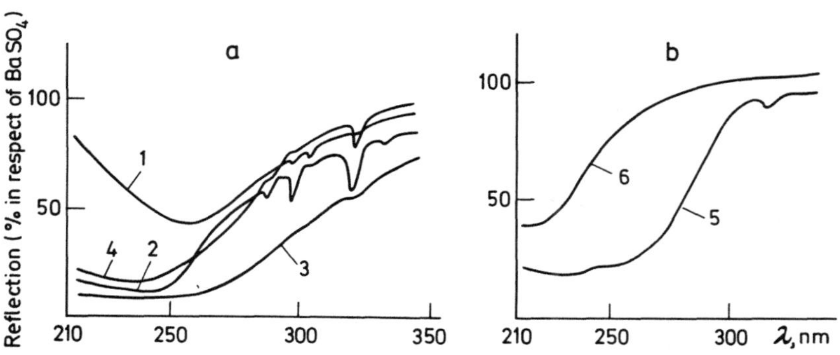

Fig. 21. a Spectra of diffuse reflection of europium germanates of A- (1), E- (2), H- (3) and G-type (4). **b** Spectra of diffuse reflection of A-type germanates $NaTbGeO_4$ (1) and $NaYbGeO_4$ (2)

G-type, 2.26 Å in the E-type, 2.28 Å in the A-types, and 2.43 Å in the H-types. The corresponding absorption band maxima corresponds to ~ 235 nm, ~ 240 nm, 255 nm, and 240–260 nm (plateau).

The reflection spectra of $NaTbGeO_4$ and $NaYbGeO_4$ are shown in Fig. 21b. A strong absorption was observed for terbium germanate in the 230–260 nm range, which is due to the interconfigurational transition $4f^8 \longrightarrow 4f^7\,5d$.

A state above the 10000-cm^{-1} level ($^4F_{5/2}$) is absent in the energy-level diagram of the Yb^{3+} ion. The absorption band in the 215 nm range for $NaYbGeO_4$ is probably due to the transfer of the charge[99].

In addition to the broad bands, Fig. 21 illustrates the narrow peaks of small intensities which correspond to the resonance absorption inside the 4f-shell of trivalent rare-earth ions (transition $4f \longrightarrow 4f$).

3.4 Electric and Magnetic Properties of Rare-Earth Germanates

The electric and magnetic properties have been investigated chiefly for the observation of germanates of the simplest compositions, Ln_2GeO_5 and $Ln_2Ge_2O_7$. These compounds can be assigned to the class of dielectrics[100]; at room temperature the samples have a high specific resistance ($\sigma = 10^{-13} - 10^{-14}$ ohm · cm) and small electric losses ($tg\delta < 0.01$).

The character of the conductivity in a range of 300–1500 °K is ionic; an anionic fraction of the conductivity was detected mostly in La_2GeO_5 and Nd_2GeO_5. The hole constituent of conductivity is also present in Er and Gd oxyorthogermanates. Cationic conductivity may occur in gadolium germanates (Table 11).

Rare-earth germanates of a much more complex composition, $A_5LnGe_4O_{12}$, are of great interest in their capacity as superionic conductors. In these compounds, which are structurally similar to the ring silicates, $Na_5LnSi_4O_{12}$, a high conductivity due to sodium ions was detected[33,101] (Table 12).

Table 11. Electrical properties of some germanates[100)]

Composition	σ, $\Omega^{-1} \cdot cm^{-1}$					ΔE, eV		Numbers of ion transfer at 1500 °	Conductivity type	T_{bend}
	300 °K	600 °K	900 °K	1200 °K	1500 °K	at low T	at high T			°K
La_2GeO_5	10^{-13}	$1.2 \cdot 10^{-3}$	$4 \cdot 10^{-5}$	$1 \cdot 10^{-3}$	$3.5 \cdot 10^{-3}$	1.33	0.87	1.00	anionic	980
Nd_2GeO_5	10^{-14}	$0.6 \cdot 10^{-10}$	$8 \cdot 10^{-4}$	$9 \cdot 10^{-6}$	$1 \cdot 10^{-4}$	1.09	1.50	0.99	anionic	950
Gd_2GeO_5	10^{-14}	$1.6 \cdot 10^{-12}$	$8.8 \cdot 10^{-9}$	$2.5 \cdot 10^{-6}$	$4.8 \cdot 10^{-5}$	1.38	2.00	0.80	anionic-hole	1025
Er_2GeO_5	10^{-14}	$0.5 \cdot 10^{-11}$	$1.8 \cdot 10^{-3}$	$0.3 \cdot 10^{-3}$	$7.2 \cdot 10^{-6}$	1.21	3.20	0.39	anionic-hole	1000–1300

Table 12. Ionic conductivity in germanates[a] with a $Na_5YSi_4O_{12}$ structure[33)]

Composition	Temperature interval	σ_{300} [b] ($\Omega^{-1}cm^{-1}$)	σ_{200} ($\Omega^{-1}cm^{-1}$)
$Na_5TmGe_4O_{12}$	150–280	$5 \cdot 10^{-2}$	$7 \cdot 10^{-3}$
$Na_5HoGe_4O_{12}$	150–280	$3 \cdot 10^{-2}$	$7 \cdot 10^{-3}$
$Na_5DyGe_4O_{12}$	140–280	$4 \cdot 10^{-2}$	$3 \cdot 10^{-3}$
$Na_5TbGe_4O_{12}$	140–280	$2 \cdot 10^{-2}$	$2 \cdot 10^{-3}$
$Na_5GdSi_{3.75}Ge_{0.25}O_{12}$	90–230	$3 \cdot 10^{-1}$	$7 \cdot 10^{-2}$
$Na_5GdSi_3GeO_{12}$	140–240	$1 \cdot 10^{-1}$	$5 \cdot 10^{-2}$
$Na_5GdSi_4O_{12}$	100–200	$3 \cdot 10^{-1}$	$8 \cdot 10^{-2}$

[a] For comparison the data for silicates are listed as well [b] Extrapolated values

$\sigma = A/T \exp(-E_a/kT)$

Table 13. Temperatures of magnetic transitions in $Ln_2Ge_2O_7$ and Ln_2GeO_5 [102)]

Composition	Temperature K	Composition	Temperature K
$Tb_2Ge_2O_7$	2.05 ± 0.05	$Er_2Ge_2O_7$	1.15 ± 0.05
$Dy_2Ge_2O_7$	2.15 ± 0.05	Dy_2GeO_5	2.5 ± 0.1
$Ho_2Ge_2O_7$	1.45 ± 0.05	Er_2GeO_5	1.25 ± 0.1

An essential peculiarity associated with a number of rare-earth germanate structures is the presence of channels which are occupied by Na, Li, and K ions in the cationic framework, and which provide mobility for these cations. This feature allows one to hope that new superionic conductors in this class of compounds may be discovered.

Magnetic properties have been discussed in[64,102)] (see Table 13). Samarium and lanthanum germanates are diamagnetic with molar susceptibilities of -14.42 and $-770.6 \cdot 10^{-6}$ respectively. Pr, Nd, and Er germanates obey the Curie-Weise law in the temperature range of 100–800 °C.

References

1. Toropov, N. A. et al.: Rare-Earth silicates and their analoques, Leningrad, Nauka, 1971, p. 230 (Russ.)
2. Kaminsky, A. A.: Laser crystals, Moscow: Nauka, 1975, p. 256 (Russ.)
3. Belov, K. P. et al.: Rare-Earth ferromagnetics and antiferromagnetics. Moscow, Nauka, 1975, p. 320 (Russ.)
4. Aleksandrova, I. T. et al.: Geology of rare element deposits. Rare-earth silicates, Part 26, Moscow, Nedra, 1966, p. 220 (Russ.)

5. Progress in the Science and Technology of the Rare-Earths (Ed. Le Roy Eyring) Vol. 1, Oxford-London, 1964
6. Les Elements des Terres Rares. Colloque Internat. du CNRS (Ed. du CNRS) N 180, Vol. 1, Paris, 1970
7. Demianets, L. N., Lobachev, A. N., Emelchenko, G. A.: Germanates of Rare-Earth Elements, Moscow; Nauka, 1980
8. Lobachev, A. N. et al.: Kristal u. Technik *5*, 425 (1970)
9. Felsche J.: Structure and Bonding *13*, 99 (1973)
10. Belov, N. V.: Essaies on structural mineralogy, Moscow, Nauka, 1976, p. 344
11. Buisson, G., Michel C.: Mat. Res. Bull. *3*, 193–194 (1968)
12. Gorbunov, Yu. A. et al.: Sov. Phys.-Kristallografiya *19*, 1081 (1974)
13. Cocbain, A. G., Smith, J. V.: Amer. Mineral. *50*, 2023 (1965)
14. Emelchenko, G. A., Ilyukhin, V. V.: Sov. Phys.-Dokl. Akad. Nauk (SSSR) *217*, 824 (1974)
15. Petrova, M. A.: Physical-chemical study of silicates of rare-earth elements. Theses. Inst. Chemi. Silicates, Acad. Sci. USSR, Leningrad, 1974, p. 20
16. Merinov, B. A. et al.: Sov. Phys.-Dokl. Akad. Nauk SSSR *241*, 353 (1978)
17. Smolin, Yu. I., Shepelev, Yu. F., Upatova, T. V.: Sov. Phys.-Dokl. Akad. Nauk SSSR *187*, 322 (1969)
18. Smolin, Yu. I., Shepelev, Yu. F., Butikova, I. K.: Sov. Phys.-Kirstallografiya *16*, 911 (1971)
19. Smolin, Yu. I.: ibid. *15*, 47 (1970)
20. Shannon, R. D., Prewitt, C. T.: J. Solid State Chem. *2*, 199 (1970)
21. Nikolsky, Yu. V. et al.: Sov. Phys.-Dokl. Akad. Nauk SSSR *233*, 362 (1977)
22. Kharakh, E. A., Chichagov, A. V., Belov, N. V.: Sov. Phys.-Kristallografiya *15*, 1064 (1970)
23. Chichagov, A. V. et al.: Dokl. Akad. Nauk SSSR *189*, 88 (1969)
24. Emelchenko, G. A. et al.: Sov. Phys.-Dokl. Akad. SSSR *222*, 343 (1975)
25. Kornev, A. N. et al.: Sov. Phys.-Dokl. Akad. Nauk SSSR *206*, 1109 (1972)
26. Zhdanova, T. A. et al.: Sov. Phys.-Dokl. Akad. Nauk SSSR *224*, 1069 (1975)
27. Christensen, A. N.: Acta Chem. Scand. *24*, 1055 (1972)
28. Gorbunov, Yu. A., Maksimov, B. A., Belov, N. V.: Sov. Phys-Dokl. Akad. Nauk SSSR *208*, 91 (1973)
29. Gorbunov, Yu. A., Maksimov, B. A., Belov, N. V.: ibid. *211*, 591 (1973)
30. Gorbunov, Yu. A. et al.: ibid. *219*, 91 (1974)
31. Klientova, G. A. et al.: Sov. Phys.-Dokl. Akad. Nauk SSSR *224*, 817 (1975)
32. Nikolsky, Yu. V., Maksimov, B. A., Belov, N. V.: ibid. *222*, 87 (1975)
33. Shannon, R. D. et al.: Inorg. Chem. *17*, 958 (1978)
34. Mill. B. V.: Sov. Phys.-Dokl. Akad. Nauk SSSR *165*, 555 (1965)
35. Roy, R.: J. Amer. Ceram. Soc. *37*, 581 (1954)
36. Lasarev, A. N.: Izv. Akad. Nauk SSSR. Neorgan. Mater. *4*, 1287 (1968)
37. Smolin, Yu. A., Tkachev, S. P.: Sov. Phys.-Kristallografiya *14*, 22 (1969)
38. Michel, C., Buisson, G., Bertaut, E. F.: Compt. Rend. Acad. Sci., Paris *264B*, 397 (1967)
39. Bondar, I. A. et al.: High temperature chemistry of silicates and oxides. (Toropov, N. A. ed.) pp. 31–36. Moscow, Nauka, 1972
40. Belokoneva, E. A. et al.: Sov. Phys.-Kristallografiya *17*, 490 (1972)
41. Shannon, R. D., Sleight, A. W.: Inorg. Chem. *7*, 1949 (1968)
42. Kornilova, E. I., Prikhod'ko, M. E., Lileev, I. S.: Izv. Akad. Nauk SSSR, Neorgan. Mater. *1*, 227 (1965)
43. Tenisheva, T. F., Lasarev, A. N., Pavlyukovich, T. M.: Izv. Akad. Nauk SSSR, Ser. khim. 1553 (1965)
44. Glushkova, V. B., Davtyan, I. A., Keler, E. K.: Izv. Akad. Nauk SSSR *3*, 119 (1967)
45. Kolesova, V. A.: Izv. Akad. Nauk SSSR, Otdel. khim. nauk *N11*, 2082 (1962)
46. Tarte, P.: Nature *191*, 1002 (1961)
47. Chichagov, A. V., Litvin, B. N., Belov, N. V.: Sov. Phys.-Kristallografiya *14*, 119 (1969)
48. Chichagov, A. V., Litvin, B. N., Belov, N. V.: Sov. Phys.-Dokl. Akad. Nauk SSSR *177*, 574 (1967)
49. Blasse, B., Brill, A.: J. Inorg. Nucl. Chem. *29*, 2231 (1967)

50. Emelchenko, G. A. et al.: Sov. Phys.-Kirstallografiya *19*, 1272 (1974)
51. Demianets, L. N., Emelchenko, G. A., Lobachev, A. N.: Zh. Koordin. khimii *1*, 1205 (1975)
52. Demianets, L. N. et al.: Izv. Akad. Nauk SSSR, Neorgan. Mater. *12*, 143 (1976)
53. Maksimov, B. A., Kharitonov, Yu. A., Belov, N. V.: Dokl. Akad. Nauk SSSR *213*, 1072 (1973)
54. Geller, S.: Z. Krist. *125*, 1 (1969)
55. Mill, B. V., Roniger, G.: Sov. Phys.-Kristallografiya *18*, 126 (1973)
56. Bauer, G.: Naturwissenschaften *58*, 622 (1971)
57. Struntz, H., Freigang, H., Contag, B.: Neues Jahrb. Mineral. Monatsch. H. 2, 47 (1960)
58. Schwarz, H.: Angew. Chem. *79*, 548 (1967)
59. Geller, S., Miller, C. E., Treuting, R. G.: Acta cryst. *13*, 179 (1960)
60. Geller, S. et al.: J. Phys. Chem. Solids *12*, 111 (1960)
61. Bondar, I. A., Petrova, M. A.: Zh. Neorgan. Khimii *14*, 2340 (1969)
62. Bondar, I. A., Petrova, M. A., Udalov, Yu. P.: Single crystal growth of refractory and rare metals, pp. 37–42, Moscow, Nauka, 1973
63. Bondar, A. I.: in: Less elements des terres rares. Colloque Internat. CNRS, Paris N180, Vol. 1, 337 (1970)
64. Wanklyn, B. M.: J. Mater. Sci. *8*, 649 (1973)
65. Wanklyn, B. M.: Rare-earth germanates. Nat. Conf. Crystal Growth, Lancaster University, p. 33 (1970)
66. Emelchenko, G. A., Demianets, L. N., Lobachev, A. N.: J. Solid State Chem. *14*, 209 (1975)
67. Chichagov, A. V., Litvin, B. N., Belov, N. V.: Geokhimiya N9, 1044 (1968)
68. Emelchenko, G. A. et al.: Sov. Phys.-Kristallografiya *19*, 436, (1974)
69. Belova, E. N. et al.: in: Growth of crystals. *12*, pp. 87–91, Erevan, EGU 1977
70. Emelchenko, G. A., Demianets, L. N., Lobachev, A. N.: Mater. Res. Bull. *9*, 1151 (1974)
71. Demianets, L. N., Emelchenko, G. A., Lobachev, A. N.: High Temperatures – High Pressures *6*, 699 (1974)
72. Emelchenko, G. A.: Izv. Akad. Nauk SSSR, Neorgan. Mater. *12*, 494 (1976)
73. Demianets, L. N., Emelchenko, G. A., Lobachev, A. N., in: Problems of crystallology, pp. 227–234, Moscow, MGU 1976
74. Kudryavtseva, O. V., Litvin, B. N., in: Problems of crystallology, pp. 171–176. Moscow, MGU 1976
75. Belova, E. N. et al., in: Mechanism and kynetics of crystal growth. IV Allunion Conf. growth of crystals, Tzakhkadzor, Akad. Nauk Arm. SSR Press, Erevan, Part 2, pp. 228–234, 1972
76. Fedosova, S. P., Gribina, I. A., Belov, N. V., in: Problems of crystallology, pp. 204–207, Moscow, MGU 1976
77. Mill, B. V.: Zh. Neorgan. Khimii N7, 1533 (1966)
78. Emelchenko, G. A. et al., in: Crystal chemistry of inorganic compounds, pp. 195–206, Kishinev, Shtiintsa 1976
79. Emelchenko, G. A., Grechushnikov, B. N., Ilyukhin, V. V., in: Spectroscopy of crystals. pp. 265–269, Moscow, Nauka 1976
80. Blasse, G., Bril, A.: Philips Res. Rept. *22*, 481 (1967)
81. El'yashevich, M. A.: Spectra of rare-earths. Moscow, GITTKh, 1953
82. Wyborne, B. G.: Spectroscopic properties of rare-earths, London, John Wiley and Sons 1965
83. Stewens, K. W. H.: Proc. Phys. Soc. *A 65*, 209 (1952)
84. Polyak, V. G., Skokov, V. A.: Standart program of minimization of the function of many variables, N IV Moscow, MGU 1967, 68 p.
85. Ikramov, Kh. D.: Standart program of calculation of own values and own vectors of complex matrix by Yakobi method, N44 Moscow, MGU 1971
86. Braumik, M. L., Nugent, L. J.: J. Chem. Phys. *43*, 1680 (1965)
87. Bril, A. et al.: Internat. Conf. Luminescence, *B. 2*, 1688, Budapest 1966

88. Bondar, I. A.: Izv. Akad. Nauk SSSR, Neorgan. Mater. *15*, 1008 (1979)
89. Emelchenko, G. A. et al.: Thermoluminescence of rare-earth germanates. Abstr. 8. Ural Conf. Spectroscopy, pp. 52–53, Sverdlovsk 1975
90. Ropp, R. C.: J. Electrochem. Soc. *115*, 841 (1968)
91. Borchardt, H. J.: Efficiency of Eu^{3+} fluorescence in oxygen-dominates host lattices. Proc. First Rare-Earth Research Conf. *1*, Ames, Iowa, pp. 65–75, 1965
92. Ropp, R. C. et al.: Electrochem. Technol. *4*, 24 (1966)
93. Blasse, G., Bril, A.: J. Chem. Phys. *45*, 3327 (1966)
94. Blasse, G.: J. Chem. Phys. *45*, 2356 (1966)
95. Bril, A., Wanmaker, W. L.: J. Electrochem. Soc. *111*, 1363 (1964)
96. Emelchenko, G. A.: Synthesis and study of rare-earth germanate single crystals. Theses on candidate degree, Moscow, Inst. Crystallography Acad. Sci. USSR 1974
97. Jorgensen, C. K.: Orbital energy difference and chemical bonding in rare-earth, Proc. Fifth Rare-Earth Conf., Ames, Iowa, *1*, 1 (1965)
98. Emelchenko, G. A., Grechushnikov, B. N.: Spectra of diffuse scattering of RE germanates. Abstr. 8 Ural Conf. Spectroscopy pp. 55–56, Sverdlovsk, 1975
99. Bryant, B. W.: J. Amer. Opt. Soc. *55*, 771 (1965)
100. Popov, V. V., Petrova, M. A.: Izv. Akad. Nauk SSSR, Neorgan. Mater. *9*, 61 (1973)
101. Hong H.Y.-P., Kafalas, J. A., Bayard, M.: Mater. Res. Bull. *13*, 757 (1978)
102. Fournier, J., Kohemuller, R.: Compt. Rend. Sci (Paris) *266*, Serie C, 1361 (1968)

Growth, Properties and Applications of Narrow-Gap Semiconductors

Horst Maier and Joachim Hesse[1]

AEG-Telefunken, Semiconductor Division, D-7100 Heilbronn, Federal Republic of Germany

This article reviews selected properties of the narrow-gap semiconductors $Pb_{1-x}Sn_xTe$, $Pb_{1-x}Sn_xSe$ and $Hg_{1-x}Cd_xTe$ which are accepted for infrared device applications, recently even in large-scale production.
The production aspect gave rise to a detailed description of the crystal growth as a first and decisive factor for device economy. Particular emphasis is placed on the important role of the thermodynamic parameters (phase diagrams) for growth, doping and physical properties of bulk crystals as well as epitaxial layers. The various growth methods are described and compared in detail.
The device aspect will be considered for photodetectors and tunable diode lasers as modern examples of applications in thermal imaging and gas spectroscopy around 10 μm.

Table of Contents

[1] Fraunhofer-Institut für Physikalische Meßtechnik, Heidenhofstraße 8, D-7800 Freiburg

1 Introduction

Optoelectronic devices have contributed to important innovations in semiconductor industry. Of decisive importance to their success was the fact that for the basic binary and ternary III-V compounds, economically viable material and device technologies could be developed.

The wavelength region from $\lambda \approx 0.5$ μm to $\lambda \approx 5$ μm can now be covered by commercially available devices on the basis of III-V compounds. For longer wavelengths, however, one has to rely on II-VI or IV-VI (narrow energy gap) semiconductive compounds. Optoelectronic devices of these semiconductors, of course, require a technology rather different from that which has been established for the III-V compounds. The positive decision as to whether the large development investment is justified depends firstly on the actual need for such devices in the longer wavelength region and secondly on the feasibility of production technologies for II-VI and IV-VI compounds. Considering the first argument, today photodetectors operating up to about 12 μm are required for thermal imaging systems in the third atmospheric window. In a still broader wavelength region, up to 20 μm, diode lasers are opening the way for improvements in gas spectroscopy including practical applications in gas analysis and air pollution monitoring. Referring to the second argument it can be stated that production oriented technologies are now available for materials such as $Hg_{1-x}Cd_xTe$, $Hg_{1-x}Cd_xSe$, $Pb_{1-x}Sn_xTe$, $Pb_{1-x}Sn_xSe$, $Pb_{1-x}Ge_xTe$, $PbS_{1-x}Se_x$ and some quaternary compounds[1)]. Single crystals can be grown in useful size and quality. Technologies for the deposition of thin epitaxial films and for producing pn-junctions or Schottky barriers are well established. Furthermore, problems of passivation, contacting and packaging, associated with these devices, have thoroughly been studied.

This article concentrates on the three most important narrow-gap materials of the IV-VI and II-VI classes: $Pb_{1-x}Sn_xTe$, $Pb_{1-x}Sn_xSe$ and $Hg_{1-x}Cd_xTe$. By varying the alloy composition they provide a very wide range of energy gaps. In fact, even zero band gap compositions are possible with these materials. This unique property is demonstrated in Figs. 1 to 3 which also represent schematic diagrams of the relevant

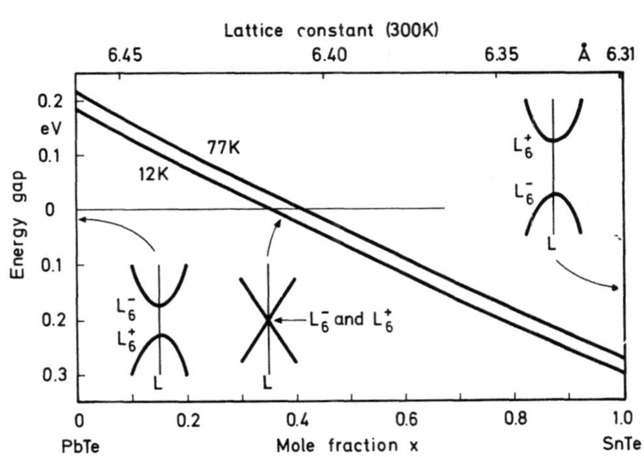

Fig. 1. Energy gap versus mole fraction x and lattice constant for $Pb_{1-x}Sn_xTe$. Schematic representation of the valence and conduction bands

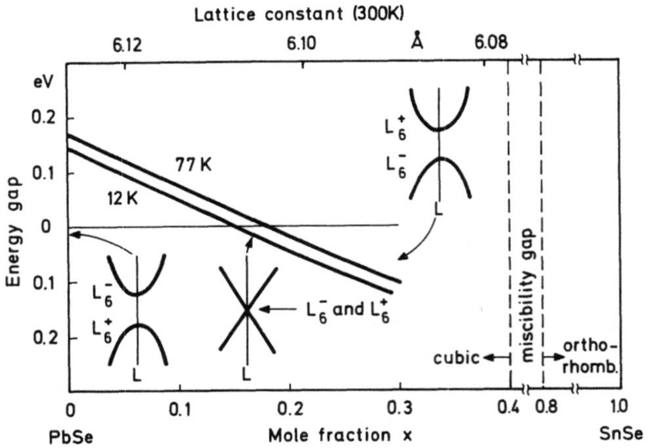

Fig. 2. Energy gap versus mole fraction x and lattice constant for $Pb_{1-x}Sn_xSe$. Schematic representation of the valence and conduction bands

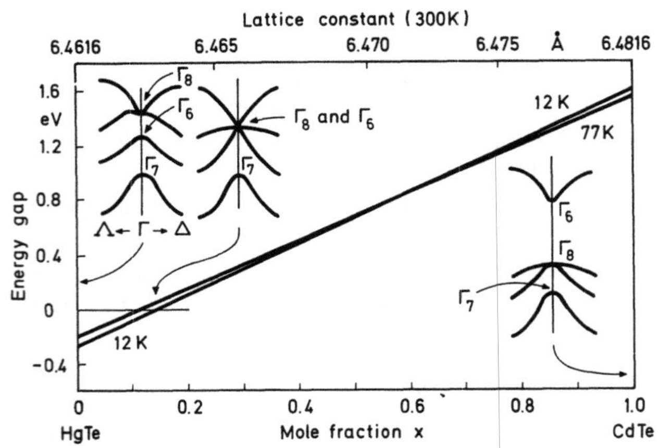

Fig. 3. Energy gap versus mole fraction x and lattice constant for $Hg_{1-x}Cd_xTe$. Schematic representation of the valence and conduction bands

bands at their direct band gap extrema, i. e. the L point in $Pb_{1-x}Sn_xTe$ and $Pb_{1-x}Sn_xSe$ and the Γ point in $Hg_{1-x}Cd_xTe$. The two former retain their semiconducting character on both sides of their band crossing compositions. HgTe, on the other hand, is a semimetal with slightly overlapping bands. By the addition of CdTe, band inversion also occurs and the alloy becomes a semiconductor. Within the range of solid solubility (which is limited only for $Pb_{1-x}Sn_xSe$) the energy gap as well as the lattice constant depend almost linearly on the composition.

Comprehensive data on the band structures and the resulting physical properties are available for these three materials and have already been reviewed[2–5]. Therefore, only their most important physical data will be presented here in tables. Our intention is, however, to complete the picture with a summary of their thermodynamic properties and their material technology. This is particularly valuable since the continuous interest in the narrow-gap materials has innovated many attractive techniques for growing single crystals and depositing epitaxial layers. Objective is their application in infrared detectors and lasers, mainly in the 10 μm wavelength region. With regard to such applications, the optimum material properties will first be discussed in Chap. 2. This is followed in Chap. 3 by a treatment of the thermodynamic aspects pertinent to the crystal growth processes. The main subjects of the paper, i. e., crystal growth itself and epitaxy, are discussed in Chap. 4 and 5, respectively. The paper will be rounded off by a short review of modern devices and applications in Chap. 6.

2 Device Requirements

2.1 Infrared Detectors

The most important parameter of an "intrinsic" quantum detector, in which a photon-induced interband transition produces the signal, is the band gap E_g. This defines the cut-off wavelength $\lambda_{co} = hc/E_g$ of the detector, i. e., the upper limit of the detectable wavelength range. The theoretical function of detectivity versus wavelength is a saw tooth which suggests to choose the alloy composition so that the desired maximum is close to the cut-off wavelength.

The detector responsivity R_λ is given by the photovoltage ΔV or photocurrent ΔI produced by the radiation power $\Delta\phi$ incident to the detector area A.

$$R_\lambda = \frac{\Delta V}{\Delta\phi} \text{ or } R_\lambda = \frac{\Delta I}{\Delta\phi} \tag{1}$$

Its detectivity $D_\lambda{}^*$ is defined as the ratio of responsivity to the total noise voltage V_N or noise current I_N in the bandwidth Δf.

$$D_\lambda{}^* = \frac{R_\lambda}{V_N}(A\Delta f)^{1/2} \text{ or } D_\lambda{}^* = \frac{R_\lambda}{I_N}(A\Delta f)^{1/2} \tag{2}$$

The first is conveniently chosen for photoconductive detectors operated under constant current conditions, i. e., with a load resistance R_L much larger than the detector resistance. The second is valid for photovoltaic detectors operated under closed circuit conditions.

Several noise sources contribute to the total detector noise I_N or V_N. The most important are the Johnson noise and the generation-recombination noise which should be small compared to the unevitable noise due to the background photon flux. Others, like the 1/f noise or noise of the signal electronics, are not inherent to the detector material itself and will not be considered here.

2.1.1 Photoconductive Detectors

For an n-type detector of area A and thickness d, operated under constant current conditions, the responsivity is

$$R_\lambda = \frac{\Delta V}{\Delta\phi} = \frac{V\eta\tau}{E_\lambda A d n} \tag{3}$$

where V represents the detector bias, E_λ the incident photon energy, η the external quantum efficiency, n the majority carrier concentration and τ the minority carrier life time. The predominant noise mechanism is a generation-recombination process if the detector bias V is sufficiently high. For the present case of a band-to-band transition photoconductor with a large mobility ratio μ_n/μ_p, as in $Hg_{1-x}Cd_xTe$, the generation-recombination noise voltage may be expressed by

$$V_{g\text{-}r} = 2\,V\left(\frac{p\,\tau\Delta f}{Adn\,(n+p)\;(1+\omega^2\tau^2)}\right)^{1/2} \tag{4}$$

and assuming a sufficiently low measuring frequency ($\omega^2\tau^2 \ll 1$), the detectivity is given by

$$D_\lambda{}^* = \frac{\eta}{2\,E_\lambda}\left(\frac{(n+p)\,\tau}{npd}\right)^{1/2} \tag{5}$$

Here n and p denote the charge carriers produced thermally and by the incident radiation ϕ, i. e.

$$n = n_o + \frac{\phi\eta\tau}{d},\; p = p_o + \frac{\phi\eta\tau}{d} \tag{6}$$

For an extrinsic n-type material with $n_o \gg p_o$ the detectivity in the thermally limited case (under reduced background radiation ϕ_B) is

$$D_\lambda{}^* = \frac{\eta}{2\,E_\lambda}\left(\frac{\tau}{p_o d}\right)^{1/2} \tag{7}$$

and in the background limited (BLIP) case $\left(\frac{\phi_B\eta\tau}{d} > \frac{n_i^2}{n_o}\right)$

$$D_\lambda{}^* = \frac{1}{2\,E_\lambda}\left(\frac{\eta}{\phi_B}\right)^{1/2} \tag{8}$$

It is seen from Eqs. (3) and (7) that the most important quantity for a photoconductive detector is the minority carrier life time τ. It has been shown[6] that in $Hg_{1-x}Cd_xTe$ the limiting band-to-band recombination mechanism in the best presently available material is an Auger process. The Auger life time is given by

$$\tau_A = \frac{2\,n_i^2\tau_{Ai}}{n_o\,(n_o + p_o)} \approx \frac{2\,n_i^2\,\tau_{Ai}}{n_o^2} \tag{9}$$

where τ_{Ai} designates the Auger life time for the intrinsic carrier concentration n_i. Substituting this expression into Eq. (7)

$$D_\lambda{}^* = \frac{\eta}{2\,E_\lambda}\left(\frac{2\,\tau_{Ai}}{n_o d}\right)^{1/2} \tag{10}$$

Thus, the second postulation for the material to fulfil highest possible detector performance is to keep the majority carrier concentration n_o as low as possible. However, care must be taken, not to enter the compensation regime, where the life time again decreases by Shockley-Read recombination processes.

Another competitive recombination process which impacts the detector performance is the surface recombination. It reduces the life time to

$$\tau = (\frac{1}{\tau_A} + \frac{2s}{d})^{-1}$$

where s is the surface recombination velocity. This, however, is rather a problem of detector fabrication and, to a lesser extent, a material property and can be overcome by proper surface treatment. The same applies to the external quantum efficiency η.

The other quantity entering Eqs. (3) and (7) is the thickness d of the photoconductor which has to be weighed against the external efficiency by considering the absorption constant of the material. The detector bias V in Eq. (3) is limited by the effect of the minority carrier sweep-out.

Above this limit, $V\tau$ is essentially constant and higher bias voltages only increase the Joule heating of the detector. It is attained when the transit time of the photo-induced carriers equals the minority carrier life time, i. e.,

$$a^2/2\,\mu_D V = \tau \tag{11}$$

where a is the detector length and μ_D designates the ambipolar drift mobility which, in extrinsic n-type material, is equal to the hole mobility $\mu_D = \mu_p$. Introducing this into Eq. (3), the sweep-out limit responsivity is given by

$$R_\lambda = \frac{e\eta R\mu_n}{2\,E_\lambda \mu_p} \tag{12}$$

where R is the detector resistance, e is the electron charge.

We are now in a position to outline the material properties desired for an instrinsic photoconductive detector:

1. High minority carrier life time
2. Low majority carrier concentration
3. High ratio of the majority to minority carrier mobilities.

These objectives can be excellently fulfilled in n-type $Hg_{1-x}Cd_xTe$ as is seen from Table 5. In the case of $Pb_{1-x}Sn_xTe$ and $Pb_{1-x}Sn_xSe$, the mobility ratio $\mu_n/\mu_p \approx 1$ is unfavorable. Moreover, it is difficult to accomplish sufficiently low carrier concentrations in the lead chalcogenides.

The foregoing treatment establishes additional postulates to the materials and detector technology:

1. Choice of composition so that optimum responsivity is achieved at the desired wavelength.
2. Homogeneity of composition in large multi-element arrays.
3. Optimization of the detector thickness in order to keep both the resistance and efficiency high.
4. Surface preparation to minimize surface recombination.
5. Contacts which minimize the 1/f noise.

6. Detector geometry to achieve high resistance in the minority carrier sweep-out mode.
7. Antireflection coating optimized for the desired wavelength range.

2.1.2 Photovoltaic Detectors

Similarly, the postulates to the material properties may be established for a photovoltaic detector. Its responsivity in a closed circuit operation mode is

$$R_\lambda = \frac{\Delta I}{\Delta \phi} = \frac{e\eta}{E_\lambda} \tag{13}$$

where e represents the electron charge, E_λ the incident photon energy, η the external quantum efficiency. Under zero bias, the Johnson noise is the only noise source to be considered besides that of the background photon flux ϕ_B, and the detectivity assumes the form

$$D_\lambda{}^* = \frac{e\eta}{E_\lambda} \left(\frac{A}{4kT/R + 2eI_B}\right)^{1/2} \tag{14}$$

which, in the thermally limited case, becomes

$$D_\lambda{}^* = \frac{e\eta}{2\,E_\lambda} \left(\frac{RA}{kT}\right)^{1/2} \tag{15}$$

and in the BLIP case, where $I_B = \eta e A \phi_B$,

$$D_\lambda{}^* = \frac{1}{E_\lambda} \left(\frac{\eta}{2\phi_B}\right)^{1/2} \tag{16}$$

The optimum properties of the photovoltaic detector are thus characterized by the diffusion limited case for which

$$R = \frac{kT}{eI_s} \tag{17}$$

where I_s is the saturation current of an abrupt junction,

$$I_s = eA \left(\frac{n_p L_n}{\tau_n} + \frac{p_n L_p}{\tau_p}\right) \tag{18}$$

Here n_p, p_n represent the minority carrier concentrations, L_n, L_p their diffusion lengths and τ_n, τ_p their life times, respectively.

Substituting this into Eq. (15) and regarding $L = \sqrt{\frac{kT}{e}\mu\tau}$, $p_n = n_i^2/n$ and $n_p = n_i^2/p$ one obtains for thermally limited detectivity (with diffusion-controlled junction current)

$$D_\lambda{}^* = \frac{\eta}{2E_\lambda n_i}\left(\frac{kT}{e}\left[\frac{1}{n}\sqrt{\frac{\mu_p}{\tau_p}} + \frac{1}{p}\sqrt{\frac{\mu_n}{\tau_n}}\right]\right)^{-1/2} \tag{19}$$

where μ_p and μ_n are the hole and electron mobilities, respectively, and k is the Boltzmann constant.

To evaluate the external quantum efficiency we consider a detector which is irradiated on its n-type side. The depth of the junction should be smaller than the minority carrier diffusion length. Then, from a general calculation[2] using

$$L_p = \sqrt{\frac{kT}{e}\mu_p\,\tau_p}$$

one obtains

$$\eta = \frac{1-r_\lambda}{1 + sl\tau_p/L_p{}^2} = \frac{1-r_\lambda}{1 + esl/\mu_p kT} \tag{20}$$

where r_λ represents the reflectivity, s the surface recombination velocity, μ_p the hole mobility, k the Boltzmann constant and l the depth of the junction.

As in the case of the photoconductive detector the optimum material properties may now be discussed for a photovoltaic detector:

1. High minority carrier life times
2. High majority carrier concentrations.

The maximum minority carrier life times are again obtained in the Auger recombination regime. Since the Auger life time is approximately proportional to $1/n^2$ or $1/p^2$, respectively [see Eq. (9)], $D_\lambda{}^*$ becomes independent of n and p. In $Hg_{1-x}Cd_xTe$ bulk material, the Auger effect has been verified to prevail down to very low carrier concentrations of $n = 5 \times 10^{14}$ cm^{-3}. However, in the pn-junction, radiative recombination may become similarly important. Moreover, generation-recombination current in the depletion region or tunneling may render our diffusion current assumption invalid, if the carrier concentrations are chosen either too low or too high, respectively[7]. In $Pb_{1-x}Sn_xTe$ and $Pb_{1-x}Sn_xSe$, radiative recombination (and the corresponding decrease of the detectivity $\propto \sqrt{n}$) was estimated to become dominant below 2×10^{15} cm^{-3} and 1×10^{16} cm^{-3}, respectively[8]. Between these concentrations and 10^{17} cm^{-3}, Auger recombination is expected whereas at higher concentrations tunneling will deteriorate the detectivity. Experimentally, a simple decay process could not be verified in $Pb_{0,8}Sn_{0,2}Te$[9, 10]. The recombination mechanism seems to be controlled by traps.

Postulates to the material and detector technology, in order to yield high detectivity, partly resemble those for the photoconductive detector:

1. Choice and homogeneity of the material composition in order to achieve highest detectivity at the desired wavelength.
2. Positioning of the junction close to the irradiated surface.

3. Contacts on p and n-type material which minimize the 1/f noise.
4. Optimal surface preparation, passivation and antireflection coating.

2.2 Infrared Diode Lasers

The major advantage of the semiconductor diode lasers is that their emission wavelength can easily be tailored to the desired range by selecting the material composition with the proper band gap. In addition, there is a certain range of tunability when the band gap is changed by varying the temperature or applying a magnetic field or a hydrostatic pressure. These features have contributed to make the lead chalcogenide and lead tin chalcogenide lasers indispensable in high-resolution IR spectroscopy for scientific applications and pollution monitoring[11–13].

In $Hg_{1-x}Cd_xTe$, on the other hand, stimulated emission could so far only be excited by optical[14, 15] and electron beam[16] pumping.

We will briefly discuss the conditions of stimulated emission in semiconductor injection lasers, some important consequences on the parameters of the material and the design of the device. The laser action is due to a first-order emission process from direct interband transitions. The condition for stimulated emission is given by[17]

$$h\nu > F_c - F_v \tag{21}$$

where $h\nu$ designates the emitted photon energy and F_c and F_v the quasi-Fermi energies. Injection into the active laser region can only be sufficiently high if one or both sides of the junction exhibit degeneracy.

In addition, the gain must be greater than the losses in the system which mainly result from free carrier absorption and penetration of light outside the active region.

$$r \exp(g - \alpha)\ a \geqslant 1 \tag{22}$$

r is the reflectivity, g the gain, α the sum of the losses and a the cavity length.

The threshold current density at which stimulated emission commences is given by

$$I_{th} = \frac{d}{\eta\beta\Gamma}\left(\alpha + \frac{1}{a}\ln\frac{1}{r}\right) \tag{23}$$

where β is the gain factor, d the thickness of the active layer, Γ the fraction of emitted light which is confined within d, and η the quantum efficiency, i. e., the ratio of the effective to the radiative minority carrier life times.

This allows to establish a few material properties which are desirable for injection lasers:

1. Effective minority carrier life time should be determined by radiative process.
2. High charge carrier densities, but optimized against tunneling, Auger life time and free carrier absorption.

However, the injection laser performance is as well critically influenced by the design and technology of the device:

1. Choice of composition to obtain the desired emission wavelength at a specified operation temperature.
2. Homogeneity of composition.
3. Low thickness of the active region.
4. High quality of the laser cavity surfaces.
5. High electrical and optical confinement, e.g., by double heterostructure design.
6. Balance between low threshold current density and emission power by choice of laser dimensions.

The characteristics of the infrared optoelectronic devices briefly compiled in this chapter may indicate the predominant role of the optimum material properties and the maintenance of these properties through device technology.

3 Phase Diagrams

A prerequisite for the fabrication of single crystals and epitaxial layers is an exact knowledge of the related phase diagrams. They exhibit information about the solid phases coexisting with the melt, the miscibility of the participating compounds and the solubility of excess metal or chalcogen in the binary or pseudobinary solid states. Particularly, the latter is important for device applications since the electronic properties of compound crystals depend on the nature and range of the homogeneity domain. The charge carrier type and concentration of a pure undoped material are determined by the deviation from stoichiometry. In order to preset and control these properties during the crystal fabrication and annealing processes, the phase diagrams offer indispensable information. Usually, the equilibrium conditions are given in multidimensional pressure-temperature-composition (p_i-T-x_j) diagrams. In practice, however, these are projected unto temperature-composition (T-x_j or liquid-solid) diagrams or unto pressure-temperature (p_i-T or vapor-solid) diagrams.

The great interest in narrow-gap lead tin chalcogenides and mercury-cadmium chalcogenides during the past decade has resulted in an extensive and reliable collection of data pertinent to the material technology, although some incompleteness and controversy still remain. These data will be discussed in the following chapter and supplemented, where necessary, by the thermodynamic properties of the binary constituents.

A survey of the solid solubility of all binary lead and tin chalcogenides is represented in Fig. 4. The compounds PbS, PbSe, PbTe and SnTe crystallize in the cubic NaCl lattice structure whereas the lattice structure of SnS and SnSe is orthorhombic. Only SnS and SnSe exhibit additional M_2C phases[18] which, however, can be avoided under proper growth conditions and will not be considered here. Mercury telluride, on the other hand, forms a complete solid solution with cadmium telluride.

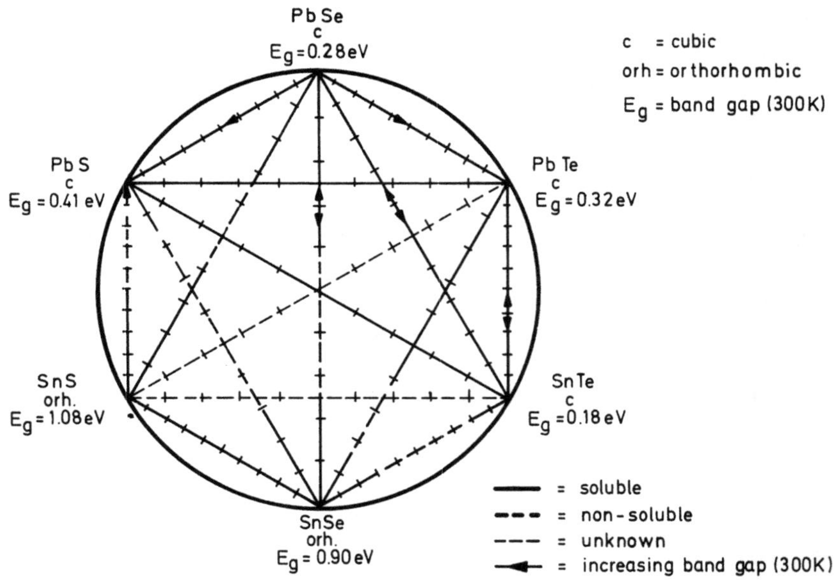

Fig. 4. Solubility of pseudo-binary lead and tin chalcogenides

3.1 Solid-Liquid Phase Diagrams

The growth of crystals from the melt is governed by the T-x_j phase diagram, describing the liquidus and solidus surfaces of the binary and pseudobinary systems as a function of temperature. Following the common nomenclature, we will designate x and (1-x) as mole fractions of the binary compounds in the solid pseudobinary alloy and 2 δ as deviation from stoichiometry. The general composition is thus represented by $(M_{1-x}N_x)_{1+\delta}\, C_{1-\delta}$. Several methods have been applied to establish the data which contribute to these diagrams, i. e., the liquidus surface and the corresponding solidus composition. One of the most important is the differential thermal analysis in a sealed crucible[19)] which indicates exothermal or endothermal reactions of the sample. Similar information may be drawn from optically inspecting the nucleation state in the melt or by weighing a seed immersed into the melt during the cooling through the liquidus surface. By rapidly quenching the melt from temperatures slightly above the liquidus surface and subsequently determining the lattice constant by X-ray analysis, gaps of miscibility have been detected[20)].

A common method of determining the solidus surface involves the growth of a crystal from the melt and the analysis of the solid composition in its first-to-freeze portion[21)]. This technique is usually applied to temperatures near the maximum melting point of the alloy.

Another point of interest is the knowledge of the tie lines connecting the liquidus surface with the solidus sheet. They govern the growth and liquid phase epitaxy from solutions at lower temperatures. Here, the question arises which liquidus composition and which appropriate liquidus temperatures, i. e., which line on the liquidus surface, corresponds to a desired composition of the solid. They may be determined experimentally by measuring the composition of a solution-grown layer or crystal, but considerable effort has also been undertaken in calculating them from thermodynamic data[22). By generalization of Vieland's equations[23) for a liquid-solid equilibrium in a ternary system $A_{1-x}B_xC$, the equilibrium equations for the solid and regular solution are

$$\gamma^{s}_{AC}(1\text{-}x) = \frac{\gamma^{l}_{A}\,\gamma^{l}_{C}}{\gamma^{sl}_{B}\,\gamma^{sl}_{C}}\, N_A N_C \exp\left[\frac{\Delta S_{AC}}{RT}(T_{AC}-T)\right] \tag{24}$$

$$\gamma^{s}_{BC}\,x = \frac{\gamma^{l}_{B}\,\gamma^{l}_{C}}{\gamma^{sl}_{B}\,\gamma^{sl}_{C}}\, N_B N_C \exp\left[\frac{\Delta S_{BC}}{RT}(T_{BC}-T)\right] \tag{25}$$

where γ_{ij} is the activity coefficient at temperature T of the indicated element in the liquid or compound in the solid, the superscripts s, l, and sl referring to the solid, liquid and stoichiometric binary liquid, respectively. N_i is the mole fraction of the indicated element in the liquid and ΔS_{ij} and T_{ij} are the entropy and temperature of fusion, respectively, of the binary compounds. The activity coefficients may be expressed in terms of the interaction parameters between the two binary solids and the interaction parameters between each of the two elements in the liquid by several models. Eq. (24) and (25) can be solved numerically to yield the full liquidus-solidus phase diagram by inserting the melting points and entropies of fusion and approximating the activity coefficients from the binary data.

Until now, non-stoichiometry in the solid has not been considered. The possibility of dissolving excess metal or chalcogen in the solid state is an important feature of the binary and pseudobinary IV–VI and II–VI compounds. An excess metal atom most likely forms a chalcogen vacancy V_C, an excess chalcogen forms a metal vacancy V_M. Incorporation of excess atoms as interstitials I_M, I_C is also possible. (The antistructure type of defect, however, can be ruled out in the rocksalt structure IV-VI compounds, and most likely also in the II-VI compounds.) These defects yield donors (metal excess) or acceptors (chalcogen excess) with zero activation energy contributing charge carriers to the bands. The carrier concentration is given by

$$n-p = z_1\ [V_C] + z_2\,[I_M] - z_3\,[V_M] - z_4\,[I_C] \tag{26}$$

where z_i indicates the charge state and the parentheses designate the concentrations of the various defects. Since the charge states z_i are not unambiguously known for all the defects (see Chap. 3.3.), it has become customary to use the carrier concentration rather than mole fractions in defining the solidus range. This has proved to be sufficient and even advantageous in device applications.

At 77 K, the intrinsic carrier concentration is so small that the minority concentration may be neglected. The carrier concentration can then be determined by Hall measurements according to n, p = r/R_He, where R_H is the Hall coefficient, r the

Hall factor which is generally near unity but may vary between 0.5 and 2 due to anisotropy of the bands, and e is the electron charge. The intersection lines through the solidus surface may now be determined by an isothermal annealing technique. This has first been applied to PbTe[24]. A crystal slab is annealed in a closed ampoule under metal-or chalcogen-rich vapor provided by a sufficiently large amount of powder with the same composition x and with metal or chalcogen excess beyond saturation. The annealing time should be long enough to equilibrate the crystal and the source, i. e., to form a homogeneous excess distribution under the influence of the infinite vapor source. Then, the crystal has to be quenched quickly to avoid precipitation[25], usually by dropping the ampoule into water. The carrier concentration of the quenched crystal is then determined by measuring the Hall coefficient at 77 K after chemically removing the surface layer which might contain a condensed second phase.

3.1.1 T-x and T-δ Diagrams of $(Pb_{1-x}Sn_x)_{1+\delta}Te_{1-\delta}$

PbTe and SnTe form a solid solution over the entire composition range and the metal atoms are randomly distributed on the sub-lattice[26]. The temperature-composition phase diagram of the ternary system Pb-Sn-Te is shown in Fig. 5 using triangular coordinates to represent the compositions. The solidus sheet of $(Pb_{1-x}Sn_x)_{1+\delta}Te_{1-\delta}$ is simplified to zero thickness (δ = O) which is only true at the highest temperatures. Furthermore, it should be slightly inclined, i. e., the maximum melting points of the alloys appear at non-stoichiometric compositions. These simplifications are nevertheless of great importance to the growth and properties of the pseudo-binary alloy and will be discussed extensively by means of the intersections through the solidus range. The solidus sheet does not touch the liquidus surface except in the melting

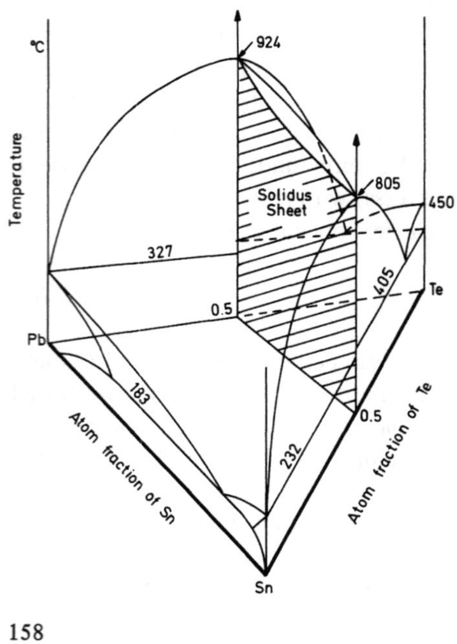

Fig. 5. Temperature-composition phase diagram for the ternary Pb-Sn-Te system[27, 28]

points of PbTe and SnTe, although the separation is small. This is shown more precisely by the solidus and liquidus lines of the pseudo-binary section in Fig. 6 which dominate the growth of $Pb_{1-x}Sn_xTe$ crystals from the melt. The liquidus line follows from melt compositions and solidification temperatures in Bridgman-type growth experiments. The solidus line is obtained from the composition of the first-to-freeze portion ot the crystal.

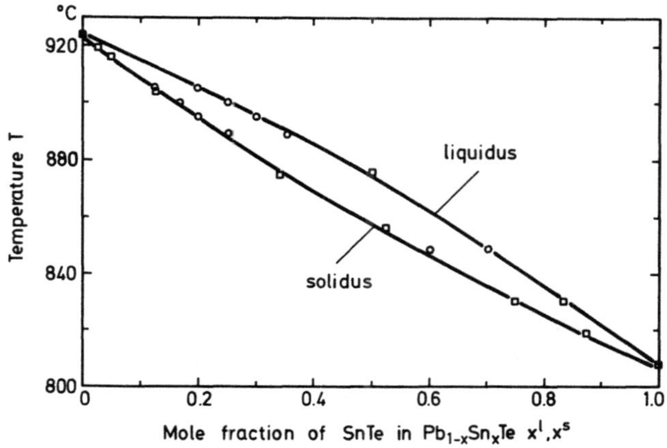

Fig. 6. Pseudo-binary T-x phase diagram for $Pb_{1-x}Sn_xTe$[21, 29]

Several attempts to calculate the phase diagramm of Fig. 5 have been undertaken. Ilegems and Pearson[22] were able to represent the liquidus curve of PbTe by the model of a regular associated solution[30, 31]. They introduced as parameters the degree of dissociation at the melting point (0.92 mole fractions were found to exist as molecular PbTe) and the interaction of the PbTe molecule with Pb or Te (depending on the branch of the liquidus curve).

Laugier et al.[32, 33] extended this model[31] to pseudo-binary compounds. However, for $Pb_{1-x}Sn_xTe$, their results show considerable disagreement with recent experimental data[34] in the low temperature and metal rich range of the phase diagram. This may be due to the neglect of ternary interactions between PbTe-Sn and SnTe-Pb in the liquid. Harris et al.[34] used a semi-empirical approach (also disregarding association in the liquid). They fitted the parameters of Eqs. (24) and (25) from low temperature data on the solubility of Te in Pb, $Pb_{0.56}Sn_{0.44}$, and Sn and Hansen's high temperature data[35] on the Pb-Te and Sn-Te liquidus. Their results, plotted in Fig. 7 are in excellent agreement with experiment. The ternary liquidus isotherms and the iso-solid composition lines exhibit several features which affect liquid phase epitaxy of $Pb_{1-x}Sn_xTe$:

a) For values of the solid composition $x < 0{,}18$, the solidus lines are nearly parallel to a line of constant Pb/Sn ratio. Thus, the solid composition within this range is almost independent of temperature.
b) A solid composition of $x = 0.22 \pm 0.02$ is achieved irrespective of the Pb/Sn ratio in the liquid within a range of $0.3 < x_l(Sn) < 0.7$ if growth temperature is sufficiently low ($T < 550$ °C).

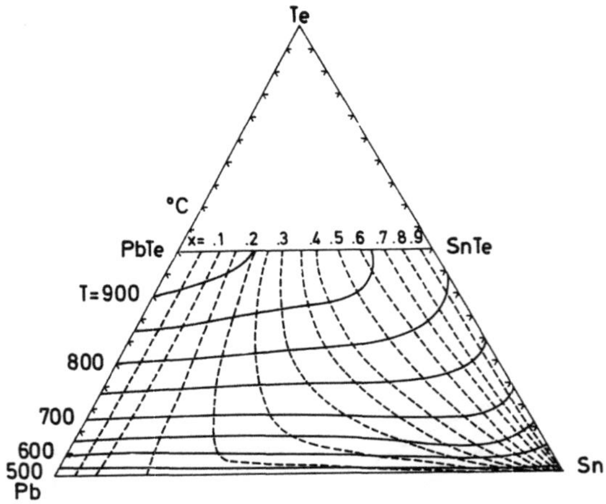

Fig. 7. Calculated iso-temperature liquidus lines (*solid*) and iso-composition solidus lines (*dashed*) for the Pb-Sn-Te system[34)]

c) Layers with high fractions of SnTe require growth at a constant temperature if large compositional grading is to be avoided.

Especially condition b) is of great advantage to liquid phase epitaxy since it presumes that $Pb_{1-x}Sn_xTe$ layers of the desired composition x = 0.2 are easily produced by liquid-phase epitaxy from metal-rich solutions. The Pb/Sn ratio of the liquid needs not to be accurately controlled and the composition of the solid is only a function of the growth temperature.

The relation between x_s(SnTe) and x_l(Sn), plotted in Fig. 8, has also been confirmed in a more conclusive way by the regular associated solution model, which takes into account all the possible ternary interactions in the liquid[22)]. It has not

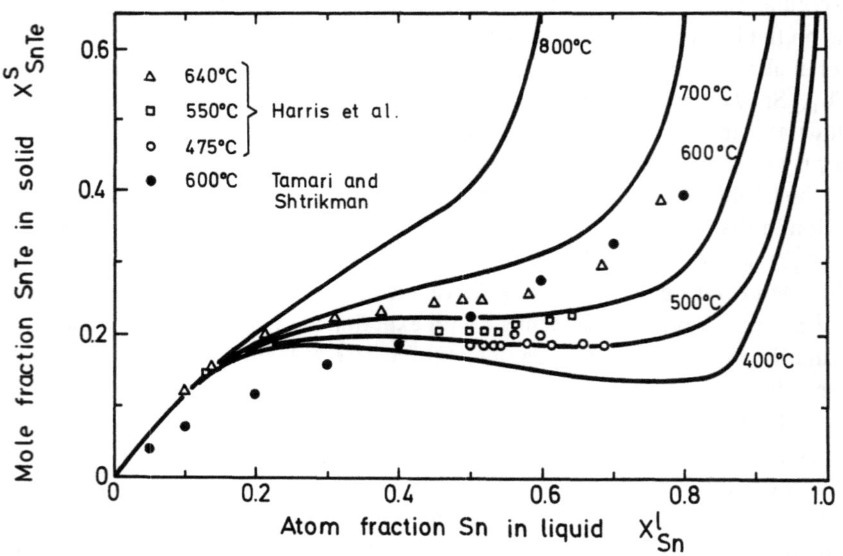

Fig. 8. Calculated solid-liquid tie lines for the Pb-Sn-Te system and experimental data[34, 37)]

been verified in recent experimental results from which a more linear behavior was derived[36, 37] (dots in Fig. 8). On the tellurium-rich part of the phase diagram (Fig. 7) a monotectic region extends to about 13% Pb in the Pb-Te system and 16% Sn in the Sn-Te system. Between this range and the stoichiometric composition of the pseudo-binary alloys, the liquidus isotherms and iso-solidus composition curves are expected to be almost straight lines[32, 33].

Intersections T-δ through the solidus domain of the phase diagram around the stoichiometric composition are expanded in Fig. 9 in terms of the equivalent carrier

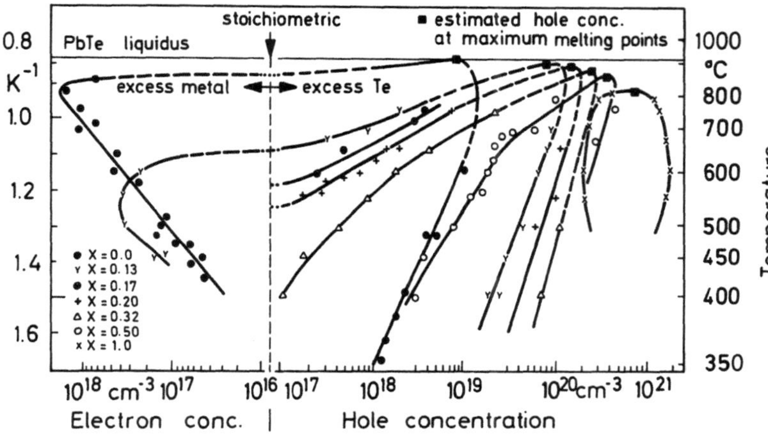

Fig. 9. Carrier concentrations at 77 K (300 K for SnTe) versus isothermal annealing temperature for several $Pb_{1-x}Sn_xTe$ compositions[28, 40, 193, 249]

concentrations. They represent the metal saturated and tellurium saturated solidus lines for several $Pb_{1-x}Sn_xTe$ alloys[28], together with the solidus lines of pure PbTe and SnTe. With increasing SnTe content the solidus field of the alloys shifts from PbTe, with electron concentrations up to $n = 1.5 \times 10^{18}$ cm^{-3} and hole concentrations of $p = 9 \times 10^{18}$ cm^{-3} (n and p values roughly twice as high as these have also been reported[38, 39]), to very high hole concentrations of 2×10^{20} cm^{-3} $\leqslant p \leqslant 2 \times 10^{21}$ cm^{-3} in SnTe. (The latter SnTe data are "apparent" hole concentrations determined at 300 K and assuming the Hall factor to be $r = 1$[40]).

It is worth mentioning that in the alloys with $0.13 < x < 0.25$, the conversion of the carrier type occurs at well defined and easily accessible temperatures. In $Pb_{0.8}Sn_{0.2}Te$ which is of particular importance in device applications the conversion temperature is 530 °C. Low electron or hole concentrations in the range of 10^{15} cm^{-3} have been obtained by simple isothermal annealing[28] or by liquid-phase epitaxy[34] at this temperature.

3.1.2 T-x and T-δ Diagrams of $(Pb_{1-x}Sn_x)_{1+\delta}Se_{1-\delta}$

The ternary temperature-composition diagram Pb-Sn-Se is shown in Fig. 10 making use of the triangular coordinate representation. The binary sections have been re-

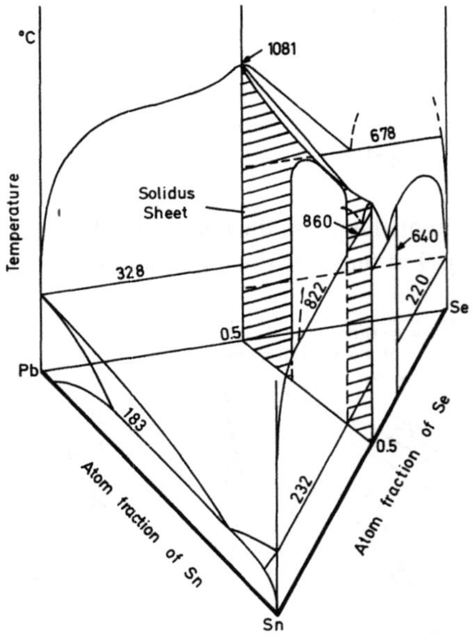

Fig. 10. Temperature-composition phase diagram for the ternary Pb-Sn-Se system[35, 41, 42]

ported by[35, 42]. Of particular interest is the solidus field between the binary compounds PbSe and SnSe. As in the diagram of Fig. 5 only its projection $\delta = 0$ is drawn and extracted in more detail in Fig. 11. Since PbSe and SnSe exhibit different crystal structures, it is more complicated than the analoguous $Pb_{1-x}Sn_xTe$ system. A gap of miscibility appears for $0.4 \leqslant x \leqslant 0.8$[20]. The peritectic point of the liquidus occurs at 870 °C and x = 0.75 and that of the solidus at x = 0.5. Concerning device application the alloys with narrow energy gaps are more interesting. They belong to the rock salt structure compositions with SnSe mole fractions of $x \leqslant 0.4$. Thus, only very few data are available about the SnSe-rich alloys with $x \geqslant 0.8$. A very wide liquidus-

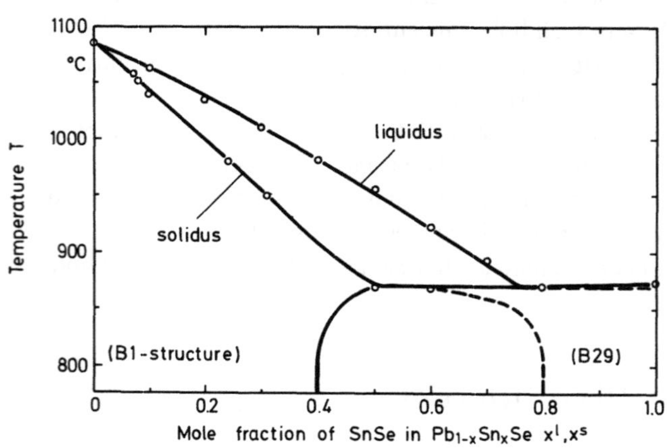

Fig. 11. Pseudo-binary T-x phase diagram for $Pb_{1-x}Sn_xSe$[28]

solidus separation is seen with segregation coefficients up to 0.5 from which considerable implications on the melt growth of $Pb_{1-x}Sn_xSe$ may be anticipated.

The liquidus isotherms and the iso-solid composition curves, i. e., liquid compositions which are in equilibrium with a particular solid composition, are shown in Fig. 12. They are experimentally verified by liquid-phase epitaxial growth[43]. The

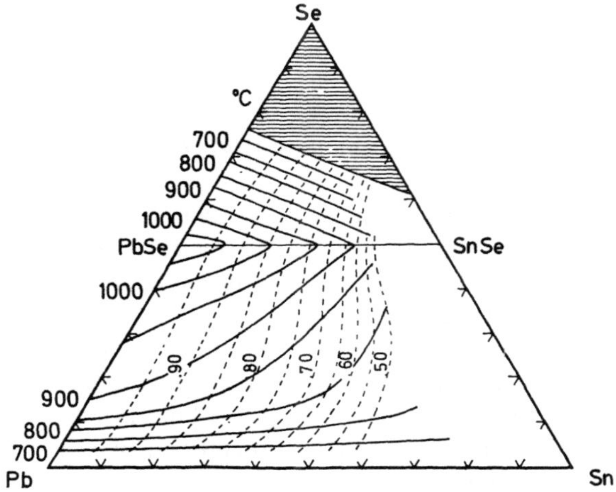

Fig. 12. Calculated iso-temperature liquidus lines (solid) and iso-composition solidus lines (dashed) for the Pb-Sn-Se system[33]

intersections with the solidus field of $(Pb_{1-x}Sn_x)_{1+\delta}Se_{1-\delta}$ with respect to δ, which represent the maximum solubilities of metal or Se, are shown in Fig. 13. They have been determined by the isothermal annealing technique described for $Pb_{1-x}Sn_xTe$.

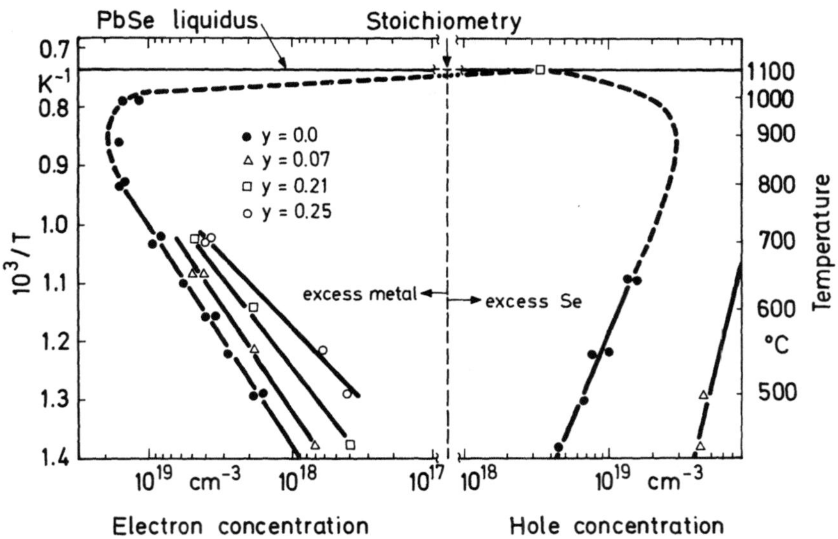

Fig. 13. Carrier concentrations at 77 K versus isothermal annealing temperature for several $Pb_{1-x}Sn_xSe$ compositions[28]

The range of homogeneous compositions of crystalline PbSe extends almost symmetrically from $n = 2 \times 10^{19}$ cm^{-3} to $p = 3 \times 10^{19}$ cm^{-3} while that of SnSe is restricted to Se-rich compositions for temperatures between 500 °C and the melting point 860 °C. It extends from $p \approx 1 \times 10^{16}$ cm^{-3} to 2×10^{18} cm^{-3} [44]. For the rock salt structure $Pb_{1-x}Sn_xSe$, the excess Se solubility increases with increasing x. However, in contrast to SnTe it is low again in orthorhombic SnSe. This may be understood from the fact that the Sn-Se bond is stronger than the Pb-Se or Sn-Te bond, as revealed from Mössbauer spectra[45].

3.1.3 T-x and T-δ Diagrams of $(Hg_{1-x}Cd_x)_{1+\delta}Te_{1-\delta}$

HgTe and CdTe form single-phase solid solutions with zinc blende structure over the entire composition range[46]. The triangular representation of this system is given in Fig. 14, with its solidus sheet extracted in more detail in Fig. 15. Several authors have contributed data to this T-x diagram[47–50]. The remarkable disagreements in their results are due to the different partial pressures of mercury above the liquid and the solidifying ingot[51]. The liquidus and solidus curves are widely separated yielding a segregation coefficient $k = x_s/x_l = 3$ in the interesting composition range of $x = 0.2$. From this fact, considerable difficulties may be anticipated for conventional melt growth techniques. No tie lines between the metal-rich or Te-rich solutions and the corresponding solidus compositions have yet been reported. However, the first results on liquid phase epitaxy from Hg-rich or Te-rich melts state, that liquid and solid compositions only slightly depend on temperature, i. e. iso-solid composition curves are expected to be straight lines.

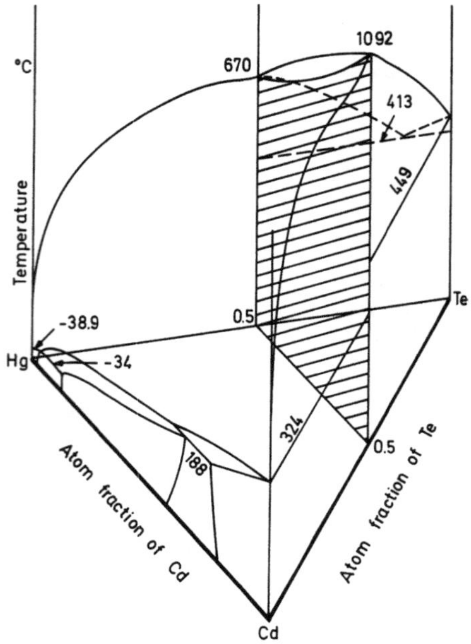

Fig. 14. Temperature-composition phase diagram for the ternary system Hg-Cd-Te[35, 52]

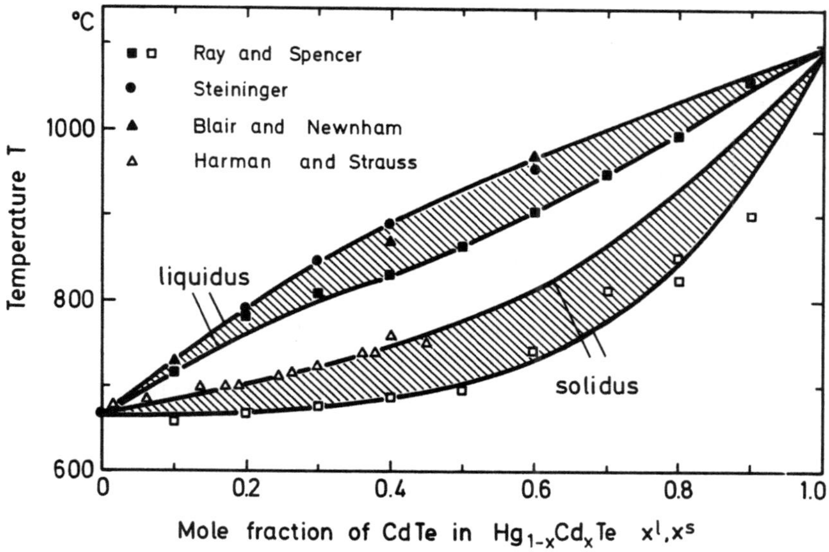

Fig. 15. Pseudo-binary T-x phase diagram for $Hg_{1-x}Cd_xTe$[47–50]

The simplified solidus sheet in Fig. 14 is again of non-zero thickness. Fig. 16 shows the solubility of excess metal and tellurium in the pure binary and in the pseudo-binary compositions x = 0.2 and x = 0.4. The data of HgTe result from measurements at 4.2 K and 100 kG which reliably exclude the contribution of highly mobile electrons to conductivity. This contribution occurs at lower hole concentra-

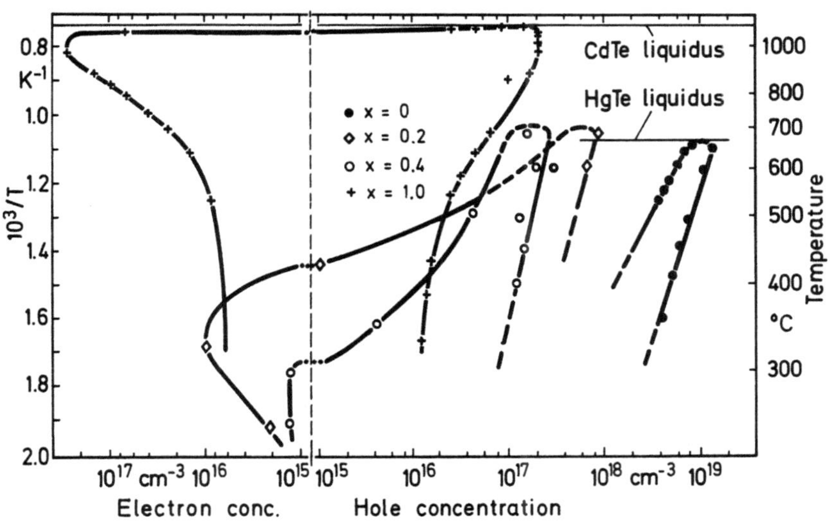

Fig. 16. Carrier concentrations at 77 K (4.2 K for HgTe) versus isothermal annealing temperature for several $Hg_{1-x}Cd_xTe$ compositions[52, 53, 74, 151]

tions and is verified by a strong dependence on the magnetic field[54]. Apparently, low-electron concentrations of $n < 10^{15}$ cm^{-3}, as desired for photoconductive detectors of $Hg_{1-x}Cd_xTe$ (x = 0.2 or 0.4), are feasible by isothermal annealing at T = 420 °C or 310 °C, respectively. But this procedure very likely produces compensated material. It is therefore advantageous to apply temperatures on the retrograde solidus branches, i. e. T < 220 °C.

3.2 Solid-Vapor Phase Diagrams

One of the most important growth methods of device-quality lead tin chalcogenides is the sublimation technique. Moreover, annealing processes under well-defined conditions are applied to achieve those electrical properties of the narrow-gap materials which are desired for device applications. This has launched considerable interest in their solid-vapor phase diagrams. Extensive studies of the vapor composition, the vapor pressure and the solid equilibrium have been performed. Although most of the available data concern the binary compounds, these elucidate the properties of the pseudobinary alloys.

A frequently used thermogravimetric method of determining the vapor pressure in equilibrium with a solid or liquid compound employs the effusion cell. The weight loss is measured either sequentially after keeping the material in the cell in a dynamic vacuum at various temperatures[55] or continuously during the heating process[56].

In the steady-state effusion, the solid acquires the congruently subliming composition if this exists within the solidus range. It corresponds to a minimum of the total pressure with partial vapor pressures P_M and P_{C_2} of the metal and chalcogen, respectively,

$$p_M = 2\, p_{C_2} = 2^{1/3}\, k_{MC}^{2/3} \tag{27}$$

where k_{MC} is defined by $p_M\,(p_{C_2})^{1/2} = k_{MC}$. (Using this definition, stoichiometric vapor is chosen for the congruently subliming composition instead of regarding the small deviation from stoichiometry, but this is negligible with respect to the large change in partial vapor pressures with a small deviation from stoichiometry in the solid). If the congruent sublimation does not exist, the thermal treatment with continuous removal of vapor leads to a solid composition on the three phase line. Here, the solid assumes a temporary steady state with constant partial vapor pressures until only the neighboring phase (a melt, a different compound or one of the components) remains.

A different method employed in the determination of the vapor pressures of chalcogenides is the optical absorption. Measurements at several wavelength bands allow partial pressures of different vapor constituents to be obtained simultaneously[57]. Furthermore, using a closed absorption cell, the composition of the solid in equilibrium with the vapor can be studied. If metal- or chalcogen-saturated compositions are chosen as a source, this procedure yields particularly important information on the three-phase line including the stability range of the compound. From the vapor pressure data, the enthalpy and entropy of sublimation may be calculated by application of the second or third law equation.

In order to obtain the partial pressures (usually that of the most volatile component) in equilibrium with compositions from within the solidus range, the experimental procedure is reversed. A crystal is annealed at a given temperature and vapor pressure and subsequently its carrier type and concentration are determined by Hall measurements. The composition may then be inferred from Eq. (26) if the predominant defect and its effective ionization state are known. The vapor pressure is varied by a liquid or solid source heated independently in a second temperature zone. Fig. 17 shows the schematic arrangements of this two-zone technique. Argon filling

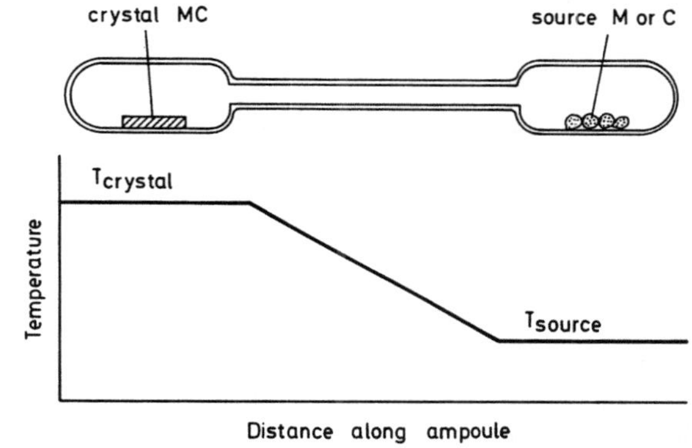

Fig. 17. Schematic diagram of the two-zone annealing

of the tube or mounting of a capillary between sample and source prevents excessive transport of the compound itself to the lower temperature region. This method has also achieved significant importance in the production of the low-carrier concentration material for device applications as shown in Chap. 4.

3.2.1 p-T-δ Diagram of $Pb_{1-x}Sn_xTe$

For PbTe a congruent sublimation behavior could be derived from effusion cell experiments[56]. The corresponding composition contains a slight excess of tellurium[39]. For SnTe, however, a remainder of pure Sn is found after the effusion runs, independently of the starting composition, indicating that a congruent sublimation does not exist within the solidus field. The steady state thus appears on the tin-rich three-phase line. The total vapor pressures were found to obey the equations[56]:

PbTe (802–1017K): $\log p[\mathrm{atm}] = 7.57 \pm 0.16 - (11.29 \pm 0.18)10^3/T[\mathrm{K}]$

SnTe (800–1026K): $\log p[\mathrm{atm}] = 7.78 \pm 0.15 - (11.36 \pm 0.18)10^3/T[\mathrm{K}]$

They agree reasonably well with earlier data. From these results, the lack of a congruent sublimation composition and a slightly preferential loss of SnTe follows for the pseudo-binary alloy $Pb_{1-x}Sn_xTe$[58]. At 700 °C a ratio of $p_{SnTe}/p_{PbTe} = 1.359$ is determined for a x = 0.5 alloy. This decreases by about 2% during complete effusion. Partial pressures of the vapor species in equilibrium with PbTe, SnTe and $Pb_{1-x}Sn_xTe$ have been measured[57–63]. In the latter material, molecules of $SnTe_2$, Sn_2Te_2, PbSnTe and $PbSnTe_2$ could be detected at 760 °C, although with partial pressures less than 1% of that of SnTe or PbTe[61]. The enthalpy of sublimation has been determined from the vapor pressure data by second- and third-law calculations[56].

PbTe: $\Delta H_{298} = 54.12 \pm 1.02$ kcal/mol (2. law)

$\Delta H_{298} = 53.77 \pm 1.35$ kcal/mol (3. law)

SnTe: $\Delta H_{298} = 54.49 \pm 1.06$ kcal/mol (2. law)

$\Delta H_{298} = 53.13 \pm 1.37$ kcal/mol (3. law)

For PbTe these values are again in good agreement with results published earlier. For SnTe, ΔH_T as low as 43.4 ± 0.9 kcal/mol has been reported[62]. The enthalpies of sublimation of the alloys coincide with those of the binary constituents within the error limits, as expected from the striking similarity of their vapor pressures[58].

The partial pressures of Pb, Sn and Te_2 over PbTe and SnTe vary greatly with deviation from stoichiometry of the solid. Or, inversely, it are these pressures which specify the solidus range and within this solidus range the composition of the solid. This is shown for PbTe in Fig. 18. The three-phase line consists of two legs meeting

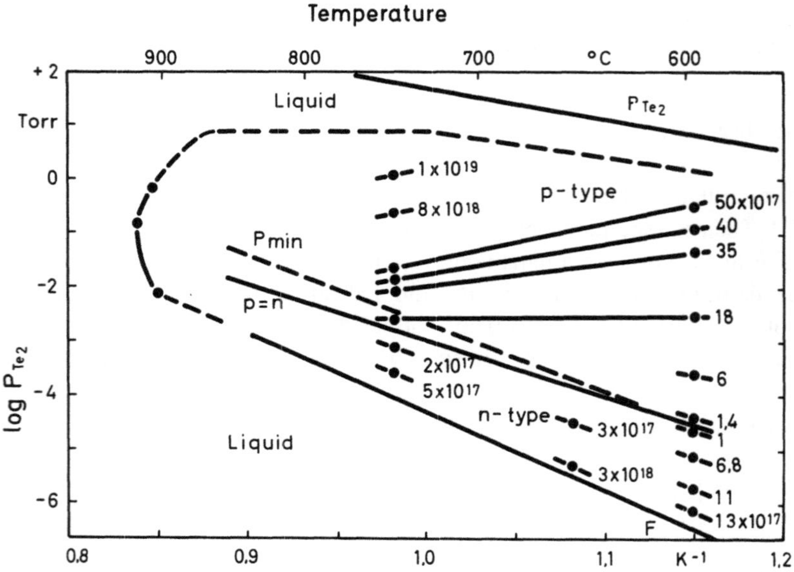

Fig. 18. p-T phase diagram for PbTe and iso-carrier concentration lines[39]

at the maximum melting point. The upper one represents the tellurium partial pressure being at equilibrium with Te-saturated PbTe. It approaches the vapor pressure of pure tellurium. The lower one represents the minimum tellurium pressure over Pb-saturated PbTe. Inside this loop only solid and vapor coexist. The line designated by p = n represents the stoichiometric composition of PbTe separating the p-type (Te-rich, upper part) from the n-type (Pb-rich, lower part) compositions. It is seen that the maximum melting point occurs at a Te-rich composition. The line denoted by p_{min} corresponds to the congruent sublimation composition. Several iso-composition lines are shown where the composition is again represented by means of electron and hole concentrations measured at low temperature. The iso-concentration lines have been determined by the two-zone annealing method reported in Chap. 3.2.

The analoguous p-T-δ diagram of SnTe is shown in Fig. 19. As expected from Fig. 6 the three-phase line only includes Te-rich compositions with a maximum melting point at 50.4 at % Te. Since the solubility of excess Te is high and covers a very wide range (0.9 at %), the optical absorption method allows to relate the partial pressure of Te_2 directly to particular compositions of the solid[63]. Furthermore, the lattice constant of SnTe crystals decreases with increasing hole concentration. This fact and comparisons between calculated and observed densities indicate that Sn vacancies are the predominant defects[64]. The data[65] available on the partial vapor pressure of Te_2 being at equilibrium with the alloy $Pb_{1-x}Sn_xTe$ ($0 < x < 1$) are not sufficient to construct p-T-δ diagrams.

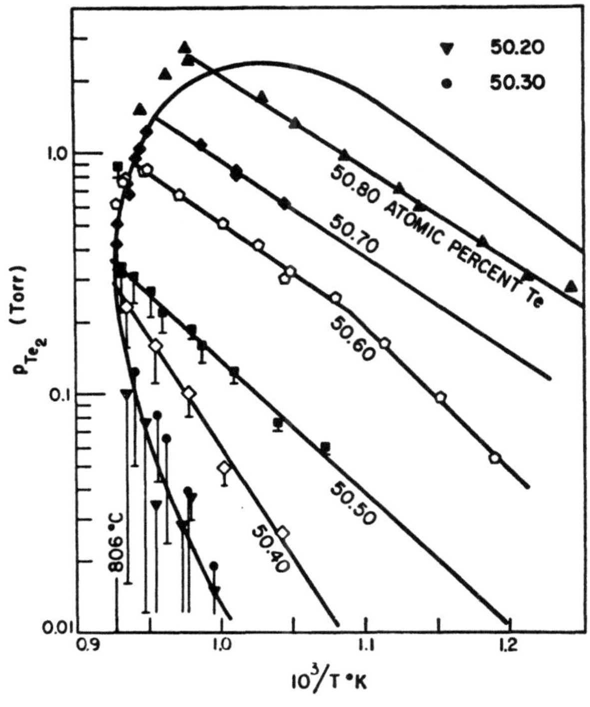

Fig. 19. p-T phase diagram for SnTe and iso-composition lines[63]

3.2.2 p-T-δ Diagram of $Pb_{1-x}Sn_xSe$

The behavior of lead and tin selenides and their alloys resemble those of the tellurides. Nevertheless, some qualitative differences exist which affect the production of alloy materials and devices:
The partial vapor pressures of PbSe and SnSe differ markedly from each other, combined with a higher partial pressure of Se_2 itself. While PbSe again possesses a congruent subliming composition (very slightly Pb-rich), this is not the case for SnSe. The total vapor pressures, as derived from effusion studies, are represented by[66, 67]

PbSe (887–1052 K): $\log p\,[\text{atm}] = 8.16 \pm 0.15 - (12.26 \pm 0.15) \times 10^3/T[K]$

SnSe (789–975 K): $\log p\,[\text{atm}] = (7.47 \pm 0.13) - (10.69 \pm 0.12) \times 10^3/T[K]$

These relations agree resonably well with earlier results[68, 69] and lead to enthalpies of sublimation on the basis of second and third law calculations[66, 67]:

PbSe: $\Delta H_{298} = 59.1 \pm 0.7$ kcal/mol (2. law)
$\Delta H_{298} = 55.66 \pm 0.2$ kcal/mol (3. law)

SnSe: $\Delta H_{298} = 51.5 \pm 0.5$ kcal/mol (2. law)
$\Delta H_{298} = 52.8 \pm 0.5$ kcal/mol (3. law)

PbSe is the only material of this group for which a p-T-phase diagram has been evaluated[18, 70, 71] (Fig. 20). The three-phase line exhibits a maximum Se_2-vapor pressure of about 4 atm and a maximum melting point of 1080 °C. A straight line

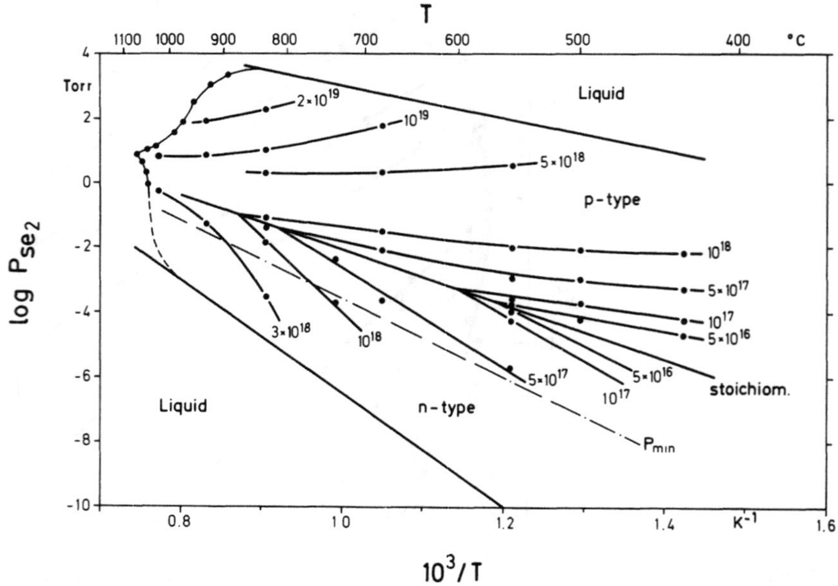

Fig. 20. p-T phase diagram for PbSe and iso-carrier concentration lines[70, 71]

again designates the stoichiometric composition; p_{min} represents the congruently subliming composition. The iso-carrier concentration curves reflect some discrepancies (for $n \geqslant 10^{18}$ cm^{-3}) between the data of different authors. No such diagram can be presented for SnSe or $Pb_{1-x}Sn_xSe$ since the available data are insufficient. However, the results of two zone annealing experiments, where hole concentrations as low as 5×10^{16} cm^{-3} could be achieved at 500 and 650 °C with Se_2-pressures of 5×10^{-7} and 1×10^{-4} Torr, respectively, have been reported[28]. These data indicate a shift of the stoichiometry line toward the lower stability limit with increasing x, as expected from the exclusively p-type character of SnSe.

3.2.3 p-T-δ Diagram of $Hg_{1-x}Cd_xTe$

The vapor phase above HgTe and CdTe was investigated by measuring its optical density[53, 72, 73]. Irrespective of composition and temperature, Hg is always the dominant vapor species over HgTe. Molecular HgTe is not detectable in the investigated range 400–700 °C and HgTe lacks a congruently subliming composition down to room temperature. Continued sublimation will form a Te-rich solid or liquid phase outside the homogeneity range of HgTe. In CdTe, the composition of the vapor phase depends on that of the solid. In equilibrium with CdTe on the Cd-rich stability limit, Cd is the dominant vapor species. Over Te-saturated CdTe, Te_2 prevails in the vapor phase. A congruently subliming composition exists within the stability range, although CdTe decomposes on evaporation. Its molecular species is not detected in the vapor within the investigated temperature range of 780 to 939 °C.

In the low-temperature region the vapor pressure of Hg over Hg-saturated HgTe is represented by[53]:

HgTe (666–735 K): $\log p_{Hg}$ [atm] $= 4.71 - 2.98 \times 10^3/T$[K]

In addition, a few distinguished vapor pressures belonging to the three-phase line of HgTe may be specified[49]:

p_{Hg} (max) over Hg-saturated HgTe (922 ± 6 K) = 19 atm

p_{Hg} over HgTe at the maximum melting point (941 K) = 13.6 atm.

The free energy of formation of solid HgTe(c) from Hg(g) and Te_2(g) is represented by[53]:

$$\Delta G^\circ \,[\mathrm{HgTe(c)}] = -41.66 + 42.71 \times 10^{-3}\,T\ [\mathrm{kcal/mol.}]$$

Analogous data for CdTe are given for the vapor pressures over congruently subliming CdTe[72]:

CdTe (1053 – 1212 K): $\log p_{Te_2}$ [atm] $= 6.346 - 10 \times 10^3/T$[K]

$$p_{Cd} = 2\, p_{Te_2}$$

and for the free energy of formation of CdTe (c) from Cd(g) and Te_2(g):

$$\Delta G^\circ \,[\mathrm{CdTe(c)}] = -68.64 + 44.94 \times 10^{-3}\ [\mathrm{kcal/mol}].$$

The Hg pressure over pseudo-binary stoichiometric melts of $Hg_{1-x}Cd_xTe$ runs parallel to that over elemental Hg with an activity a_{Hg} = (p over $Hg_{1-x}Cd_xTe$

melt)/p_{Hg}) = 0.345 ± 0.020 independently of x and T in the investigated range. It follows that over pseudo-binary stoichiometric melts with $0 \leqslant x \leqslant 0.6$ [49]:

$Hg_{1-x}Cd_xTe$ (943–1238 K): $\log p_{Hg}$ [atm] = 4.43 – 3.10 x 10^3/T[K]

If, on the other hand, the melt composition of $(Hg_{1-x}Cd_x)_{1+\delta}Te_{1-\delta}$ deviates from stoichiometry, i. e., if $\delta \neq 0$, a marked influence on the activity coefficient is observed[49].

The p-T-δ phase diagram of HgTe and $Hg_{0.8}Cd_{0.2}Te$ representing their existence range versus the partial pressure of Hg and the inverse temperature, is given in Fig. 21 [74]. The upper three-phase curves of both materials run into the vapor pressure line of pure Hg, the lower ones show significant differences. For $Hg_{0.8}Cd_{0.2}Te$, some data on the composition are included, i. e., the carrier type and concentration as revealed from isothermal and two-zone annealing experiments. The hatched area designates stoichiometry and the transition from p to n. Thus, high temperature processes always produce p-type crystals and very low isothermal annealing temperatures are required to achieve low-electron concentrations.

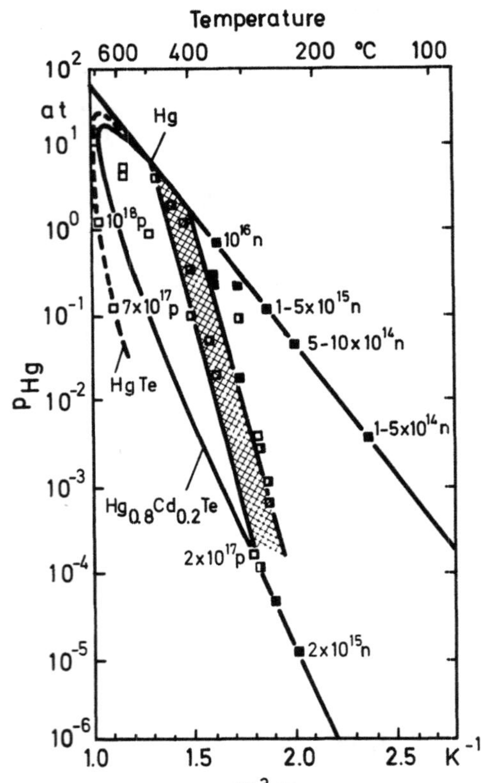

Fig. 21. p-T phase diagram for $Hg_{1-x}Cd_xTe$ (x = 0, x = 0.2) and some carrier concentrations at 77 K for x = 0.2. ▪ n-type, ▫ p-type, ▣ compensated material[74]

3.3 Properties of Intrinsic Defects and Foreign Impurities

It has already been mentioned in previous chapters that excess metal produces n-type, excess chalcogen p-type conductivity in both the IV-VI and II-VI-compounds. The physical properties of these defects have been the subject of many investigations which will be briefly reviewed here. Some results on foreign impurities are added although these are known to a lesser extent.

3.3.1 Native Defects in $Pb_{1-x}Sn_xTe$ and $Pb_{1-x}Sn_xSe$

Although there is a general agreement concerning the doping character of native defects in lead-tin chalcogenides, i. e., vacancies and interstitials of either type of ions, the energetic levels and the ionic charge states associated with these defects are still in discussion. The latter is important to the extent that it relates the deviation from stoichiometry to the carrier concentration. In a theoretical approach the change of the electronic levels in the vicinity of vacancies and interstitials in PbTe was calculated[75, 76] using the Greens-function method of Koster and Slater. The strong perturbations around the defects cause valence band states to shift to the conduction band. As a result, a Te vacancy provides two electrons for the conduction band and a Pb vacancy two holes for the valence bands. The Pb interstitial yields a single electron whereas the Te interstitial is found to be neutral. Later, a cluster calculation was performed[77] by the scattered-wave method for PbTe and SnTe which confirmed the position of the metal vacancy, but suggested a Te vacancy level about 14 meV below the conduction band edge.

Experimental studies on defect energy levels in these materials within the band gap have often been undertaken[78, 79]. On the basis of a recent thorough study on PbTe, epitaxial layers with carrier concentrations as low as p, n = 2×10^{16} cm^{-3}, at T = 1.8 K and with magnetic fields up to B = 15 T deep levels of lattice defects or impurities could be effectively excluded[80]. Shallow hydrogen-like impurity donor or acceptor levels, however, may still form impurity bands under these conditions. The states in fact seem to exist and merge into the conduction band at temperatures above 170 K[81].

Several attempts have been made to relate both the carrier type and concentration to the type and concentration of lattice point defects in the lead-tin chalcogenides. In SnTe, the very large deviation from stoichiometry (Fig. 9) could be verified directly by density measurements. It was found that the prevailing defect is a doubly ionized Sn vacancy[64].

In PbTe, a step-by-step ion implantation of Pb and Te into p- and n-type epitaxial layers followed by low temperature annealing processes was performed[82]. Comparing the changes of carrier concentration and implantation doses, the effective state of ionization of the defects [z_i in Eq. (26)] could be determined: $z_1(V_{Te}) = 2$, $z_2(I_{Pb}) = 1$, $z_3(V_{Pb}) = 2$ and $z_4(I_{Te}) = 0$. This result provides a plausible explanation for the well-known fact that lattice damages, produced by ion-, proton- or electron-bombardment, yield n-type samples. If an approximately equal number of Pb and Te Frenkel defects is produced, the net effect will be the appearance of excess electrons. Recently, these observations have been supported for

PbSe[83] and $Pb_{1-x}Sn_xSe$[84]. Crystals of these materials were grown from sources of exactly defined composition to which several small quantities of Pb and Se were added. The difference in carrier concentration could be related to the additive elements and yielded $z/r = 2$ for n- and p-type material where r is the Hall factor which is expected to be near unity. Since in PbSe and $Pb_{1-x}Sn_xSe$ (at least for sufficiently low Sn content) metal or Se vacancies are believed to be the predominant defects, the state of ionization refers to these vacancies [i. e., $z_1 = z_3 = 2$ in Eq. (26)]. However, contrary results have also been reported. If the results of Fujimoto et al.[39, 85] on the annealing of PbTe samples at various Te pressures (see Chap. 3.2.1) are considered at a fixed temperature (Fig. 22) conclusions may be drawn on the ionization

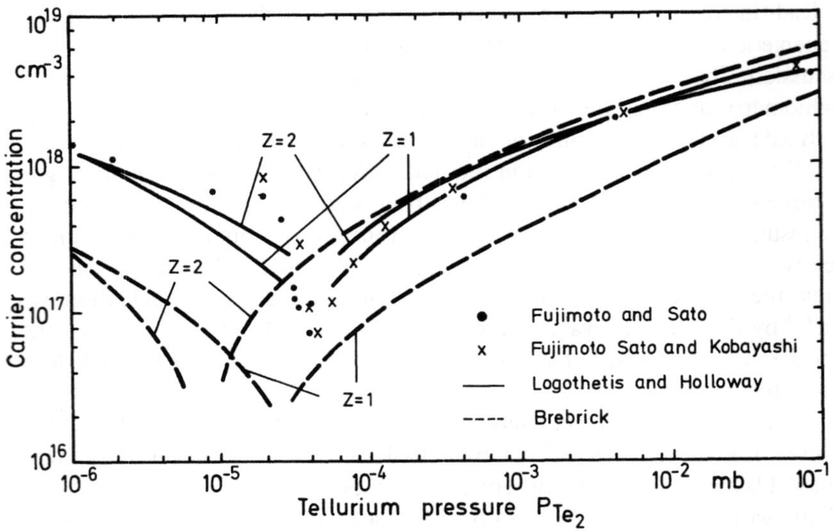

Fig. 22. Experimental and calculated carrier concentrations of PbTe after annealing at 600 °C versus tellurium pressure; z denotes the ionization of the defects[39, 85–87]

state of the predominant lattice defects. These authors found $z_1 = z_3 = 1$ whereas Brebrick's[86] more conclusive analysis reveals some ambiguity in this approach (Fig. 22). This ambiguity is verified by a different model of Logothetis and Holloway[87]. The authors take into account a high degree of compensation (Schottky constant $k_s^2 = [V_M][V_C] \approx 10^{35}\ cm^{-3}$ to $10^{37}\ cm^{-3}$ for T = 600 °C and 750 °C, respectively) and show that with both, z = 1 and 2, reasonable fits into the experimental results can be attained.

3.3.2 Foreign Impurities in $Pb_{1-x}Sn_xTe$ and $Pb_{1-x}Sn_xSe$

If high purity elements are synthesized and crystal growth is performed under proper purity precautions, foreign impurities may usually be neglected in crystals with carrier concentrations above 10^{17} cm^{-3}. Below this carrier concentration, foreign impurities can play a role by compensating lattice defects.

The doping characteristics of specific impurities in the pseudo-binary alloys $Pb_{1-x}Sn_xTe$ and $Pb_{1-x}Sn_xSe$ and their binary constituents has been investigated[88–91]. Crystals were grown from intentionally doped material and the change of carrier type with respect to undoped crystals was analyzed. Although most data stem from experiments with PbTe, the results may be generalized with respect to the other IV-VI compounds and their alloys at least in a qualitative description, disregarding the specific solubilities, the influence of metal or chalcogen saturation and of impurity pairs or clusters. However, the different positions of foreign atoms in the lattice, either interstitial or substitutional on metal or chalcogen sites complicate unambiguous statements. Table 1 summarizes the results. In general, the following characteristics can be assumed: Group Ia and Ib impurities tend to be either

Table 1. Properties of dopants in PbTe

Atom	Substituting Me/Te	Doping character	Max. attained carr. conc. (cm^{-3})	(Pb, Te)-Anneal. temp. (°C)	sat.	Ref.
Na	Me	Acc	$p = 2.5 \times 10^{20}$	924	(melt)	91)
Cu	Me	Acc	$p = 2 \times 10^{16}$	800	(Te)	88)
	interstitial	Don	$n = 2 \times 10^{19}$	500–800	(Pb)	
Ag	Me	Acc	$p = 3 \times 10^{19}$	800	(Te)	88)
			$p = 1 \times 10^{19}$	350	(Te)	
	interstitial	Don	$n = 7 \times 10^{18}$	800	(Pb)	
			$n = 6 \times 10^{18}$	500	(Pb)	
Au	Me	Acc				89)
Zn	Me	Don				89, 90)
Cd	Me	Don	$n = 1.2 \times 10^{18}$	400	(Te)	89, 90, 213)
In	Me	Don	$n = 7 \times 10^{18}$	800, 500	(Pb)	88)
			$n = 0.2 - 5 \times 10^{18}$	800, 350	(Te)	
Tl	Me	Acc	$p = 1 - 2 \times 10^{19}$	350–800	(Pb, Te)	88)
As	Te	Acc	$p = 6 \times 10^{18}$	800	(Pb)	88)
			$p = 7 \times 10^{18}$	500	(Pb)	
	Me	Don	$p = 3 \times 10^{18}$	800	(Te)	
			$n = 0.7 - 2 \times 10^{17}$	800	(Te)	
Sb	Te	Acc	$n = 2 - 7 \times 10^{17}$	800	(Pb)	88)
			$n = 1 \times 10^{16}$	500	(Pb)	
	Me	Don	$p = 3 \times 10^{18}$	800	(Te)	
			$n = 1 \times 10^{18}$	350	(Te)	
Bi	Me	Don	$n = 2 \times 10^{19}$	800, 500	(Pb)	88)
			$n = 3 \times 10^{18}$	350	(Te)	
Cl, Br	Te	Don	$n = 3 \times 10^{19}$	924	(melt)	89)
undoped			$n = 2 \times 10^{18}$	800	(Pb)	
			$n = 2 \times 10^{17}$	500	(Pb)	
			$p = 6 \times 10^{18}$	800	(Te)	
			$p = 1 \times 10^{18}$	350	(Te)	

acceptors if substitutional or donors if interstitial. Group III impurities are found to be donors except for Tl which is an acceptor, group IV elements are isoelectronic and therefore electrically neutral. Group V impurities have to be donors on the chalcogen site and acceptors on the metal site. Group VI elements are isoelectronic and group VII impurities donors.

3.3.3 Native Defects in $Hg_{1-x}Cd_xTe$

Three types of native defects are believed to exist in $Hg_{1-x}Cd_xTe$, namely Hg and Te vacancies and Hg interstitials, the first being an acceptor and the latter two donors.

The Hg vacancy corresponds to an acceptor level whose energy depends on the composition x. For $x < 0.16$ it becomes resonant with the conduction band[92].

In $Hg_{1-x}Cd_xTe$ ($0.21 \leqslant x \leqslant 0.5$) an acceptor state with about 20 meV above the valence band has been detected by carrier freeze-out and photoluminescence[93]. In $Hg_{1-x}Cd_xTe$ ($x = 0.15$), measurements of transport properties as a function of pressure revealed two different energy states. In annealed samples a level about 9 meV above the valence band is detected whereas in as-grown p-type samples a level at 20 meV is found which causes impurity-band conduction[94]. Both levels have also been detected in semi-metallic HgTe and are attributed to a double-acceptor Hg vacancy[95]. A resonant donor state 7.5 meV above the conduction band, observed in transverse magneto-resistance and in far-infrared transmission, is attributed to the Te vacancy[96].

3.3.4 Foreign Impurities in $Hg_{1-x}Cd_xTe$

Since $Hg_{1-x}Cd_xTe$ is used in photoconductive as well as in photovoltaic detectors, foreign impurities are considered from two points of view. Firstly, as residual impurities whose concentrations define the lowest achievable carrier concentration and the onset of compensation; secondly, as intentionally introduced dopants to obtain pn-junctions. Whereas the highest possible purity is desirable in the first case, high and well-defined concentrations of foreign atoms are intended in the second case.

The doping properties of foreign impurities in II-VI compounds are better understood than in IV-VI compounds. Group I and group III elements are expected to substitute on the metal sub-lattice and group V and group VII elements on the chalcogen sub-lattice. Consequently, group I and group V atoms act as acceptors while group III and group VII atoms act as donors. Group IV elements are believed to behave nearly amphoteric. Si is still a donor, Sn is neutral. In contrast to HgTe where diffusion constants have been determined for several metal impurities[97], only qualitative data on their solubility and diffusion properties in $Hg_{1-x}Cd_xTe$[98] are available. They have been derived either by adding elements to the melt prior to crystal growth or annealing crystals with proper sources up to 300 °C in open systems or up to 650 °C in closed ampoules. Hall measurements have been performed with In-contacts on n-type materials or Au/Cr/In-contacts on p-type materials[99]. Table 2 compiles the properties of some important dopants.

Table 2. Properties of dopants in $Hg_{1-x}Cd_xTe$ (x = 0.4)[98]

Atom	Substituting Me/Te	Doping character	Max. attained carr. conc. (cm^{-3})	Annealing temp. (°C)	Diffusion
Li	Me	Acc	10^{18}–10^{19}	250	very fast
Na	Me	Acc	–	–	fast
Cu	Me	Acc	1.7×10^{18}	290	fast
Ag	Me	Acc	1.5×10^{17}	290	fast
Al	Me	Don	3.5×10^{18}	650	slow
Ga	Me	Don	–	300	fast
In	Me	Don	–	–	–
Si	–	Don	1.7×10^{16}	650	slow
Sn	–	neutral	–	300	slow
P	Te	Acc	1×10^{18}	300	slow
As	Te	Acc	1.6×10^{19}	300	slow
Cl	Te	Don	–	300	–
Br	Te	Don	8×10^{17}	300	slow

The energy levels of these impurities in the band gap are unknown. Especially, the donors are expected to produce shallow hydrogen-like states with binding energies of the order of < 1 meV and strongly overlapping wave functions even at low concentrations. In the semi-metallic range of the alloy system (see Fig. 3) an energy level of about 1 meV above the Γ_8 band edge was detected for the Cu-acceptor[92].

4 Growth of Single Crystals

Optoelectronic device applications require a very high crystal quality of the alloys. First of all, this comprises metallurgical properties such as homogeneity of composition, absence of precipitates and a minimal number of dislocations and point defects. But it also includes electrical properties such as carrier mobility and excess carrier life time which are affected by the purity, the deviation from stoichiometry and the degree of compensation.

It is obvious from the phase diagrams that the carrier type and concentration are often determined by the growth process. Optimum data for a particular device may then only be obtained by a subsequent heat treatment, but the success of crystal fabrication nevertheless clearly depends on the correct choice of the cation/anion composition.

Pseudo-binary lead and tin chalcogenide crystals have been grown successfully from the melt. Some attempts have also been undertaken to develop growth procedures involving three phases. The best results, however, are obtained through growth from the vapor phase.

Mercury cadmium telluride crystals have been generated by several melt and solution growth techniques, although the solid-state recrystallization is considered superior. Vapour-phase growth is impaired by the highly dissimilar partial pressures of the constituents. Single crystals have so far only been obtained with the binary HgTe[100].

4.1 Growth of $Pb_{1-x}Sn_xTe$ and $Pb_{1-x}Sn_xSe$

4.1.1 Growth from the Melt

A Bridgman-Stockbarger technique has early been applied to produce crystals of the binary lead chalcogenides PbS, PbSe and PbTe[101, 102]. These experiments have already revealed the crucial role of the purity of the constituents. Care was especially exercised to remove residual oxygen by reducing the elements in a hydrogen flow before mixing them in the growth ampoule and sealing off the latter. The problem of purity seemed to be particularly severe in the case of PbS where local reactions with the quartz ampoule lead to internal cracks in the crystals during the cooling down process or even to a fracture of the ampoule itself. Depositing thin graphite layers on the inside of the ampoule by high temperature pyrolysis of methanol[103] or using a graphite crucible within the ampoule[104] could prevent these disadvantages. In the growth of pseudo-binary alloys from the melt, the T-x phase diagram with segregation coefficients different from unity generally imposes additional problems concerning the homogeneity and morphology of the crystals. This is expected in $Pb_{1-x}Sn_xTe$ which, for example, at $x = 0.2$ and at the corresponding maximum melting point $T_{max} = 895\ °C$, shows $x_s/x_l = 0.66$, and even more for $Pb_{1-x}Sn_xSe$ which, at $x = 0.07$ and $T_{max} = 1050\ °C$, exhibits $x_s/x_l = 0.57$. Despite this disadvantage, crystals with a surprisingly high homogeneity have been grown from both materials[21] although not completely free of imperfections such as inclusions, low-angle grain boundaries or cellular sub-structures[105]. Along the total crystal length of 5 cm, the composition varies by about 10%; in transverse cross sections a composition gradient could not be detected within the accuracy $\Delta x = \pm 0.005$ of the electron microprobe analysis. The growth procedure is illustrated in Fig. 23. 400 g

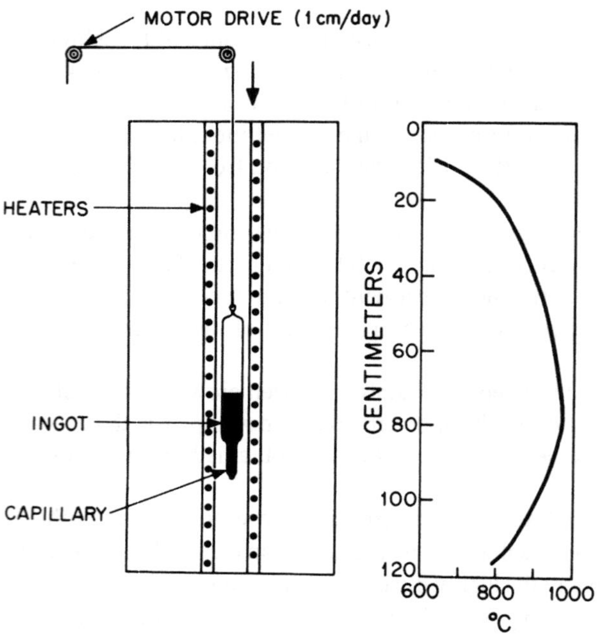

Fig. 23. Bridgman growth apparatus for $Pb_{1-x}Sn_xTe$ and $Pb_{1-x}Sn_xSe$ crystals. Temperature profile for $Pb_{0.8}Sn_{0.2}Te$[21]

charges containing stoichiometric proportions of the elements are placed in a fused silica ampoule coated with pyrolitically produced graphite. The inner diameter of the ampoule is 25 mm ending in a 15 mm tube with a tapered end at the bottom. Since the ampoules crack upon cooling, in the case of $Pb_{1-x}Sn_xSe$ another protection tube filled with 0.25 atm argon is used. During the reaction and homogenization at about 50 °C above the liquidus temperature, the furnace is horizontal. It is rotated to a vertical position before the ampoule is lowered into a temperature gradient of 10 K cm^{-1} at a rate of about 1 cm/day. The method has been improved[28] by using a charge composed of binary PbTe and SnTe which have previously been zone-refined. This procedure ensures a composition near the maximum melting point composition $(Pb_{1-x}Sn_x)_{1+\delta}Te_{1-\delta}$, required for a minimum infliction by constitutional supercooling effects due to excess metal or chalcogen. The crystals thus prepared were free of metallurgical imperfections which had been observed earlier[105]. They exhibited p-type conductivity with hole concentrations of the order of $p = 10^{19} cm^{-3}$, as expected from Fig. 9. Similar results were obtained[106] with smaller charges of 50–100 g. A higher growth rate of 3.4 cm/day was applied without formation of cellular structures or precipitates using a temperature gradient > 50 K cm^{-1}.

Recently, the parameters inherent to the Bridgman growth of $Pb_{1-x}Sn_xTe$ have been investigated[107]. Combining temperature gradients of G = 20 K cm^{-1} and 60 K cm^{-1} and growth rates of V = 1 mm h^{-1} and 5 mm h^{-1}, the effects of convection, interface shape and constitutional supercooling could be studied[107]. The characteristic thickness d of the interface layer between the solid and the homogeneous melt was approximated. Its values 1×10^{-2} cm < d < 2.5×10^{-2} cm are one order of magnitude smaller than expected from melt diffusion coefficient and

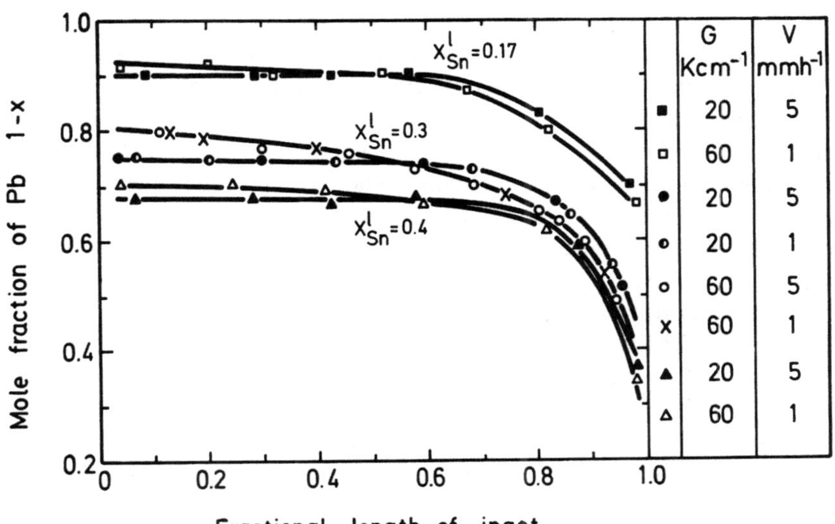

Fig. 24. Longitudinal composition of $Pb_{1-x}Sn_xTe$ crystals generated by the Bridgman technique using various growth velocities (V) and temperature gradients in the melt (G)[107]

growth rate, indicating that convectional transport dominates the interface composition even with the high growth rates of 5 mm h^{-1}. Figure 24 shows the distribution of the composition along crystals of 5 cm length grown under various conditions[107].

A modified melt growth technique for $Pb_{1-x}Sn_xTe$ has been suggested[108] which avoids the necessity of high growth rates and large temperature gradients (see Fig. 25). The crystal is grown from a liquid droplet continuously fed via the vapor

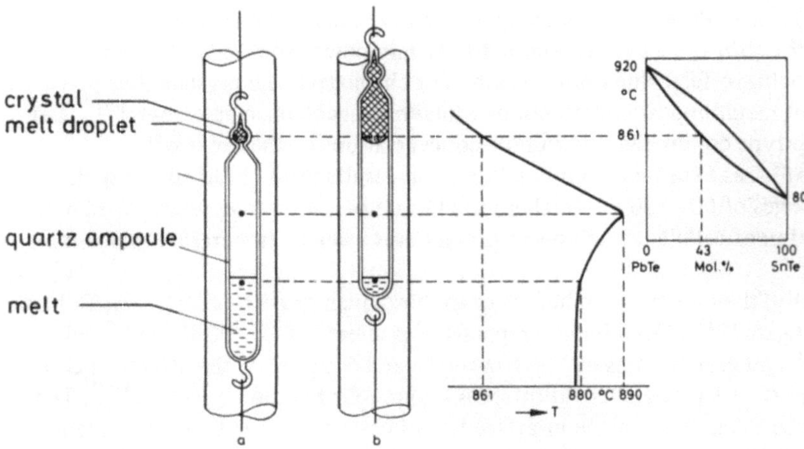

Fig. 25. Three-phase growth technique for $Pb_{1-x}Sn_xTe$ crystals. Temperature profile for $x = 0.43$[108]

phase from the melt source in the lower part of the ampoule. The velocity of pulling the ampoule through the temperature gradient has to be controlled very accurately in order to maintain the system in equilibrium (i. e., not to freeze the droplet or increase it beyond the size borne by its surface tension). Crystals of 9 mm diameter and of 60 mm length have been obtained whose compositional homogeneity and structural quality exceed those of crystals grown by the Bridgman technique for comparison. This method, in principle, offers a valuable alternative to the Bridgman technique. However, the equilibrium state is reported to be difficult to maintain even by continuous visual inspection through an eye port in the furnace.

The Czochralski technique has been applied to the fabrication of PbTe and $Pb_{1-x}Sn_xTe$[109]. Figure 26 schematically describes a growth apparatus. The high volatility of the compounds (see Chap. 3.2.1) must be suppressed to maintain the melt composition and the visibility in the system. It can be achieved by liquid encapsulation with B_2O_3. This material is less dense than the charge, has a low melting point of $T_m = 450$ °C, does not react with the PbTe or SnTe, and is of excellent transparency in the visible spectral range. In experiments of liquid-encapsulated zone refining B_2O_3 has even been observed to extract some impurities from the liquid charge for which it appears to exhibit higher solubility.

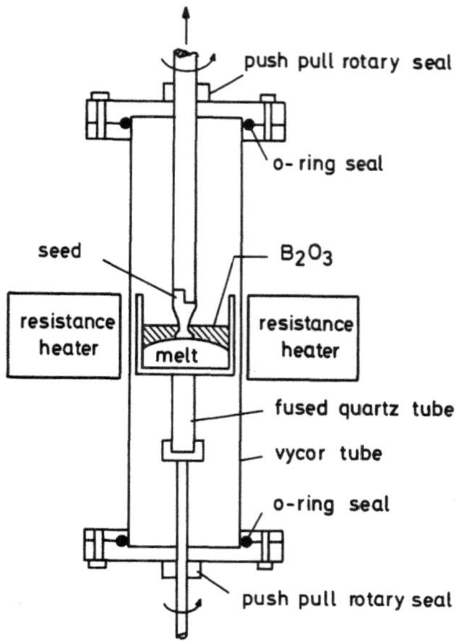

Fig. 26. Czochralski growth apparatus with liquid B_2O_3 encapsulation[29)]

60 g charges of $Pb_{1-x}Sn_xTe$ in fused silica crucibles covered by 25 g B_2O_3 under a nitrogen pressure of 1.4 to 1.8 atm have been used[109)]. Losses of PbTe and SnTe could thus be reduced to negligible amounts. A tantalum capillary has been employed as a seed rather than a crystal of the same material. Growth rates range from 1 mm h^{-1} to 5 cm h^{-1} at a rotation of 22 rev. min^{-1}. The higher pulling speeds result in strong cellular growth with compositional scatter of about 5 mol % whereas growth rates between 1 and 3 mm h^{-1} yield good crystals. The compositional gradients inherent to the growth method and the structural quality, however, have not been specified.

The Czochralski growth from melts containing different metal/tellurium compositions have been investigated[29)]. While no problems arise with metal excesses up to 5 %, a tellurium excess $>$ 1 %, however, causes volatilization which prevents a regular crystal growth. Crystals of 10 cm length (80 g) have been produced at typical growth rates of 1 cm h^{-1} using PbTe seeds which were preferentially oriented to a [111] – growth direction. Their compositional variation from top to base amounts to $\Delta x = 0.007$ (in the case of the first-to-freeze composition $x = 0.131$). They contain dislocation densities of 2 to 4 x 10^5 cm^{-2}.

More recent experiments[110)], however, have revealed a serious disadvantage of the Czochralski technique. The composition of $Pb_{1-x}Sn_xTe$ crystals grown under conditions very similar to those described above (8 mm h^{-1}, 10 rev. min^{-1}) has been carefully analyzed after mechanical and electrochemical polishing treatment. This procedure envisages a strong striational structure across the ingot. Striations are well-known in doped semiconductors pulled from the melt[111)]. They can be attributed to the rotational type corresponding precisely to the ratio of rotational to pulling speed and are due to an asymmetric temperature gradient in the melt, or they are of the non-rotational type and then caused by destabilizing thermal gradients. The striations in $Pb_{1-x}Sn_xTe$ have been identified to belong to the second type, their period (200 μm) being an order of magnitude higher than expected from the rotation/pulling ratio. They consist of a periodic compositional variation in the Pb/Sn ratio, as verified from electron microprobe analysis and from extra frequencies in Shubnikov-de Haas spectra. Furthermore, a monotonic increase in the Pb/Te ratio is observed along the length of the crystal. It exceeds the solidus range and suggests the presence of eletrically inactive microprecipitates.

Inhomogeneities as revealed in the latter careful analysis, of course, confine the device applicability of the Czochralski grown material. Similarly, the residual variation of the metal/chalcogen composition in Bridgman grown crystals, although smaller than may be anticipated from the effective growth conditions, restricts the yield of device quality crystal wafers. Besides that, the dislocation densities of 10^5 to 10^6 cm^{-2} in melt grown crystals are inferior to those achieved, for instance, with the vapor-phase growth technique to be described below. The three-phase growth technique seems to overcome these disadvantages, but it is based on a thermal equilibrium which is difficult to control. To our knowledge, melt growth is presently no longer used in the production of pseudo-binary $Pb_{1-x}Sn_xTe$ and $Pb_{1-x}Sn_xSe$ crystals for devices. It is, however, very valuable in rapidly produced binary lead chalogenides as substrate material for high quality pseudo-binary epitaxial layers.

4.1.2 Growth from the Solution

Recently, some efforts have also been undertaken to grow $Pb_{1-x}Sn_xTe$ bulk crystals from solution. In this technique, a solvent is driven through a polycrystalline ingot. At one end of the solvent zone the crystal is deposited while at its other end it is resaturated continuously from the source. Compared to the gradient-freeze Bridgman technique this method offers interesting advantages such as higher homogeneity in composition and lower growth temperatures leading to lower concentrations of lattice defects and impurities. If, moreover, $Pb_{0.8}Sn_{0.2}Te$ is grown from a PbSn solution at temperatures close to the cross-over point (see Fig. 9), then crystals with low carrier concentrations should be attained. Jäger and Seipp[112] used the travelling-heater procedure with a PbSn solvent at 750 °C to grow $Pb_{1-x}Sn_xTe$ (x = 0.18) crystals of 7.6 mm diameter and without detectable variation in the composition over several cm length (accuracy of their electron microprobe analysis $\Delta x \leqslant \pm$ 0.002).

In the traveling solvent technique, a small stationary temperature gradient causes the solution zone to move through the charge. From small sandwich charges of a [100]-oriented Bridgman grown seed crystal, a $Pb_{0.8}Sn_{0.2}$ foil of 200 μm thickness and a platelet of source material crystals of 1.5 mm thickness have been grown by this method[113]. They exhibited hole concentrations as low as $p = 6 \times 10^{16}$ cm^{-3}. Their carrier mobility and structural quality was superior to those of melt grown crystals. However, difficulties may be anticipated in extending this method to the production of much larger crystals.

A growth method which utilizes the same principle in an isothermal recrystallization process has been described[114]. A polycrystalline $Pb_{1-x}Sn_xTe$ (x = 0.2) ingot either stoichiometric or with a slight metal excess is placed into an evacuated sealed quartz ampoule and heated to 860 °C for 5 to 14 days. Recrystallization takes place for which the small amount of liquid metal (which is also present in stoichiometricly composed charges, Fig. 9) is believed to provide the main transport medium and facilitates equilibration. It migrates through the crystal and is collected at its lower, usually polycrystalline end. Uniform crystals up to the order of 1 cm^3 could be produced by this method, their final shape, of course, being random.

4.1.3 Growth from the Vapor

The best results with respect to crystalline structure as well as homogeneity and control of composition have been achieved with the sublimation growth technique. It is particularly suitable for lead and tin chalcogenides since these sublime as molecules, i. e., the dissociation in the vapor phase is negligible. Moreover, if pseudobinary crystals are sublimation-grown in closed ampoules and close to thermal equilibrium, any difference in the partial pressures of the molecular components is irrelevant. However, a successful growth process with a reasonable growth rate is only realized if these molecular components constitute the major part of the total pressure in the ampoule. This requires a very high purity of the source material and a carefully adjusted composition of its metal/chalcogen ratio. Crystals are usally grown on the metal-rich solidus line[115)], but growth within the narrow solidus range has also been accomplished[83, 84, 116, 117)].

4.1.3.1 Synthesis of the Source Material

In the following, the procedure of preparing the source material and carrying out the growth is described in more detail. The first step involves the synthesis of the binary compounds. High purity elements (usually 6 N grade) are weighed in stoichiometric portions and sealed in quartz ampoules previously etched and baked under 10^{-6} mb. The ampoule is heated to about 50 K above the melting point of the compound, held at this temperature for about 4 h, and then quickly quenched in water.

The binary compounds are subsequently purified either by horizontal liquid encapsulated zone refining or by a sublimation process. In the former technique the homogenized ingot is placed into another ampoule and covered with B_2O_3[44, 117, 118)]. In a horizontal furnace, one or several narrow molten zones are repeatedly passed along the length of the ingot (about 30 cm) (Fig. 27). This procedure not only effectively purifies the compound but also adjusts very accurately the metal/chalcogen composition to a value given by its maximum melting point. It is obvious that the B_2O_3 has a double function: firstly, to prevent the volatile impurities and secondly the eventual excess chalcogen from diffusing back into the solid material. The sublimation purification[83)] utilizes the separation of compound material and impurities by condensation in different temperature zones of the furnace when sublimation occurs at 800 °C. Most volatile impurities, particularly disturbing O_2, H_2O, CO_2, are extracted by a liquid nitrogen pump at the low temperature end of the ampoule. If a congruent sublimation composition exists in the binary compound (PbTe, PbSe), the material condensed under dynamic vacuum conditions (at temperatures between 670 and 750 °C with PbSe, 480 and 550 °C with PbTe), assumes that

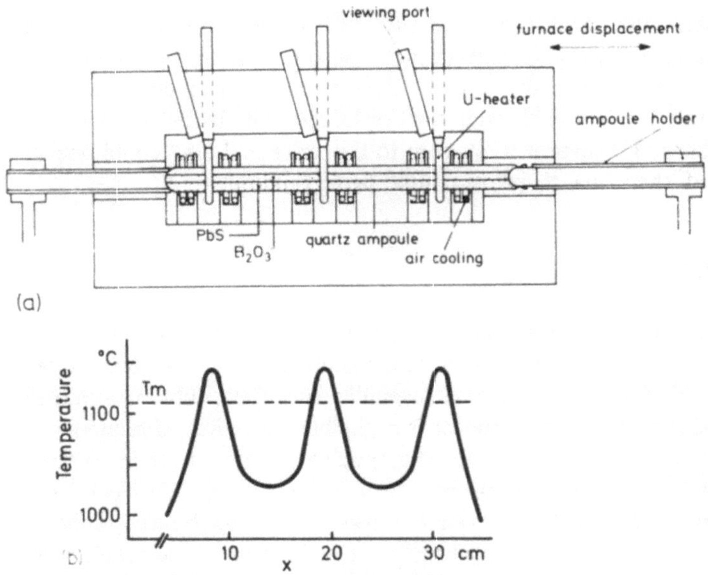

Fig. 27 a and b. Liquid encapsulation zone refining apparatus. Temperature profile for PbS[118]

composition. (This is exactly true only for equilibrium, but a comparison with effusion cell experiments indicates that equilibrium is closely approached at the condensation temperatures.)

Van der Pauw-Hall measurements on crystals grown from the binary materials yield their composition in terms of carrier concentration. Good agreements are found for PbSe with the p_{min} curve of Ohashi and Igaki[71] (see Fig. 20). For PbTe, hole concentrations of $p = 1 \times 10^{18}$ cm^{-3} to 1.5×10^{18} cm^{-3} at condensation (and effusion cell) temperatures between 480 and 820 °C suggest a p_{min} curve lying somewhat higher[119] than that given by Fujimoto and Sato[39]. If a congruent sublimation composition does not exist (SnSe, SnTe, see Chap. 3.2) the condensed material acquires a composition outside the solidus field (Sn-rich). A second sublimation transport[84] then separates the excess Sn to yield a composition very close to the solidus range, making use of the large difference in vapor pressure between Sn and SnSe or SnTe.

Pseudo-binary source material of $Pb_{1-x}Sn_xTe$ and $Pb_{1-x}Sn_xSe$ is then produced by annealing and homogeneizing the appropriate portions of the purified binary compounds in a closed quartz ampoule at 800 °C for several days. In order to adjust the desired deviation from stoichiometry (i. e., carrier type and concentration) an exactly calculable amount of metal or chalcogen is added to the charge in this stage. This procedure has been very successfully applied to produce single crystals with a predetermined carrier concentration necessary for device application in the case of binary PbSe, pseudo-binary $Pb_{1-x}Sn_xSe$ (Fig. 28) and PbTe[117]. With PbSe it furthermore gives evidence[83] that the ionic state of the metal and chalcogen vacancies is $z = 2$.

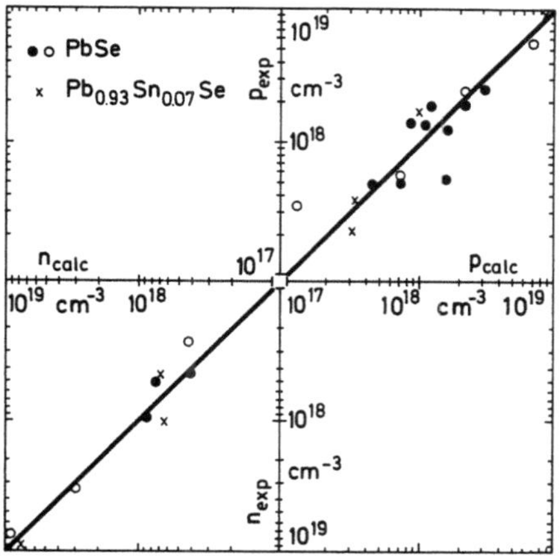

Fig. 28. Experimental carrier concentrations at 77 K of vapor-grown $Pb_{1-x}Sn_xSe$ crystals (x = 0, x = 0.07) versus calculated carrier concentrations of source materials to which various amounts of Pb or Se have been added. ● crystals furnace-cooled to 500 °C; ○, x crystals quenched in water[83, 84)]

4.1.3.2 Sublimation Growth Procedure

For the actual closed-tube vapor growth process two different procedures are applied and developed to high state of the art: a horizontal unseeded growth and a vertical seeded growth technique. They are equally applicable to all the binary and pseudobinary lead and tin chalcogenides.

For the unseeded vapor growth[117, 120)] about 20–40 g of the source material, prepared as described above, is ground to coarse grains. These are then sealed under a vacuum of at least 10^{-6} mb in a quartz ampoule of 25 mm diameter and 15 cm length which was previously thoroughly etched and outgassed. The ampoule is placed into the isothermal zone of a horizontal furnace at 780–850 °C. The growth process takes about 5 to 10 days. Its success is crucially dependent on several parameters: The grain size of the source material should not be too small; crystallites with 1 to 3 mm edge length appear to be optimal[83, 121)]. The largest crystals can be obtained with stoichiometric or slightly metal-rich source material. A high excess of metal (i. e. $\delta > 0.01$) leads to metal precipitates whereas an excess of chalcogen reduces the material transport and thus the growth rate due to its higher partial vapor pressure. Despite this fact satisfactory crystals have also been grown from a chalcogen-rich source[122)]. The vapor pressure established by volatile impurities is equally critical, as revealed by a direct correlation between the crystal size and the residual total pressure determined after the growth period. A pressure < 1 mb was found to be a prerequisite to a reasonable growth process[117)]. This makes a careful purification of the material and ampoule obligatory. The absolute growth temperature is not found to be a parameter which sensibly governs the growth. It may be chosen between 780 and 850 °C. However, the temperature profile strongly affects the growth process. It should be flat to within 0.5 °C in the longitudinal direction and exhibit a gradient of 1 K cm^{-1} in the direction transverse to the ampoule axis, the bottom being hotter[83)]. Under these conditions, the crystals grow on the source material with the [100]-facets pointing upward. A minimal contact with the quartz walls of the ampoule is achieved which leads to an excellent metallurgical structure. The total charge is collected into a small number (usually 1 to 3) of large crystals

Fig. 29. Ampoule for the unseeded vapor growth of lead-tin chalcogenides before and after the 5 day growth run

with facets up to 2.5 cm^2. An ampoule before and after the growth period is shown in Fig. 29. If the ampoule is slowly cooled down within the furnace, dislocation densities of 10^2 to 10^3 cm^{-2} are obtained. The unsurpassed structural quality demonstrates the advantage of this growth procedure; its disadvantage is the non-reproducible shape of the crystals which implies problems to large-scale device production. This can be overcome in the seeded growth described below. Before that, we shall briefly mention two modifications of the horizontal unseeded growth which make use of a controlled atmosphere within the ampoule. In the first one, applied to PbSe,

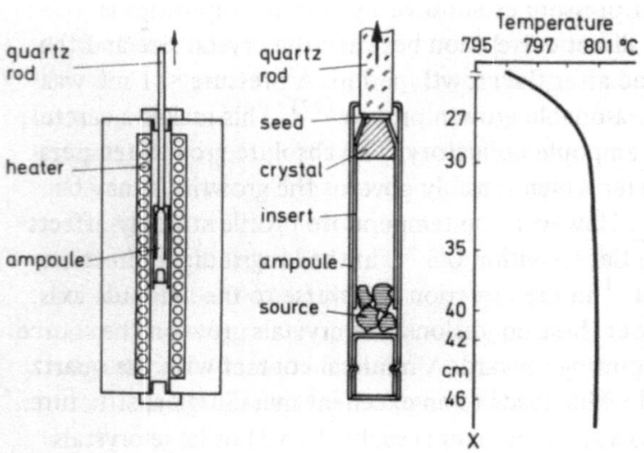

Fig. 30. Growth apparatus, ampoule and temperature profile for the seeded vapor growth of lead tin chalcogenide crystals

the chalcogen pressure during growth is controlled by a second lower temperature reservoir containing pure chalcogen[116]. In the second one, applied to PbTe, a capillary monitors the flow of the compound vapor from the molten source into the growth region of the ampoule[123]. Both procedures are more elaborate and less reproducible than the one described above.

The ampoule used for the seeded growth technique which so far has been mainly applied to $Pb_{1-x}Sn_xTe$[119, 124] is schematically described in Fig. 30. It consists of a source region containing a few lumps of polycrystalline material and a growth region in which a seed is attached to the polished end of a quartz rod extending into the

Table 3. Physical properties of $Pb_{1-x}Sn_xTe$ (x = 0; 0.2; 1)

	PbTe	$Pb_{0.8}Sn_{0.2}Te$	SnTe
Lattice structure	cub. (NaCl)	cub. (NaCl)	cub. (NaCl)
Lattice constant a (300 K) [Å]	6.460	6.4321 ± 0.003	6.327 Sn-sat. 6.302 Te-sat.
Density γ (300 K) [gcm^{-3}]	8.242	7.91	6.445
Opt. phonons (300 K)			
LO [cm^{-1}]	114	120	
TO [cm^{-1}]	32		
Dielectr. constant			
ϵ_s (300 K)	400	570	
ϵ_∞ (300 K)	32.8	38	
Band gap			
Eg (300 K) [eV]	0.32	0.20	0.18
Eg (77 K) [eV]	0.22	0.10	
Eg (0 K) [eV]	0.190	0.081	
dEg/dT (300) [$10^{-4}eVK^{-1}$]	4.85	4.85	
Effective masses (4 K)			
m_c^*/m_o	0.034		
m_v^*/m_o	0.032		
Anisotropy K	12		
Mobilities (77 K)			
μ_n [$cm^2V^{-1}s^{-1}$]	4×10^4	2.5×10^4	
μ_p [$cm^2V^{-1}s^{-1}$]	3×10^4	2.5×10^4	1×10^3
Intrins. carr. conc. n_i (77 K) [cm^{-3}]		3.2×10^{13}	
Melting point T_m [K]	1197	1168 (sol.) 1178 (liq.)	1078
Heat cap. (300 K) C_p [cal $mol^{-1}K^{-1}$]	12.1		12
Therm. exp. coeff.			
α (300 K) [$10^{-6}K^{-1}$]	19.8	20	
α (77 K) [$10^{-6}K^{-1}$]	15.9		

ampoule. The seed may be a [111]-oriented BaF_2 crystal whose lattice constant a = 6.200 Å and thermal expansion coefficient $\alpha = 19 \times 10^{-6}\ K^{-1}$ reasonably well match the values of the lead tin chalcogenides (see Tables 3 and 4) and which can easily be obtained by cleaving slices from commercially available crystal rods[125]. It is found, however, that often small-angle grain boundaries result from this choice of seed material[119]. In another method[126] a quartz tip is used and the first deposit of material acts as a non-oriented seed in the growth. Good results are obtained by use of PbTe or $Pb_{1-x}Sn_xTe$ seed crystals cut from a single crystal and carefully

Table 4. Physical properties of $Pb_{1-x}Sn_xSe$ (x = 0; 0.07; 1)

	PbSe	$Pb_{0.93}Sn_{0.07}Se$	SnSe
Lattice structure	cub. (NaCl)	cub. (NaCl)	orthorhombic
Lattice const. (300 K) a [Å]	6.1265	6.118	a = 4.46 b = 4.14 c = 11.47
Density (300 K) γ [gcm^{-3}]	8.274	8.13	6.179
Opt. phonons (300 K)			
LO_Γ [cm^{-1}]	133		
TO_Γ [cm^{-1}]	44		
Dielectr. const.			
ϵ_s (300 K)	200	227	
ϵ_∞ (300 K)	22.9	25	
Band gap			
Eg (300 K) [eV]	0.28		0.9(ind.); 1.2(dir)
Eg (77 K) [eV]	0.17	0.10	0.95(ind.)
Eg (0 K) [eV]	0.15	0.07	
dEg/dT (300) [$10^{-4}eVK^{-1}$]	5		– 4.3
Effective masses (4 K)			
m^*_c/m_o	0.047		
m^*_v/m_o	0.041		0.5
Anisotropy K	2		1 : 1 : 4
Mobilities (77 K)			
μ_n [$cm^2V^{-1}s^{-1}$]	3×10^4	3×10^4	
μ_p [$cm^2V^{-1}s^{-1}$]	3×10^4	3×10^4	7×10^3(c)
Intrins. carr. conc. n_i (77 K) [cm^{-3}]		3.3×10^{13}	
Melting point T_m [K]	1354	1325 (sol.) 1340 (liq.)	1134
Heat cap. (300 K) C_p [cal $mol^{-1}K^{-1}$]	12.0		12.7
Therm. exp. coeff.			
α (300) [$10^{-6}K^{-1}$]	19.4		
α (77) [$10^{-6}K^{-1}$]	16.0		16.9

polished and etched before inserting them into the ampoule. Their orientation should be [111] since this has shown to cause less strain and structural defects such as voids and small angle grain boundaries in the crystal[124, 127, 128]. Pressing the seed crystal against the quartz rod can cause detrimental strain or scratches which is avoided by the inverse technique where the seed is situated on the bottom of the ampoule[119]. The growth temperature may be chosen between 800 and 850 °C whereas the temperature profile is again crucial for the structural quality of the growing crystal. A continuous decrease by 2–3 K from source to seed with a steep gradient immediately behind the seed could result in a material transport in a stationary ampoule. In this case, the growth rate is expected to change with time. Its average value has to be below 0.7 mm/day and the growth time amounts to 4 to 6 weeks[127]. A flat temperature profile and a gradient of 1 to 2 K cm^{-1} in the seed region through which the ampoule is pulled allows, on the other hand, a careful control of the growth rate and to increase the latter to 2.2 mm/day. The growth time is then reduced to 3 weeks[119].

Large crystals (Fig. 31) with excellent homogeneity and structural quality have been grown with this technique. Their dislocation density depending strongly on the cooling rate after the growth period is of the order of 10^3 to 10^4 cm^{-2}. It is inferior to that of crystals from unseeded horizontal growth due to the contact with the

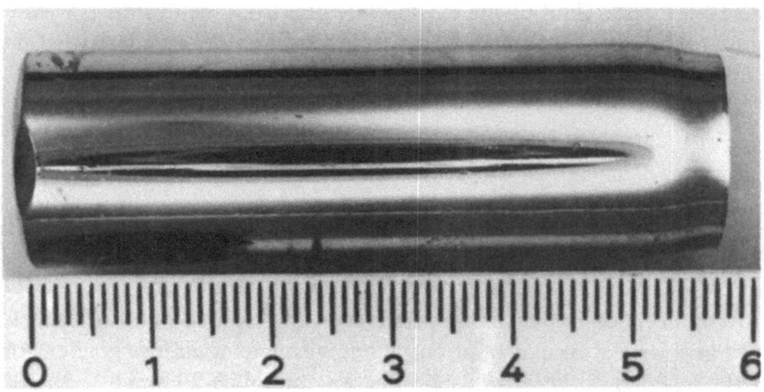

Fig. 31. $Pb_{0.8}Sn_{0.2}Te$ crystal grown in <111> direction by the seeded vapor-growth technique

quartz walls, but is still suitable for device applications. Lead-tin chalcogenide crystals of this perfection are soft and have to be handled with great care. Dislocations and slip lines (plastic deformation) are easily introduced by mechanical force or thermal shock[129]. Sawing and lapping inevitably cause additional damages of several hundred microns depth which have to be removed by chemical etching. The delicate structure of the crystals must be observed in all technological steps of device fabrication, like glueing slices on a substrate, curing photoresists or covering it with layers of different thermal expansion. The difficulty of handling is one of the disadvantages of the lead tin chalcogenides, as compared to mercury cadmium telluride.

4.2 Growth of $Hg_{1-x}Cd_xTe$

4.2.1 Growth from the Melt

Compared to lead chalcogenides, the melt growth of $Hg_{1-x}Cd_xTe$ is hampered by the wider separation of the liquidus and the solidus. The segregation coefficient $k_{CdTe} = x_s/x_l \approx 3$ for the composition x = 0.2 of the solid which is of particular importance to infrared detectors (Fig. 15). Unfortunately, compared to $Pb_{1-x}Sn_xTe$ and $Pb_{1-x}Sn_xSe$, the band gap depends more distinctly on composition variations (Fig. 3). Thus, homogeneity of composition is considered as the major problem in melt growth of device grade $Hg_{1-x}Cd_xTe$. If the propagation of the solid-liquid interface is slow, the longitudinal gradient in x is inevitable. Fast propagation (say > 1 mm h^{-1}) requires tremendously large temperature gradients in the liquid ahead of the interface to avoid constitutional supercooling. Since these are not achievable homogeneously across the interface, strong longitudinal and transverse composition variations must arise[130)]. However, not only CdTe shows an unfavorable segregation coefficient with respect to HgTe. This is also observed for excess Te relative to Hg and Cd. A segregation coefficient $k_{Te} = \delta_s/\delta_l \approx 0.1$ has been reported[131)]. Excess Te is strongly enriched in the melt and induces cellular growth and the formation of inclusions. More recently, an electron microscopy study of as-grown and annealed samples has been performed[132)]. Te precipitates were observed in samples grown on the Te-saturated limit of the existence region from melts with 1 % excess Te, sometimes even from stoichiometric melts. Precipitates of an ordered HgCd phase were observed in samples grown or annealed on the metal-saturated limit.

4.2.1.1 The Bridgman Growth

Bridgman growth processes using a variety of parameters have been described by several authors[50, 130, 131, 133)]. The most comprehensive study on the influence of the growth rate on the compositional variation parallel and perpendicular to the growth direction has been published recently[134)]. This publication describes growth runs in a vertical Bridgman arrangement with a temperature gradient of 100 K cm^{-1} at the freezing isotherm and solidification rates of 0.25 mm h^{-1} to 3000 mm h^{-1}. The ingots have a length of 60 mm and a diameter of 12.5 mm. The composition has been investigated by optical absorption. At high growth rates good axial homogeneity (with an effective segregation coefficient $k_e = 1$) is attained except in the very first range of the ingot, but a strong radial variation in x is still present (Fig. 32). At low growth rates the axial homogeneity is poor, and even the radial composition is not uniform except, occasionally, in the narrow regions of the ingot where the growth interface is believed to undergo an unstable change from convex to concave shape. The high compositional inhomogeneity at low growth rates suggests an interface thickness of several mm. Interruption of the growth run reveals, however, that the interface curvature extends only over approximately 1 mm. Density-driven convective flow is assumed, leading to an interface which no longer follows an isotherm. This effect slowly decreases when the increasing concentration of mercury telluride in the melt levels out the density variations. Then, growth on the isotherm may be-

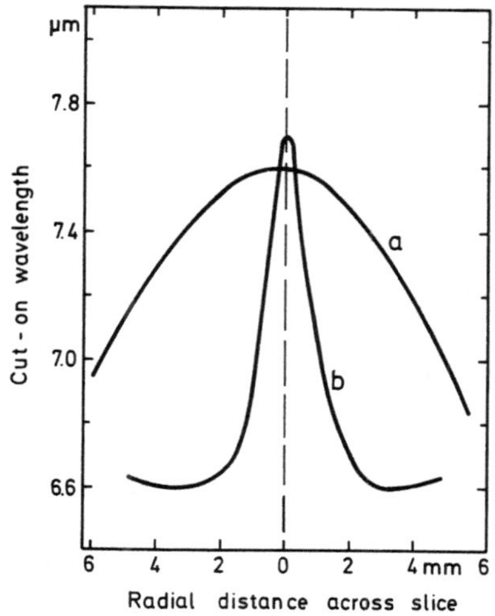

Fig. 32. Transverse variations of cut-on wavelength of transmission with melt-grown $Hg_{1-x}Cd_xTe$ crystals. *a* growth velocity 0.5 mm h^{-1}, *b* growth velocity 3000 mm h^{-1} [134)]

come dominant. The authors conclude that it is not possible to grow crystals of $Hg_{1-x}Cd_xTe$ which are both axially and radially completely uniform. If this statement is accepted, the fast growth is preferable and the highest possible homogeneity is to be sought for by additional precautions. This will be treated in Chap. 4.2.3.

An X-ray topographic investigation[135)] has recently revealed additional disadvantages of the Bridgman growth technique. A high density of grains has been found; the grains exhibiting a size of 0.3 to 0.5 mm in the center of the wafers and of 0.05 to 0.2 mm at the circumference. Their misorientation is 2 to 3 min and 6 to 9 min in the center and at the circumference, respectively. These structures have not been observed in comparable crystals grown by the solid-state recrystallization method.

4.2.1.2 *Zone Melting and Travelling Solvent Growth*

The above consideration elucidates the problems connected with melt growth in the directional freeze or Bridgman technique. Fast solidification necessarily yields unstable growth conditions because of constitutional supercooling; slow solidification produces very large axial inhomogeneities (besides residual radial inhomogeneities) due to diffusional and convectional exchange in the melt. It is therefore reasonable to employ slow growth processes from a narrow zone of the liquid phase which is continuously fed by a homogeneous polycrystalline solid source when driven through the ingot.

In fact, the first attempt to grow $Hg_{1-x}Cd_xTe$ crystals has been performed using this scheme of zone melting[46)]. The binary HgTe and CdTe were first zone refined in a horizontal furnace and, after smelting them together, zone levelled in the same apparatus. An analysis of the homogeneity has not been reported at that time. Later, the zone melting technique was applied to grow $Hg_{0.8}Cd_{0.2}Te$ crystals[136)]. Propagation rates of 0.17 mm h^{-1} and 0.27 mm h^{-1} have been used in a vertical furnace with a temperature gradient of 40 K cm^{-1} in the liquid-solid interface and a molten zone width of about 30 mm. By interrupting the growth process, the influence of the interface curvature and of the melt composition on the homogeneity of the single crystal could be studied. Exceptionally large radial and axial inhomogeneities were detected with the faster zone propagation rates. Referring to the foregoing discussion of the Bridgman growth process, this anomaly may be understood in terms of density-driven convections. It should then be easier to grow homogeneous crystals with higher CdTe content. In fact, $Hg_{0.1}Cd_{0.9}Te$ crystals with a compositional variation of less than $\Delta x = \pm 0.005$ over half of the ingot length and with no detectable radial inhomogeneity have been fabricated by the zone melting technique[137].

A variation of this method has been suggested[138], which also involves a vertical system, but introduces a gap of a few mm between the small amounts of liquid from which crystallization occurs and the polycrisalline source (Fig. 33). Material is transported via vapor phase and liquid droplets. The melt composition should not change if the source ingot is homogeneous and diffusion in the solid is negligible. Large grain crystals are obtained with a reasonably axial homogeneity. Residual variations in x are attributed to the difficulty to control the often discontinuous material flow.

The zone levelling technique has also been applied using a Te solvent. Triboulet et al. produced high quality binary HgTe crystals[139] and, in an extension of the above method, attempts have been made[140] to grow pseudo-binary $Hg_{1-x}Cd_xTe$ crystals with a travelling solvent technique. The radial compositional uniformity is found to improve by reducing the length of the liquid zone. However, a 10% scatter in x across the ingot diameter is still observed.

In principle, the zone melting or travelling solvent growth offer some important advantages which can be further exploited. Low growth rate is possible under continuous stable feeding of the liquid zone. A flat and narrow zone under proper temperature profiling of the furnace should provide the radial homogeneity. The

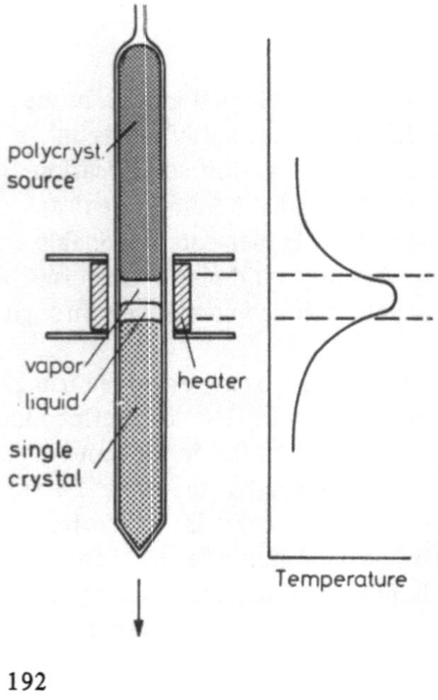

Fig. 33. Vertical zone melting or travelling solvent apparatus for $Hg_{1-x}Cd_xTe$ crystals. The liquid is fed from the polycrystalline source via the vapor phase[137, 139]

effect of purification is a useful by-product for achieving pure, low-carrier concentration crystals.

4.2.2 Growth and Recrystallization from Slush

A further technique involving a continuous source of material in the growth interface has been investigated[141]. A melt of composition x/k where x is the required composition of the crystal and k the segregation coefficient is homogenized and quickly dropped into a temperature gradient of 10 K cm^{-1}. That part of the ingot, whose temperature is below the solidus point, freezes rapidly. In the adjacent range at temperatures between the solidus and liquidus, a slush composed of solid (CdTe-rich) and liquid (HgTe-rich) material is formed. Above this range the melt exhibits the original composition. The ampoule is kept at fixed position for a long period (44 d). A crystal is formed in the solid-slush interface and the decrease in the CdTe content is compensated by diffusion from the melt above the interface until this itself depletes of CdTe; the furnace power is then shut off. The region covered by the growth front during this time is narrow (about 3 mm) but of high structural quality and of a rather uniform composition ($\Delta x \leqslant \pm 0.005$) over its entire 1.8 cm diameter (Fig. 34). The preferential growth direction is found to be [100].

A recent modification is explained in Fig. 35[142]. Here, the total melt of the composition x_o is quickly quenched. The homogeneous ingot thus formed is heated to a temperature between the corresponding solidus and liquidus points. This procedure generates a slush of solid and liquid phases with the compositions x_s and x_l,

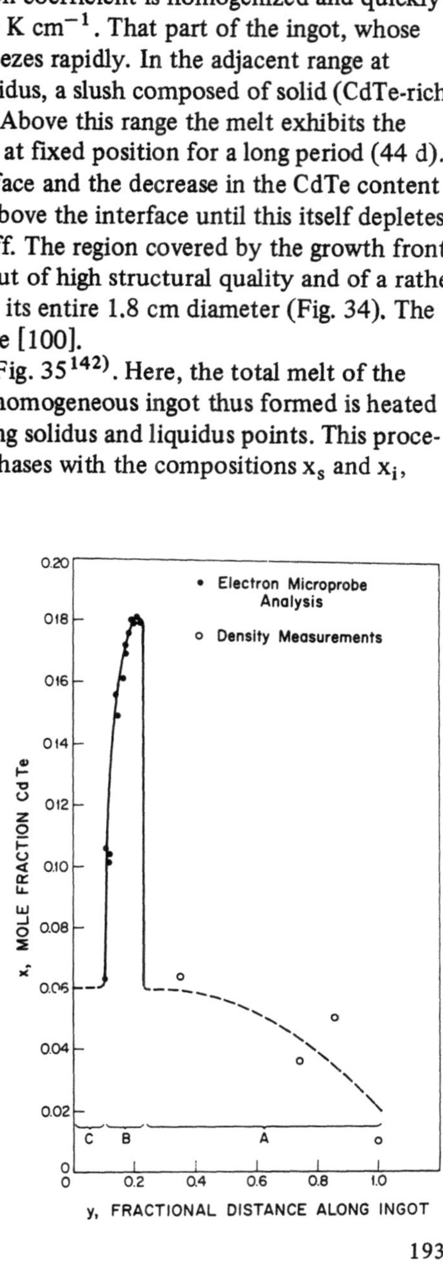

Fig. 34. Composition profile of an ingot of $Hg_{1-x}Cd_xTe$ after slush recrystallization[141]

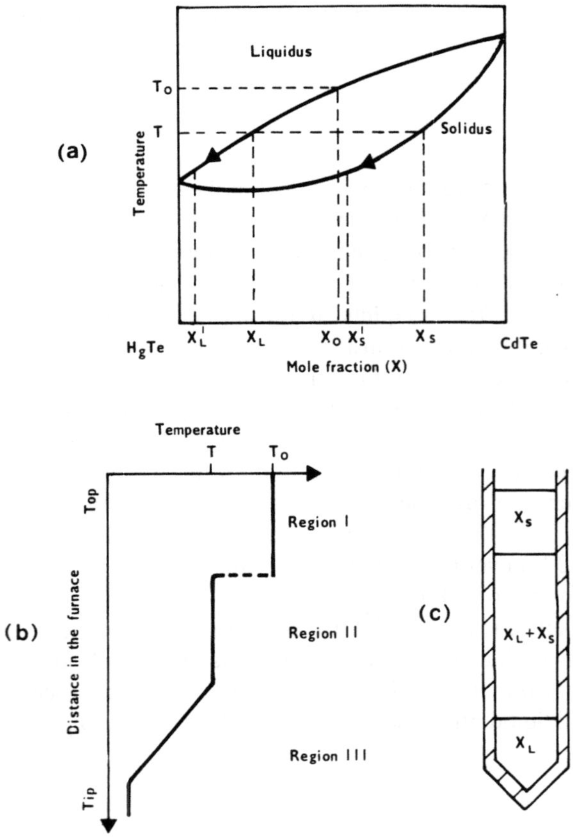

Fig. 35 a–c. Diagram of the slush growth technique[142)]

respectively. If the ampoule is then slowly (0.5 mm h^{-1}) lowered into a temperature gradient, a crystal starts growing from the tip. Its CdTe content is initially too high ($x_s = kx_1$) but decreases rapidly to a value x_s' close to x_0. This value can be maintained over the majority of the crystal since the solid CdTe-rich phase of the slush dissolves in the neighborhood of the growth interface and acts as a continuous source for CdTe.

It has been found that not all the CdTe-rich crystallites dissolve before reaching the interface and might form undesired seeds. To achieve a sufficient separation between the liquid zone ahead of the growth interface and the slush zone, the ampoule has been inverted by 180° after quenching and before re-inserting it into the furnace. The melt (with x_l) now penetrates into the originally empty part of the ampoule. Above it, zones are formed consisting of a slush ($x_l + x_s$) and a predominantly solid (x_s) phase. The width of the liquid zone, free from the solid, is an adjustable parameter, given by the empty volume of the ampoule. Crystals of promising homogeneity have been obtained by this method which, however, certainly deserves further development.

4.2.3 Solid-State Recrystallization

Swink and Brau[143] have first reported on the growth of $Hg_{1-x}Cd_xTe$ crystals by recrystallization from the solid state. Today, it is the most frequently employed technique in industrial laboratories and production facilities since it provides crystals of the highest presently achievable homogeneity. Its particular features will therefore be considered in more detail in this chapter.

The method involves two steps: A melt synthesis process of the elements followed by rapid quenching and a recrystallization process just below the solidus temperature. Although the first step is very similar to that described in Chap. 4.2.1, some specific aspects may still be noted. Since the additional purification effect of the directional freezing procedure is lost, higher precaution with respect to purity of the starting material is essential. Usually, multiply zone-refined Te and Cd and multiply distilled Hg are used. Special care must be taken to avoid contamination by oxygen since the amalgam of Hg and Cd readily oxidizes and its oxides attack the quartz of the ampoule[131]. In choosing the proper melt compositions, the Hg amount lost into the empty space of the ampoule and the deviation from stoichiometry at the maximum melting point (see Fig. 16) must be considered. Ampoules with an inner diameter between 8 mm[143] and 13 mm[144] are used with wall thicknesses of 2 to 3 mm to avoid explosion at the high Hg pressure used. The ampoule is closed under vacuum and heated to above the liquidus point of the chosen composition (e. g., to about 800 °C for $x = 0.2$) for 24 h. Continuous rocking of the furnace avoids gravitational segregation of HgTe and CdTe.

The parameters of the subsequent quenching process very critically determine the homogeneity of the polycrystalline ingot and the success of the subsequent recrystallization process. Different propagation rates and directions of the solid-liquid interface are employed. A very fast radial solidification is effected if the total ampoule or a large part of it is immersed in oil. The eruption of the innermost melt during the sudden breakdown of the Hg pressure can be avoided by decreasing the free ampoule

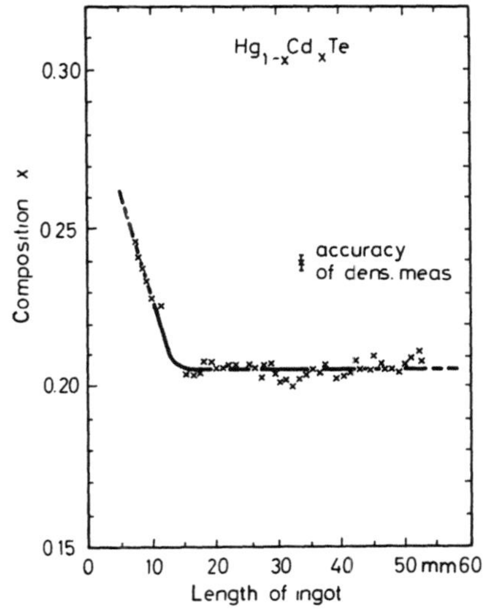

Fig. 36. Axial composition profile from density analysis of a $Hg_{1-x}Cd_xTe$ crystal grown by solid-state recrystallization after air quenching[147]

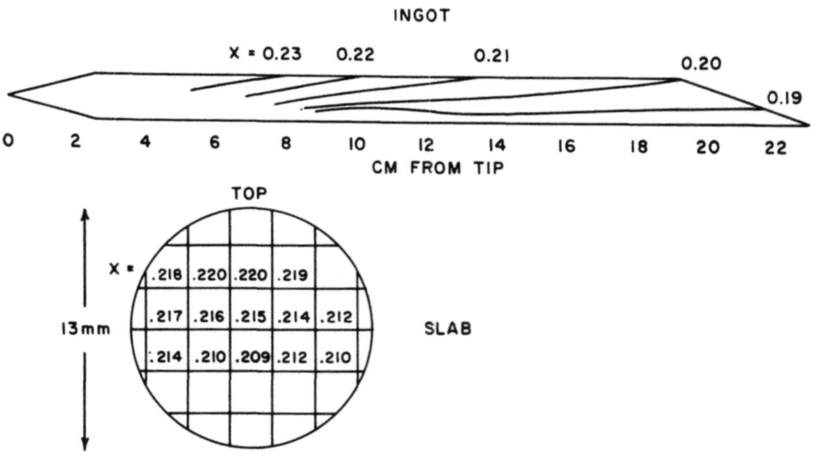

Fig. 37. Composition profile of a $Hg_{1-x}Cd_xTe$ crystal grown by solid-state recrystallization after sequential furnace zone cooling[144)]

volume to only a few mm^3 [145)]. Slower and preferentially axial solidification happens if only the tip of the ampoule is immersed in oil or cooled by an air flow[146)]. The longitudinal composition of an ingot obtained by air cooling is given in Fig. 36. A still slower and also directional type of solidification occurs if the heating zones of the furnace are sequentially cooled down[144)]. The composition profile of an ingot prepared in this manner is shown in Fig. 37. It shows that, if the furnace and the ampoule are tilted during the quenching process, the effect of convection mentioned in Chap. 4.2.1.1 loses its deleterious nature. Longitudinally cut planks of such a crystal may exhibit the required homogeneity over areas suitable for device applications.

An interesting modification of the synthesis and solidification procedure has been suggested[148)]. The high-pressure reflux technique is schematically represented in Fig. 38. A combination of the inert gas pressure and a steep negative temperature

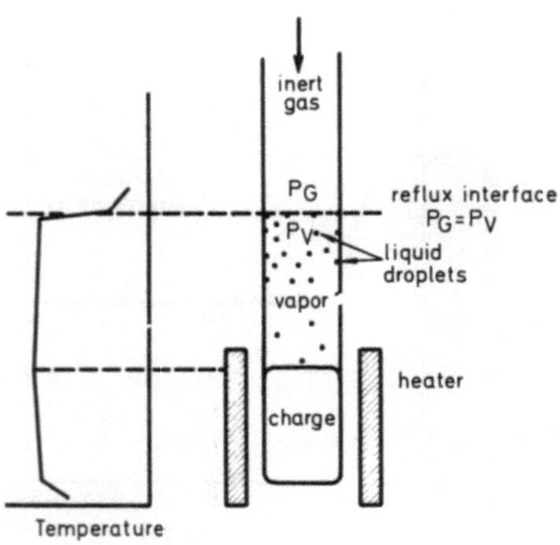

Fig. 38. Diagram of the high-pressure reflux growth technique. P_G and P_V denote the pressures of inert gas and refluxing vapor, respectively[148)]

gradient prevents the escape of the volatile species. Losses are reported to be lower than 0.1 g even after days of high temperature operation. It is even possible to control the vapor pressure and hence the stoichiometry of the melt. The major advantage, however, is that very large ingots (diameter 2.5 cm, length 6 cm) can be produced since the pressures within and outside the ampoule are balanced. The melt is cooled down within 10 to 20 min to provide a large central section of sufficient homogeneity, e. g., $x = 0.210 \pm 0.015$.

The various solidification processes described above yield ingots consisting of grains and dendritic structures, but exhibiting sufficient integral homogeneity. An example is presented in Fig. 39. The subsequent procedure of recrystallization is performed to produce single crystals and to level out the remaining local inhomogeneities. A secondary recrystallization[149)] is desirable in which selected grains grow due to their preferred orientation. This is supported by the directional solidification in the quenching process.

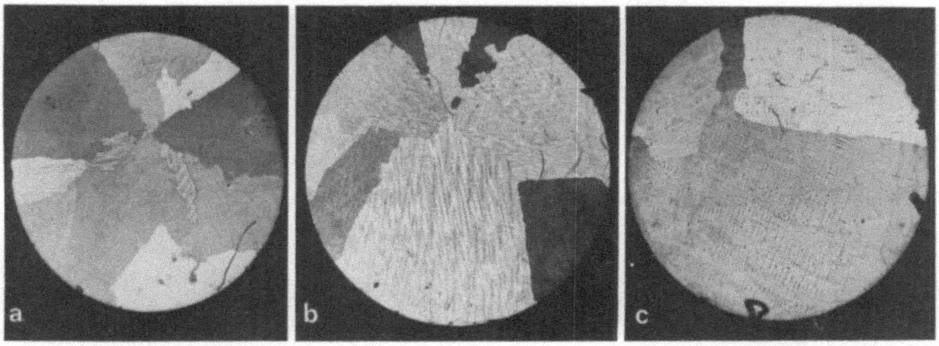

Fig. 39 a–c. Grains and dendritic structure of an air quenched ingot; **a** tip, **b** 5 mm from tip, **c** 25 mm from tip

The recrystallization process may be performed directly in the synthesis ampoule. However, to assure a better control of the mercury pressure, it is advantageous to either reencapsulate the ingot into another ampoule with a proper amount of excess mercury[143)] or to use a two-zone procedure with an indepedently controlled Hg vapor source. The optimum temperature and pressure for the recrystallization is chosen just below the solidus data of the particular composition x; e. g., for $x = 0.2$, about 670 °C and 10 to 15 atm are desirable according to Figs. 15 and 21. The temperature profile is either flat over the total ingot length or exhibits a gradient in which the first-to-freeze part of the ingot is positioned at higher temperatures. A combination of the recrystallization procedure and Harman's slush method, in which the uppermost tip of the ingot is re-melted, is advantageous[150)]. Although the liquid-solid interface then no longer exhibits the desired composition, and diffusion of Cd may render too Cd-rich an additional adjacent region of the solid, the crystals growing in the interface act as seeds which accelerate recrystallization for the remaining ingot.

The recrystallization procedure is usually applied for 3 to 4 weeks. An X-ray topography which verifies the single crystallinity after a 4-week duration is shown in Fig. 40; a typical crystal is represented in Fig. 41.

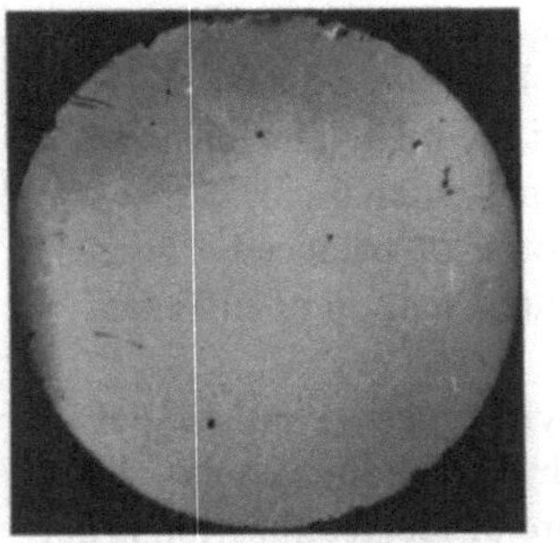

Fig. 40. Reflection X-ray topography of a slice of the ingot from Fig. 39 after recrystallization

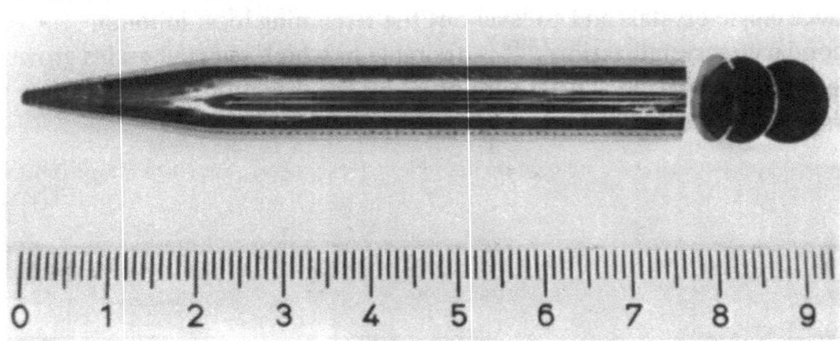

Fig. 41. $Hg_{1-x}Cd_xTe$ crystal grown by solid-state recrystallization

4.2.4 Stoichiometry Annealing

Due to the position of the stoichiometry line in the p-T-δ diagram (see Fig. 21), the as-grown or recrystallized ingot is p-type with hole concentrations of $p = 10^{16}$ to 10^{18} cm^{-3}. If other carrier concentrations or n-type conduction are requested, an additional annealing procedure has to be performed. This particularly applies to photoconductive detectors which demand electron concentrations of $n < 7 \times 10^{14}$ cm^{-3} at 77 K. The phase diagram offers two possibilities, an isothermal annealing with parameters of the Hg saturated three-phase line or a two-zone annealing with parameters from within the existence region (see Fig. 42). Since the

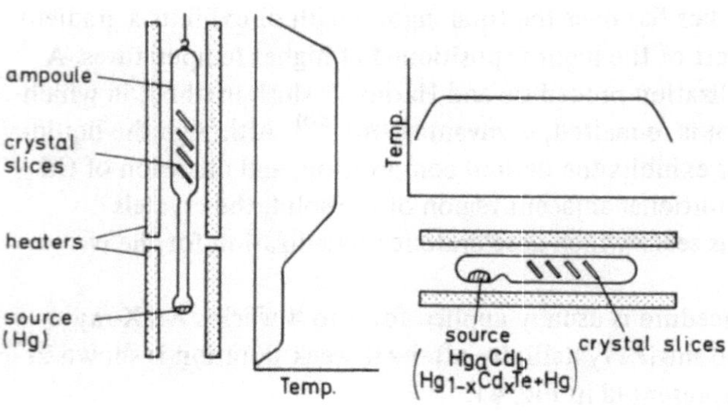

Fig. 42. Diagram of the two-zone and isothermal annealing technique for $Hg_{1-x}Cd_xTe$ crystals[147)]

annealing temperatures have to be low, unduely long annealing is expected for total ingots. Therefore, slices of thickness $\leqslant 1$ mm are cut and polished prior to annealing. The source for the appropriate Hg vapor pressure is either a $Hg_{1-x}Cd_xTe$ powder with additional Hg[151] or an alloy of Hg_aCd_b [147]. Since the composition a/b determines the vapor pressure of Hg at a certain temperature according to its activity coefficient, the latter source even permits an isothermal annealing process with data from within the existence region. The maximum life time of photoexcited minority carriers so far reported in n-type $Hg_{0.8}Cd_{0.2}Te$ is 7 μs[7]. The effect of partial compensation by residual impurities or by coexisting native defects of both types usually limits the highest attainable carrier mobility and life time. This effect is reflected in the anomalous dependence of the carrier concentrations and mobility on the magnetic field and temperature and in the decreasing mobility with decreasing net carrier concentration. This is attributed to charged-impurity scattering[54, 95, 99, 152], which will not be considered here in more detail.

5 Growth of Epitaxial Layers

In principle, the quality of the bulk crystals grown by the above techniques is sufficient for device fabrication. Diffusion or ion implantation processes provide the p-n junctions for infrared lasers or detectors. Highest performance of modern devices, however, often require more sophisticated structures. For lasers operating under moderate cooling conditions, single or double hetero-structures reduce the threshold current by electrical and optical confinements. For high efficiency detectors or arrays with very high element density, hetero-structures minimize the surface recombination and allow back-side illumination. These structures involve thin layers of different compositions x and/or δ which may only be achieved with epitaxial growth techniques.

5.1 Epitaxy of $Pb_{1-x}Sn_xTe$ and $Pb_{1-x}Sn_xSe$

Standard methods such as liquid, gas phase and molecular beam epitaxy are well-known in optoelectronic device fabrication, particularly from III-V compounds. However, intensive studies and specific modifications were necessary in order to apply these techniques to pseudo-binary lead tin chalcogenides. This is particularly true for the hot-wall epitaxy, an interesting development of the molecular beam method (Chap. 5.1.3), that causes layer growth under quasi-equilibrium conditions.

5.1.1 Liquid Phase Epitaxy

The physics and experimental procedures of liquid phase epitaxy have been extensively reviewed in recent general articles[153, 154]. Again, the knowledge of the phase diagrams is of considerable aid and almost inevitable in the understanding of the growth process. It is seen from Chap. 3.1.1 that for $Pb_{1-x}Sn_xTe$ the phase diagrams are very suitable for liquid phase epitaxy. Layers with the properties desired for optoelectronic devices may be grown from metal-rich solutions. The composition $x = 0.2$ is easily achievable and only slightly dependent on the growth temperature and the Pb/Sn-composition of the solution. A growth temperature around 530 °C may be chosen which coincides with the conversion temperature of the charge carrier type (Fig. 9) and allows very low carrier concentrations to be produced. This yields optimum conditions for IR detectors and active layers of heterostructure lasers. Uncertainty exists on how critically constitutional supercooling influences the morphology of the growing layer of $Pb_{1-x}Sn_xTe$. The need of a strong temperature gradient in the solution/layer interface has been emphasized[155]. Thus, a gradient of 5 K cm^{-1} was obtained and very slow cooling rates of the order of $dT/dt =$ 0.1 K min^{-1} at $T_{initial} = 580$ °C were used. On the other hand, it has been found that a temperature gradient is unnecessary and high cooling rates ($\geqslant 1$ K min^{-1}) are rather advantageous[156, 157].

Nelson's dipping method[158] has early been used to produce epitaxial $Pb_{1-x}Sn_xTe$ layers[159] but is difficult to remove the residual melt from the specimen. The slider technique, although a slightly more complicated set up, eliminates this difficulty and provides a good wipe off. It offers the additional advantage of allowing layers of different compositions x and δ to be grown successively without opening the system. For instance, the layer sequence in double-heterostructure lasers has been chosen as n-PbTe (LPE)/n-$Pb_{1-x}Sn_xTe$ (LPE)/p-PbTe (substrate)[156] and in hetero-junction detectors as n-PbTe (LPE)/p-$Pb_{1-x}Sn_xTe$ (LPE)/p-$Pb_{1-x}Sn_xTe$ (substrate)[160]. Especially in the latter case, the objective is a low hole concentration $Pb_{1-x}Sn_xTe$ layer. Layers of $p < 10^{16}$ cm^{-3} have been achieved utilizing the favorable phase diagram of this alloy. Often a diffusion between these layers of different composition happens even at the relatively low temperatures of liquid-phase epitaxy. The heterostructure then weakens up and the position of the pn-junction is shifted. To avoid this implication, doping of epitaxial $Pb_{1-x}Sn_xTe$ layers has successfully been applied[161]. Tl-doping was used to produce as-grown p-type layers, In-doping yielded n-type layers. The carrier type of these layers cannot be inverted by annealing under metal or chalcogen rich vapor, respectively.

Generally, [100]-oriented PbTe- or $Pb_{1-x}Sn_xTe$-substrates are employed. Their surface preparation, of course, critically effects the layer quality. The dislocation density of the substrate is preserved in homoepitaxial layers and considerably increased in heteroepitaxial layers even with small lattice mismatch[162]. Very few results are reported on $Pb_{1-x}Sn_xSe$ alloys[43] since their phase diagrams do not exhibit the attractivity of the liquid phase epitaxy technique as with $Pb_{1-x}Sn_xTe$ (see Fig. 13). The large separation of the solidus and liquidus lines makes a constitutional supercooling effect probable. A temperature gradient of about 12 K cm^{-1} perpendicular to the growing surface was generated by air cooling of the substrate holder and slow growth applying a temperature decrease of 0.1 to 0.2 K min^{-1} appeared to be necessary. n-Type $Pb_{0.93}Sn_{0.07}Se$ layers with a carrier concentration of $n = 6 \times 10^{18}$ cm^{-3} were obtained from $Pb_{0.9}Sn_{0.1}$ solutions in fair agreement with Fig. 12.

5.1.2 Vapor Phase Epitaxy

The epitaxy from the vapor phase using reactive or passive transport gases is well-known[163]. It has, however, not received the attention in the case of lead-tin chalcogenides as, for instance did the liquid phase epitaxy. Nevertheless, some results on the vapor phase deposition of PbTe and $Pb_{1-x}Sn_xTe$, which demonstrate the efficiency of this technique, have been reported.

A vertical open-tube vapor transport epitaxy system has been designed and described in detail[164] (Fig. 43). The substrate is located at about 600 °C in a steep temperature gradient. The transport gas is hydrogen and care is taken to maintain a symmetrical flow around the substrate. A similar apparatus has been successfully used for producing epitaxial n-PbTe/p-$Pb_{0.8}Sn_{0.2}Te$ hetero-structures on PbTe substrates[165, 166, 167].

More recently, a horizontal vapor transport epitaxy set up has been described and the conditions of depositing low carrier concentration $Pb_{0.79}Sn_{0.21}Te$ have been studied[168]. Although the layers are grown at temperatures $425 \leqslant T_{subst} \leqslant 550$ °C (close to or below the cross-over point of the phase diagram (Fig. 9)), they are all of the p-type with hole concentrations as low as $p = 3 \times 10^{16}$ cm^{-3}. The optimum charge composition is the stoichiometric one while 1% of either metal or tellurium excess deteriorates the structural quality of the layers. In all vapor phase experiments the purity of the source material and transport gas (Pd diffused H_2) is crucial for a uniform and reproducible growth. Much work must also be invested in producing well-oriented, damage-free and clean substrate surfaces.

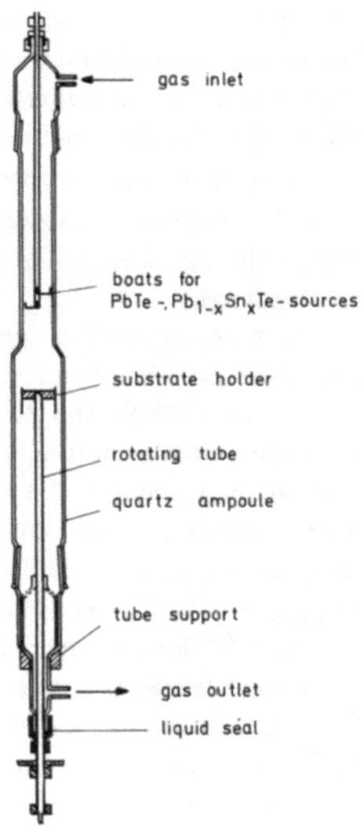

Fig. 43. Vertical open-tube system for vapor phase epitaxy of lead-tin chalcogenides[164]

5.1.3 Vacuum Evaporation Epitaxy

One of the objectives in the growth of epitaxial layers is to keep the growth condition as close as possible to thermodynamic equilibrium. The liquid phase and vapor phase epitaxy usually fulfil this requirement. The techniques described in this chapter closely resemble each other but differ essentially in just this condition.

In the molecular beam epitaxy utilizing thermal evaporation in a vacuum, the kinetics on the substrate surface (i. e., impingement rate, sticking coefficients and re-evaporation) govern the growth rather than thermal equilibrium. In hot-wall epitaxy, the system is closed within dimensions much smaller than the free path length so that equilibrium is almost restored. Consequently, the composition of the growing layer has to be controlled very carefully in the first method but less in the second one.

5.1.3.1 Molecular Beam Epitaxy

A review of the molecular beam epitaxy technique has been presented very recently in these series[169]. The beam is produced by a Knudsen effusion cell. Additional cells are necessary to grow alloys from their compounds, to compensate for the loss of the more volatile component in partially dissociating compounds or if dopants are required. The beams are periodically recorded by a mass spectrometer whose signal is used to control the cell temperatures. The structural quality and the chemical composition of the growing layer are analyzed in situ by low- or high-energy electron diffraction and Auger electron spectroscopy, respectively. An electron sputtering gun provides a cleaning of the substrate surface before epitaxy. The substrate temperature has to be low to minimize re-evaporation from the growing layer. It follows that the impingement rate has to be low, too, to assure high crystalline quality. Then, however, the background vapor already contributes a considerable concentration of impurities (the inpingement rate at a pressure of 10^{-6} mb corresponds to 1 monolayer per second) so that ultra-high vacuum of the order of 10^{-10} mb is required.

In spite of these exceptional difficulties the method is very successfully applied to III-V compounds, especially to GaAs and $Ga_{1-x}Al_xAs$, where super-lattice structures with an individual layer thickness of 50 Å[170] and multi-layer structures[171] are obtained.

It is also applied to deposit IV-VI compound alloys where, due to the lower dissociation rate of the molecules, the effort can fortunately be kept somewhat lower.

Layers of PbSe and $Pb_{0.94}Sn_{0.06}Se$ have been grown[172] on BaF_2 and SrF_2 substrates cleaved in the [111]-direction, which equally well match the thermal expansion coefficient of the selenides (difference $\Delta\alpha = 2 \times 10^{-6} K^{-1}$) and possess lattice constants either slightly above (a_{BaF_2} = 6.20 Å) or below (a_{SrF_2} = 5.80 Å) that of $Pb_{1-x}Sn_xSe$ (see Table 4). Two Knudsen cells at the same temperature ($T_{source} \approx 700$ °C) have been employed for PbSe and SnSe whose orifices determine the ratio of the impingement rates. This has been preferred to using two separately controlled furnaces or one cell containing both compounds[173] since temperature fluctuations and unavoidable compositional changes of the source, respectively, can no longer influence the compositions of the growing layer. A third cell containing pure selenium has been used to compensate for the loss of Se by re-evaporation and dissociation and to control the carrier type and concentration. Thus, at substrate temperatures between 250 and 400 °C, p- and n-type PbSe layers with carrier concentrations in the range n, p = (0.3–6) x 10^{17} cm^{-3}, and $Pb_{0.94}Sn_{0.06}Se$ layers with p = (1–6) x 10^{17} cm^{-3} or n = (0.3–2) x 10^{17} cm^{-3} have been produced. Without the additional Se source, the achievable carrier concentrations would be restricted to the metal-rich solidus line (Fig. 13) and metal precipitates due to their retrograde character could not be avoided. This happens since the materials assume either the congruently subliming (PbSe) or the metal-rich two-phase (SnSe) composition at the source temperature of 700 °C which are far outside the homogeneity range at the substrate temperature (Fig. 20). Mobilities up to $\mu_{77K} = 3 \times 10^4$ $cm^2V^{-1}s^{-1}$ and structural investigations verify the quality of the layers although their dislocation density is estimated to be 10^7–10^8 cm^{-2}. The layers have been used to produce Schottky barrier diodes for the 3–7 μm[174)] and 8–12 μm[175)] wavelength range which are illuminated through the IR transparent substrate.

Similar techniques have been applied to PbTe[176)] and to $Pb_{1-x}Sn_xTe$ in order to produce infrared detectors[177, 178)] and double hetero-structure lasers with conventional Perot-Fabry geometry[179)] or distributed feedback operation[180)]. The deposition of $Pb_{0.77}Sn_{0.23}Te$ from separate Knudsen cell sources containing the elements Pb, Sn and Te have been compared with that from sources with binary compounds and elemental Pb[177)]. The layers have been deposited at rates of 1 μm h^{-1} on [100]-oriented BaF_2 substrates and controlled in situ by low-energy electron diffraction and Auger spectroscopy. In the first case, nearly stoichiometric conditions can only be achieved with excess Te_2 which is obviously re-evaporated unless it is incorporated into the layer. p-Type layers with hole concentrations between p = 4 x 10^{17} cm^{-3} and 3 x 10^{18} cm^{-3} are obtained. In the second case, using binary sources and additional Pb, conditions closer to stoichiometry are achievable at 380 °C yielding carrier concentrations between p = 7 x 10^{17} cm^{-3} and n = 2 x 10^{17} cm^{-3} at 77 K. Layers grown in the neighborhood of the p-n cross-over point, however, appear to be inhomogeneous.

The growth condition under vacuum evaporation from pseudo-binary sources of $(Pb_{1-x}Sn_x)_{1+\delta}Te_{1-\delta}$ ($x = 0.2$, $\delta = -0.005$ to -0.01) on [111]-oriented BaF_2 has been studied by Chan[178)]. The films exhibit compositions x = 0.18 to 0.205. Fig. 44 shows how the epitaxial growth as well as the carrier type and concentration depend on the substrate temperature and on the deposition rate. Above 400 °C re-evaporation becomes significant.

5.1.3.2 Hot-Wall Epitaxy

In molecular beam epitaxy the large difference between source and substrate temperatures, i. e., the non-equilibrium growth conditions, renders controlling of the layer composition difficult. The hot-wall epitaxy technique has been developed to overcome this disadvantage and has recently been reviewed[181)].

The basic design is demonstrated in Fig. 45. The source cell is extended to the substrate and either almost totally closes the system (compared to the mean free path length at the pressures corresponding to the temperatures in question) or keeps only a narrow distance. In both cases, it increases the molecular flux to the substrate, so that the source temperature can be decreased and/or the substrate temperature

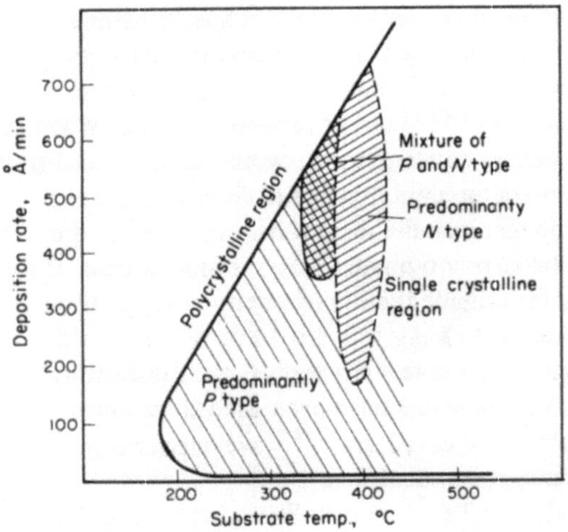

Fig. 44. Region of single-crystalline growth and carrier types versus deposition rate and substrate temperature in molecular beam epitaxy of $Pb_{0.8}Sn_{0.2}Te$[178)]

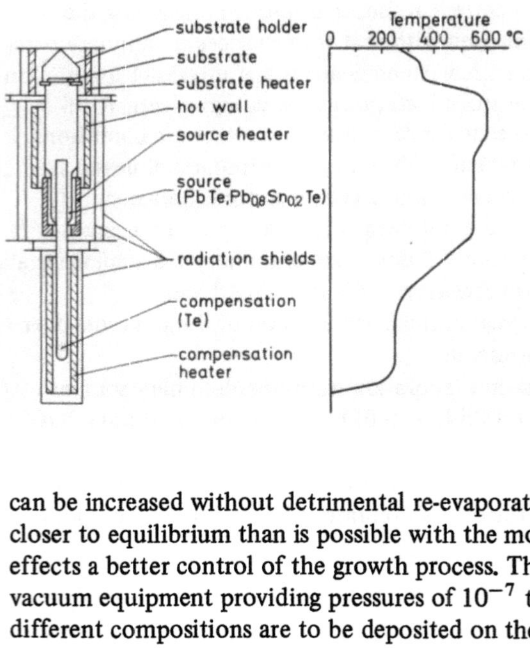

Fig. 45. Hot-wall apparatus for epitaxial deposition of lead-tin chalcogenide films. Temperature profile for PbTe and $Pb_{0.8}Sn_{0.2}Te$[193)]

can be increased without detrimental re-evaporation. This produces conditions much closer to equilibrium than is possible with the molecular beam epitaxy technique and effects a better control of the growth process. The furnace system is inserted into a vacuum equipment providing pressures of 10^{-7} to 10^{-8} mb. If consecutive layers of different compositions are to be deposited on the same substrate, this may be achieved by moving the specimen to other equivalent furnace systems without interrupting the vacuum.

The hot-wall method has extensively been applied to PbS, PbSe and $PbS_{1-x}Se_x$ and to the hetero-structures of these compounds for laser and detector fabrication[182–186)], which will not be reviewed here (see for example Ref. 13). It has also been very successful in producing PbTe[187–189)], $Pb_{1-x}Sn_xTe$[189–191)] and hetero-structures of both[192, 193)]. Very few data have hitherto been reported on the hot-wall epitaxy of $Pb_{1-x}Sn_xSe$[195, 196)]. Usually, [100]-oriented NaCl, [111]-oriented BaF_2, either cleaved or polished[194)], are chosen as substrates. They are transparent to IR radiation and can be used for backside illuminated detectors. Besides, these insulating substrates are favorable for studying the electrical properties of the epitaxial layers by Hall measurements. To reduce strain effects and dislocations originating from the lattice misfit, $Pb_{1-x}Sn_xTe$ single crystal substrates are preferred. Hetero-structures are preferably composed of $Pb_{1-x}Sn_xTe$ and a lattice-matched composition of $PbTe_{1-y}Se_y$ where, for instance, $y = 0.08$ fits to $x = 0.20$[193].

Similar to the molecular beam epitaxy technique, an additional elemental source is crucial for the film growth by hot-wall epitaxy. Its temperature is controlled separately and allows to adjust the metal/chalcogen composition to values within the homogeneity range. With hot-wall epitaxy this composition can even be adjusted to the desired carrier type and concentrations[181] as shown in Fig. 46. The carrier concentrations of layers grown at a constant source temperature of 560 °C and a rate of 2 $\mu m\ h^{-1}$ on BaF_2 and NaCl are plotted versus the nominal Te_2 partial pressure, i. e., the pressure corresponding to the Te_2 reservoir temperature. At low pressures the layers grow n-type with carrier concentrations corresponding to the metal-rich boundary of the homogeneity range. With increasing pressure a transition from n- to p-type occurs and the hole concentration increases $\propto p^{1/4}$ (see Chap. 3.3.1). It is interesting to note that the transition takes place at somewhat higher Te_2 pressures than expected from the phase diagram in Fig. 18. This is explained by the assumption that congruently subliming PbTe acquires a composition close to the metal-rich boundary[181], in contrast to Fig. 18. A second feature of interest is that carrier concentrations of layers grown on BaF_2 are consistently higher than those of layers grown on NaCl. The reason may be the different growth direction.

The similarity of the vapor pressures of PbTe and SnTe favors the growth of $Pb_{1-x}Sn_xTe$ alloys since only slight differences in the compositions of source and growing layers occur. This has been thoroughly investigated[189]. If the temperature

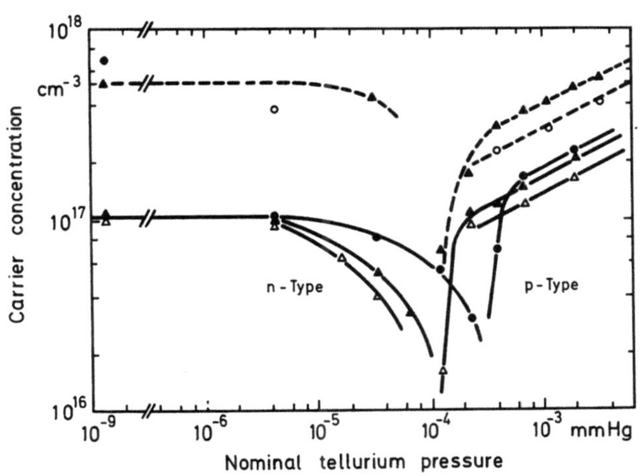

Fig. 46. Carrier concentration of PbTe films grown by hot-wall epitaxy as a function of the substrate temperature and the nominal Te_2 pressure in the growth apparatus.
● 400 °C; ▲ 380 °C, ○ 360 °C, △ 340 °C; --- BaF_2 substrate; — NaCl substrate.
The growth rate is 2 $\mu m\ h^{-1}$ for all experiments[181]

of source and substrate are chosen correctly, the source material is unchanged even after multiple use. Employment of a feed lock and vacuum substrate manipulators provide stable conditions for several hundred-hour epitaxy time, making the hot-wall technique a valuable tool for large-scale production[193].

The electrical properties at 77 K and 4 K and the investigation of the optically excited excess carrier life time[9] and the Shubnikow-de Haas effect[197] confirm the high quality of the epitaxial layers grown by this technique. A few examples may be presented[193]: PbTe layers are reproducibly grown with electron concentration as low as $n \leqslant 3 \times 10^{16}$ cm^{-3} and mobilities $\mu = 3 \times 10^4$ cm^{2}V^{-1}s^{-1} at 77 K. $Pb_{0.8}Sn_{0.2}Te$ layers reproducibly exhibit hole concentrations $p \leqslant 2 \times 10^{17}$ cm^{-3}, and mobilities $\mu = 2 \times 10^4$ cm^{2}V^{-1}s^{-1}. Much lower carrier concentrations of 10^{15} cm^{-3} and 10^{16}cm^{-3}, respectively, have also been obtained. However, these usually result from compensation effects and are less accurately predetermined by the epitaxial parameters.

5.1.4 Rf Multicathode Sputtering

The inert gas (Ar) and reactive (Ar + O_2) rf-sputtering technique has also been successfully used to deposit $Pb_{1-x}Sn_xTe$ layers for photoconductive IR detectors[198]. In early experiments three elemental targets of Pb, Sn and Te were employed simultaneously, but more reliable results follow from compounded PbTe and

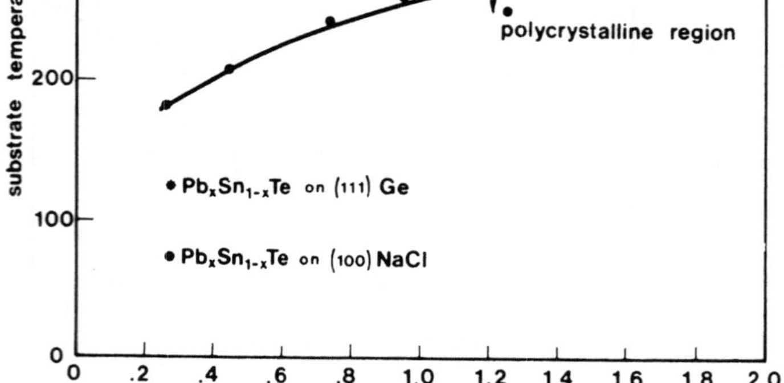

Fig. 47. Effect of growth rate and substrate temperature on the structure of $Pb_{1-x}Sn_xTe$ films sputtered onto (111)-Ge and (100)-NaCl substrates[198]

$Pb_{1-x}Sn_xTe$ targets with an additional Te target to control stoichiometry. The substrates, NaCl, BaF_2[199] or [100]- or [111]-oriented Ge[200], are mounted on a rotating platform. Highest crystallinity was obtained with layers deposited on Ge in spite of the large lattice mismatch (a_{Ge} = 5.675 Å). Fig. 47 represents the limit of single crystalline growth as a function of deposition rate and temperature[198]. Layers on BaF_2 and Ge exhibit carrier concentrations $p, n < 10^{17}$ cm^{-3} and mobilities $\mu > 10^4$ $cm^2 V^{-1} s^{-1}$. Infrared detectors have been fabricated with these layers which, however, have not yet reached the performance of those produced by the other epitaxy techniques.

5.2 Epitaxy of $Hg_{1-x}Cd_xTe$

Photoconductive detectors of $Hg_{1-x}Cd_xTe$ require crystals of 10 to 15 μm thickness (see Chap. 2.1.1). Up to now they have been exclusively produced by lapping and polishing bulk material to this final thickness, although films grown by epitaxial methods would save efforts and costs. The reason is that attempts to produce epitaxial layers of the required purity have failed so far. Photovoltaic detectors, on the other hand, for which carrier concentration of $n, p \geqslant 10^{16}$ cm^{-3} are sufficient, are frequently produced from epitaxial material. Liquid phase and vapor phase techniques are employed to produce this material. CdTe is used as a substrate which is attainable in adequate crystalline structure and size.

5.2.1 Liquid Phase Epitaxy

Although this method appears to be superior to the vapor phase epitaxy and several groups are engaged in its development, very few details have been published until now. Growth from both metal-rich and tellurium-rich solutions is possible, the latter being more convenient since higher temperatures, i. e., higher solubilities, are accessible at moderate operation pressures. An open system slider technique has been applied with growth temperatures of about 500 °C (Hg vapor pressures of the order of 0.1 atm). Since the tie lines between the liquidus and solidus surfaces are only inaccurately known, $Hg_{1-y}Te_y$ solutions ($0.76 \leqslant y \leqslant 0.815$), which are saturated over a CdTe source prior to deposition, have been used[15]. The layers then exhibit mole fractions of CdTe between x = 0.41 and 0.73 depending on saturation and growth temperatures. $Hg_{0.6}Cd_{0.4}Te$ layers from $(Hg_{0.9}Cd_{0.1})_{0.175}Te_{0.825}$ solutions at 500 °C have been obtained[201]. Substrates are usually CdTe crystals[201, 202], but the use of lattice-matched $CdTe_{1-x}Se_x$ substrates improves the layer quality[15]. The optimal orientation is found to be [111] B. Very small misorientations already produce a triangularly cross-hatched surface structure on the film. The layers grow p-type with hole concentrations of $p = 4 \times 10^{16}$ cm^{-3} to 10^{17} cm^{-3}. They may be used directly to produce pn-junctions or converted to electron concentrations of as low as $n = 4 \times 10^{15}$ cm^{-3}. Their mobilities, however, do not yet reach those of high-quality bulk material.

Table 5. Physical properties of $Hg_{1-x}Cd_xTe$ (x = 0; 0.2; 1)

	HgTe	$Hg_{0.8}Cd_{0.2}Te$	CdTe
Lattice structure	cub. (ZnS)	cub. (ZnS)	cub. (ZnS)
Lattice constant a (300 K) [Å]	6.463	6.4645	6.482
Density γ (300 K) [gcm^{-3}]	8.076	7.63	5.85
Knoop-hardness	37		45
Opt. phonons (77 K)			
LO [cm^{-1}]	138	137, 157	170
TO [cm^{-1}]	118	120, 150	142
Dielectr. constant			
ϵ_s (300 K)	20.5	18.5	10.5
ϵ_∞ (300 K)	15	13	7.2
Band gap			
Eg (300 K) [eV]	− 0.15	+ 0.165	1.505
Eg (77 K) [eV]	− 0.25	+ 0.09	1.60
Eg (0 K) [eV]	− 0.30	+ 0.064	1.605
Effective masses (77 K)			
m^*_c/m_o	0.029	0.005	0.096
m^*_v/m_o	0.35–0.7	0.2–0.6	0.66
Mobilities (77 K)			
μ_n [$cm^2V^{-1}s^{-1}$]		2.5×10^5	4×10^4
μ_p [$cm^2V^{-1}s^{-1}$]		7×10^2	3.8×10^4
Intrins. carr. conc.			
n_i (300 K) [cm^{-3}]		2×10^{16}	
n_i (77 K) [cm^{-3}]		2×10^{13}	
Melting point T_m [K]	943	940 (sol.) 1050 (liq.)	1365
Therm. exp. coeff.			
a (300 K) [$10^{-6}K^{-1}$]	4	4.3	5.5
a (77 K)	2		
Therm. conductivity			
C (300 K) [$Wcm^{-1}K^{-1}s^{-1}$]	0.04		0.07
C (77 K)	0.15		0.4

5.2.2 Vapor Phase Epitaxy

The growth of epitaxial layers of $Hg_{1-x}Cd_xTe$ has very early been attempted[203]. The method consists of an isothermal evaporation-diffusion process using a HgTe source. This technique has been further developed and refined[204–208].

Figure 48 schematically describes some of the experimental arrangements employed. A close spacing between source and substrate is always maintained. The growth temperature of 600 °C is most often applied and appears to be superior to

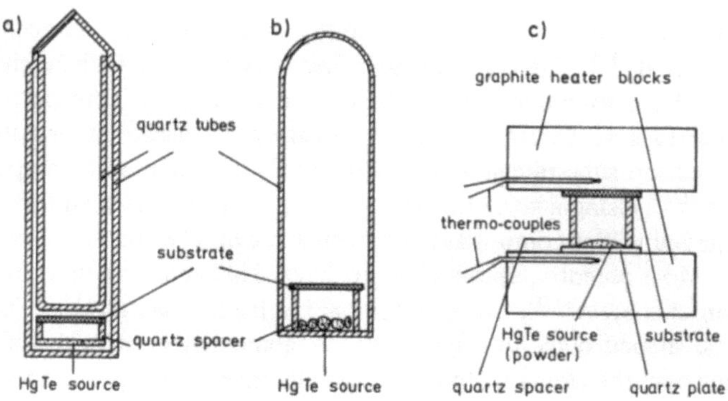

Fig. 48 a–c. Various experimental arrangements used for close-spaced epitaxial growth of HgTe ($Hg_{1-x}Cd_xTe$) layers on CdTe substrates. **a** Ref.[206]; **b** Ref.[207]; **c** Ref.[208]

lower temperatures since it provides smooth, mirror-like surfaces. Although a pure HgTe source is used, the epitaxial surface usually contains a high amount of CdTe whereas the $Hg_{1-x}Cd_xTe$ extends below the original substrate surface. The thickness of the layer, measured from the CdTe/$Hg_{1-x}Cd_xTe$ interface, is proportional to the square-root of the deposition time. An unexpected characteristic of the process is that the Cd content x in the surface is independent of the deposition time within a large investigated range (41 to 176 h)[206] but is strongly influenced by the Hg pressure during growth. On the other hand, the Hg pressure affects the deposition rate as soon as it exceeds the vapor pressure over HgTe (Fig. 49). Thus, the control of surface composition and its homogeneity becomes merely a problem of controlling

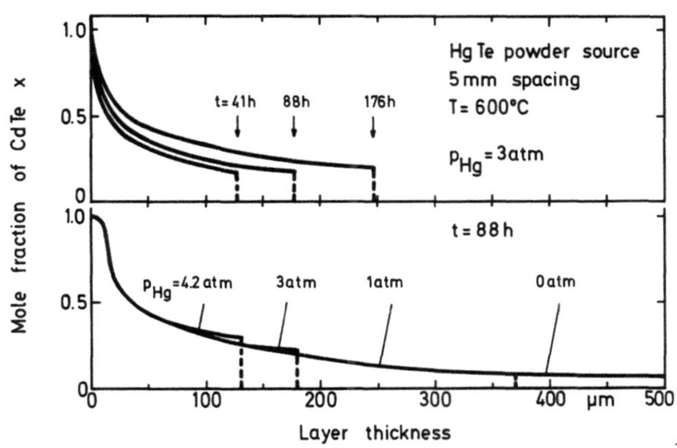

Fig. 49. Compositional profiles of epitaxial layers grown from a HgTe source on CdTe substrates.
Top: Equal isothermal conditions but different deposition times
Bottom: Equal isothermal conditions but different added mercury pressures (different amounts of excess mercury)[206]

the layer thickness and its uniformity. These features have been theoretically investigated[209]. The behavior described above is only slightly changed if $Hg_{1-x}Cd_xTe$ is used as a source instead of HgTe since some CdTe is transported under isothermal conditions. However, it differs completely if a large temperature gradient between source and substrate and an additional H_2 flow is used to support the vapor transport[207]. Using $x = 0.3$ in the source, epitaxial layers with $0.2 < x < 0.3$ are obtained with flat composition profiles through most of the layer thickness.

More recently, epitaxial layers have also been deposited from separated elemental sources[210]. An open system has been designed (Fig. 50) consisting of a cross-shaped silica tube with four independently controlled furnaces. Hydrogen flows transport the elements to a reaction chamber in which the substrate is positioned. The tube diameters, the flow rates and the relative vapor pressures critically affect the composition of the layer. The rates are adjusted to achieve a simultaneous deposition of CdTe and HgTe according to the desired composition x which can cover the entire range $0 \leqslant x \leqslant 1$. Layers with a uniform composition to about one half of its thickness and with a radial scatter of $\Delta x \leqslant 0.004$ have been grown.

Vapor-phase grown layers are generally p-type; n-type layers have only been found in the growth from elemental sources. The carrier type may, however, be changed by an after-growth annealing procedure. The electrical properties of graded gap epitaxial $Hg_{1-x}Cd_xTe$ layers and their applicability to infrared detectors have recently been reviewed[207].

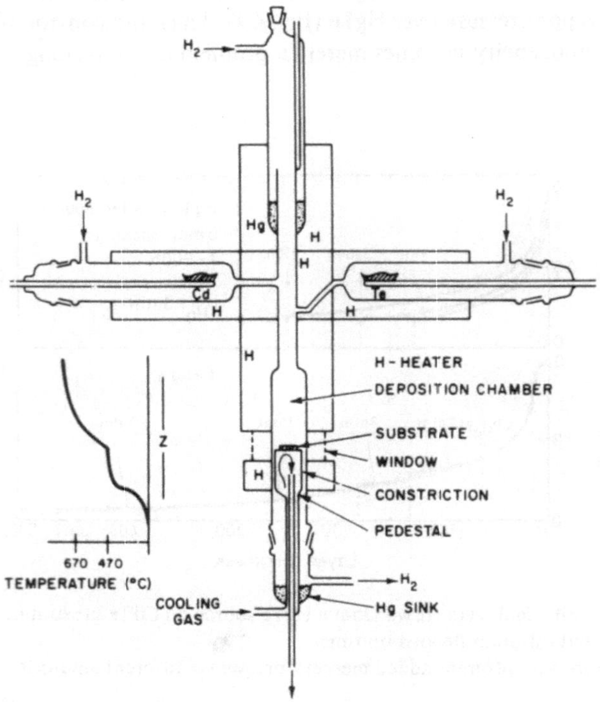

Fig. 50. Diagram of the reactor and temperature profile in the deposition chamber for the vapor-phase epitaxy of $Hg_{1-x}Cd_xTe$ films[210]

6 Modern Device Structures and Technologies

$Pb_{1-x}Sn_xTe$, $Pb_{1-x}Sn_xSe$ and $Hg_{1-x}Cd_xTe$ have early been applied in detector and laser devices. In recent years, advances in the state-of-the-art of these devices brought about very sophisticated device structures and further progress is still being expected. Some of the fabrication techniques and device designs will briefly be reviewed in the following chapter.

6.1 Infrared Detectors

The narrow-gap semiconductors treated in this article are predominantly applied to detectors for the 8 to 14 μm wavelength range, i. e., the third atmospheric window. Due to the stringent requirements for optimum performance (see Chap. 2), only $Hg_{1-x}Cd_xTe$ could be used successfully for photoconductive IR detectors. Attempts with polycrystalline rf-sputtered $Pb_{1-x}Sn_xTe$ films have been made[198)], but do not yet show results comparable to those of $Hg_{1-x}Cd_xTe$.

The technological steps leading to photoconductive detectors such as polishing to optimal thickness, contacting, passivation, packaging etc. are straightforward although they undoubtedly impose tough problems to the manufacturer. Ultimate performance and stability of the device is especially limited by the surface preparation and by the passivation technique which usually is kept proprietory by the producer.

For photovoltaic detectors all the materials treated here are intensively studied although recently a tendency toward the exclusive use of $Hg_{1-x}Cd_xTe$ is noticeable. Many different techniques have proved to yield n-p junctions with high resistance-area products in lead tin chalcogenides. Common methods involve intrinsic doping by isothermal or two-zone annealing[211)] and impurity doping by In or Cd diffusion[212, 215)]. Proton bombardment or ion implantation have been successful in PbTe[216)] but appear to be difficult in narrower gap compositions. Only very recently has implantation of In been reported to produce good diode characteristics[217)]. In $Hg_{1-x}Cd_xTe$, isothermal annealing in Hg vapor[218)], diffusion[219)] as well as ion implantation are very valuable techniques. Several ions such as Al[220)], Hg[221)] and B[222)] have been implanted into p-type crystals, Au and P[223)] into n-type crystals. Schottky barrier detectors were most intensively studied in $Pb_{1-x}Sn_xSe$ ($x = 0.07$) where Pb-contacts on p-type epitaxial layers provide surface inversion[175, 195)]. The device is designed for backside illumination through the IR transparent BaF_2 substrate, exploiting interference for increasing the spectral responsivity[224, 225)]. In p-$Pb_{1-x}Sn_xTe$[226)] and in p-$Hg_{1-x}Cd_xTe$[227)], Schottky barriers have recently been studied, too.

The most modern detectors on the basis of lead tin chalcogenides are epitaxially grown hetero-structures of p-$Pb_{1-x}Sn_xTe$/n-$Pb_{1-y}Sn_yTe$ ($x > y$) or p-$Pb_{1-x}Sn_xTe$/n-$PbTe_{1-y}Se_y$ both on PbTe or $Pb_{1-x}Sn_xTe$ substrates. Liquid-phase epitaxy[228, 229)], vapor-phase epitaxy[165)] and hot-wall epitaxy[193)] are employed. Recently, the hetero-structure design has also been applied to photovoltaic detectors of $Hg_{1-x}Cd_xTe$[202, 207)] grown on CdTe by liquid-phase epitaxy. p-n junctions have not yet been obtained epitaxially but are manufactured by ion implantation or diffusion. In these hetero-structure devices, illumination is effected by the wider band-gap material, in order to minimize surface recombination. They are used in thermal imaging systems with high sensitivity to small temperature differences and high spatial resolution. This requires a large number of small elements which are ordered in one column or in two rows interlacing each other. An example is shown in Fig. 51. A rotating mirror scans the image across the linear detector array. This parallel scanning mechanism has recently been extended to a serial-parallel mechanism which employs a two-dimensional detector array. The N elements of a line subsequently provide signals from the same image point which can be added coherently. Since noise is added incoherently, the detectivity may thus be increased by a factor $\sqrt{N}$. The detector technology, of course, is pointing toward very large two-dimensional arrays, the staring focal plane arrays having element numbers $> 10^5$. The signals will have to be read out by either charge-coupled devices (CCD) or charge injection devices (CID). At the output of the multiplexer, a serial video signal will be obtained.

Solutions to this complex problem are being intensively investigated by:

a) A hybrid structure of a photovoltaic detector array from narrow-gap material and a silicon read-out device;
b) an integrated structure in narrow-gap material;

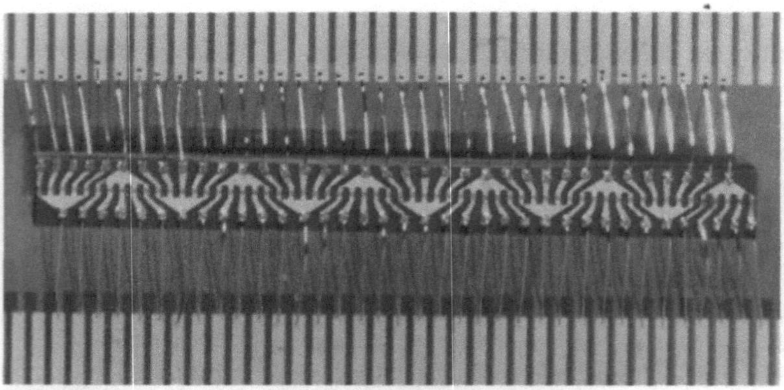

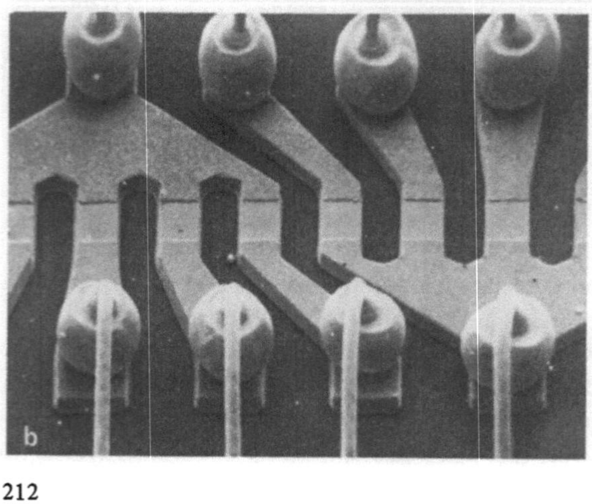

Fig. 51 a and b. Photoconductive 60-element $Hg_{1-x}Cd_xTe$ detector array mounted on a sapphire substrate (elements 50 x 50 μm^2); a general view, **b** detail, (AEG-Telefunken, Semiconductor Division, Heilbronn, Germany)

c) an integrated structure in silicon consisting of a photoconductive extrinsic silicon detector and a silicon read-out device. A description of the latter device exceeds the scope of this article. (Concerning pertinent review articles see e. g., Ref.[230]).

Up to now, the hybrid-structure performance has been demonstrated with arrays of up to 32 x 32 elements on backside illuminated $PbTe/Pb_{0.8}Sn_{0.2}Te$ hetero-structures in a solder bump technique using a Si multiplexer[231] and similarly with 32 element linear arrays of $Hg_{0.6}Cd_{0.4}Te$ connected with a Si multiplexer[232].

Impressive results have also been resported on monolithic $Hg_{1-x}Cd_xTe$ devices[233, 234]. A MIS technique using native oxides was developed and 32 stage CCD shift registers in $Hg_{0.7}Cd_{0.3}Te$ have already been achieved. They exhibit a charge transfer efficiency of up to 0.9994[235]. These results suggest that large monolithic detector arrays with electronic read-out multiplexing may become feasible at least within the 3–5 μm wavelength region of $Hg_{0.7}Cd_{0.3}Te$. Their higher operation temperature offers considerable advantages compared to analogous extrinsic silicon devices.

6.2 Infrared Diode Lasers

In contrast to its extensive use in detectors, $Hg_{1-x}Cd_xTe$ has not gained significance in laser applications. Only externally pumped lasers[15] have hitherto been accomplished. However, advanced diode laser technologies utilizing pseudo-binary lead and lead tin chalcogenides are being developed. Their capability to cover a wide wavelength range from 3 to 30 μm render these materials very attractive. Some reviews on lead salt lasers to which we refer in this article have already been published[4, 13, 236]. There are two directions of development, which are dictated by application requirements: If a rugged, high-power device with high reproducibility is required and low-operation temperature is not a serious obstacle, then diffused p-n junction lasers are employed. Up to 10 mW they are obtained with Cd diffusion in $Pb_{1-x}Sn_xTe$[237] and $Pb_{1-x}Sn_xSe$[238]. If, on the other hand, low-threshold currents and high operation temperature are desired, double hetero-structure lasers, e. g., $PbTe/Pb_{1-x}Sn_xTe/PbTe$, are preferred. They utilize optical and electrical confinements due to the narrower gap and higher refraction index of the active layer embedded between wider band-gap layers. These devices have been fabricated by molecular beam epitaxy[239], hot-wall epitaxy[240, 241] and liquid phase epitaxy[36, 156, 242, 243]. In the latter technique, foreign impurity doping with Tl and In has successfully been applied in order to reduce interdiffusion effects[161, 244].

Stripe geometries of double hetero-structure lasers provide additional lateral confinements. They are realized by means of narrow contact stripes, mesa configurations or buried active layers[13]. The end faces of the resonator are usually cleaved parallel to (100) crystal planes, but the mode structure may be considerably improved by an additional polishing process and proper antireflection and mirror coatings[245]. Proper packaging plays a dominant role in laser technology. It has to provide effective and homogeneous cooling of the device without damaging the latter upon repeated temperature cycling. Fig. 52 shows a modern packaging design[13].

The high performance of modern diode lasers has contributed to a better understanding of the hetero-junction such as recombination velocity[246] and threshold-

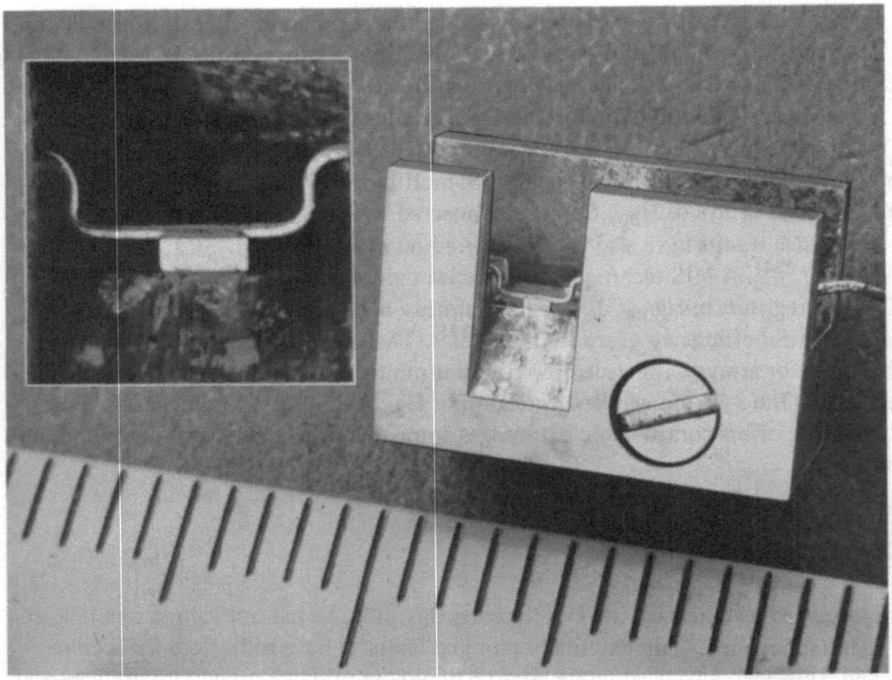

Fig. 52. $Pb_{1-x}Sn_xSe$ diode laser mounted in laser package (AEG-Telefunken, Research Institute, Frankfurt/Main, Germany)

current density and its dependences on the operation temperature[247, 248]. The practical application of these lasers is divided into two fields: One is the heterodyne detection of very low-intensity infrared radiation where diode lasers supply the local oscillator. The other one is high resolution gas spectroscopy which has recently gained considerable attention in pollution monitoring and process control[12]. Due to the very narrow linewidth of $\Delta\nu \leqslant 10^{-4}$ cm^{-1} traces of specific gases are detected without cross sensitivity to other constituents of the atmosphere. The tunability range of 1 to 2 cm^{-1} for a single mode by far exceeds the natural pressure-broadened line width of molecular absorption lines (0.1 cm^{-1}). If the laser is pulse-operated, the emission can be chirped over this range by the temperature increase during the pulse[185]. Signal integration over the pulse duration determines the area under the absorption line which is proportional to the gas concentration but independent of the pressure and composition of the atmospheric matrix. Future development of infrared laser devices manufactured from narrow-gap materials will tend toward improved (mono-) mode structures and higher tunability ranges of the emission. Here, the first successful use of distributed feed-back operation by surface corrugation of a double hetero-structure $Pb_{1-x}Sn_xTe$ laser may have opened new ways[180]. Higher operation temperatures, as recently realized with $PbS_{1-x}Se_x$ hetero-structure lasers[185, 241], will ease their applicability in industrial and pollution monitoring processes.

7 References

1. Hesse, J.: Japan, J. Appl. Phys. *16/1*, 297 (1977)
2. Melngailis, I., Harman, T. C. in: Semiconductors and Semimetals, Vol. 5 (ed. Willardson, R. K., Beer, A. C.), p. 111, New York: Academic Press 1970
3. Long, D., Schmit, J. L.: ibid., p. 175
4. Harman, T. C., Melngailis, I. in: Applied Sol. State Science, p. 1, New York: Academic Press 1974
5. Dornhaus, R., Nimtz, G. in: Springer Tracts in Modern Physics, Vol. 78, p. 1, Heidelberg: Springer 1976
6. Kinch, M. A., Borrello, S. R., Simmons, A.: Infrared Phys. *17*, 127 (1977)
7. Kinch, M. A., Borrello, S. R.: Infrared Phys. *15*, 111 (1975)
8. Preier, H.: Infrared Phys. *18*, 43 (1978)
9. Schlicht, B., et al.: Sol. State Electron. *21*, 1481 (1978)
10. Sizov, F. F., Prokopchuk, L. F., Sal'kow, E. A.: Sov. Phys.-Sol. State *19*, 1078 (1977)
11. Calawa, A. R.: J. Luminescence *7*, 477 (1973)
12. Hinkley, E. D., Nill, K. W., Blum, F. A. in: Topics in Applied Physics, Vol. 2. (ed. Walther, H.), p. 125, Berlin, Heidelberg: Springer 1976
13. Preier, H.: Applied Phys. *20*, 189 (1979)
14. Melngailis, I., Strauss, A. J.: Appl. Phys. Lett. *8*, 179 (1966)
15. Harman, T. C.: J. Electron. Mater. *8*, 191 (1979)
16. Verié, C., Granger, R.: Compt. Rend. Acad. Sci. *261*, 3349 (1965)
17. Bernard, M. G. A., Duraffourg, G.: Phys. Stat. Sol. *1*, 699 (1961)
18. Novoselova, A. V., et al.: Progr. in Sol.-State Chem. *7*, 85 (1972)
19. Elwell, D., Neate, B. W., Smith, S. H.: J. Thermal Analysis *1*, 319 (1969)
20. Krebs, H., Grün, K., Kallen, D.: Z. Anorg. Allg. Chem. *312*, 307 (1961)
21. Calawa, A. R., et al.: Trans. Met. Soc. AIME *242*, 374 (1968)
22. Ilegems, M., Pearson, G. L.: Ann. Rev. Mater. Sci. *5*, 345 (1975)
23. Vieland, L. J.: Acta Met. *11*, 137 (1963)
24. Brebrick, R. F., Allgaier, R. S.: J. Chem. Phys. *32*, 1826 (1960)
25. Albers, W., Haas, C., Vink, H. J.: Philips. Res. Rept. *18*, 372 (1963)
26. Reti, A. M., Jena, A. K., Bever, M. B.: Trans. Met. Soc. AIME *242*, 371 (1968)
27. Linden, K. J., Kennedy, C. A.: J. Appl. Phys. *40*, 2595 (1969)
28. Harman, T. C.: J. Nonmetals *1*, 183 (1973)
29. Wagner, J. W., Willardson, R. K.: Trans. Met. Soc. AIME *242*, 366 (1968)
30. Jordan, A. S.: Met. Trans. *1*, 239 (1970)
31. Jordan, A. S.: J. Electrochem. Soc. *119*, 123 (1972)
32. Laugier, A.: Rev. Phys. Appl. *8*, 259 (1973)
33. Laugier, A., et al.: J. Cryst. Growth *21*, 235 (1974)
34. Harris, J. S., et al.: J. Cryst. Growth *28*, 334 (1975)
35. Hansen, M.: Constitution of Binary Alloys, Mc Graw-Hill, New York 1958
36. Groves, S. H.: J. Electron. Mater. *6*, 195 (1977)
37. Tamari, N., Shtrikman, H.: J. Cryst. Growth *43*, 653 (1978)
38. Chou, N., Komarek, K., Miller, E.: Trans. Met. Soc. AIME *245*, 1553 (1963)
39. Fujimoto, M., Sato, Y.: Japan. J. Appl. Phys. *5*, 128 (1966)
40. Brebrick, R. F.: J. Phys. Chem. Sol. *24*, 27 (1963)
41. Matthes, H.: unpubl.
42. Seidmann, D. N.: Trans. Met. Soc. AIME *236*, 1361 (1966)
43. Kasai, I., Bassett, D. W.: J. Cryst. Growth *27*, 215 (1974)
44. Maier, H., Daniel, D. R.: J. Electron. Mater. *6*, 693 (1977)
45. Bekker, A. A., et al.: Inorg. Mater. *9*, 757 (1973)
46. Lawson, W. D., et al.: J. Phys. Chem. Sol. *9*, 325 (1959)
47. Ray, B., Spencer, P. M.: Phys. Stat. Sol. *22*, 371 (1967)
48. Harman, T. C., Strauss, A. J. cited in: Harman, T. C.: Physics and Chemistry of II-VI Compounds (ed. Aven, M., Prener, J. S.), p. 767, Amsterdam: North Holland 1967

49. Steininger, J.: J. Electron. Mater. *5,* 299 (1976)
50. Blair, J., Newnham, R.: Metallurgy of Elemental and Compound Semiconductors, Vol. 12, New York: Wiley Interscience 1961
51. Schmit, J. L., Speerschneider, C. J.: Infrared Phys. *8,* 247 (1968)
52. Zanio, K. in: Semiconductors and Semimetals, (ed. Willardson, R. K., Beer, A. C.), p. 1, New York: Academic Press 1978
53. Brebrick, R. F., Strauss, A. J.: J. Phys. Chem. Sol. *26,* 989 (1965), ibid. *31,* 2293 (1970)
54. Nishizawa, J., et al.: J. Phys. Chem. Sol. *37,* 33 (1976)
55. Pashinkin, A. S., Novoselova, A. V.: Russ. J. Inorg. Chem. *4,* 1229 (1959)
56. Northrop, D. A.: J. Phys. Chem. *75,* 118 (1971)
57. Brebrick, R. F., Strauss, A. J.: J. Chem. Phys. *40,* 3230 (1964)
58. Northrop, D. A.: J. Electrochem. Soc. *118,* 1365 (1971)
59. Colin, R., Drowart, J.: Trans. Faraday Soc. *60,* 673 (1964)
60. Winchell, P.: Adv. Energy Conversion *8,* 81 (1968)
61. Sokolov, V. V., et al.: Russ. J. Phys. Chem. *40,* 885 (1966)
62. Sokolov, V. V., et al.: Izv. Akad. Nauk SSR, Neorg. Mater. *5,* 15 (1969)
63. Brebrick, R. F., Strauss, A. J.: J. Chem. Phys. *41,* 197 (1964)
64. Brebrick, R. F.: J. Phys. Chem. Sol. *32,* 551 (1971)
65. Sugiyama, K.: Japan. J. Appl. Phys. *5,* 128 (1966)
66. Hansen, E. E., Munir, Z. A., Mitchell, M. J.: J. Am. Ceram. Soc. *52,* 610 (1969)
67. Blair, R. C., Munir, Z. A.: J. Amer. Ceram. Soc. *53,* 301 (1970)
68. Hirayama, C., Ichikawa, Y., de Roo, A. M.: J. Phys. Chem. *67,*1039 (1963)
69. Zlomanov, V. P., Popovkin, B. A., Novoselova, A. V.: Russ. J. Inorg. Chem. *4,* 1231 (1959)
70. Igaki, K., Ohashi, N.: J. Phys. Soc. Japan *18,* Suppl. II, 143 (1963)
71. Ohashi, N., Igaki, K.: Trans. Japan. Inst. Metals *5,* 94 (1964)
72. Brebrick, R. F., Strauss, A. J.: J. Phys. Chem. Sol. *25,* 1441 (1964)
73. Schwartz, H.: Thesis, Marquette Univ., Milwaukee, Wisc. 1977
74. Farrar, R. A., et al.: J. Mater. Sci. *12,* 837 (1977)
75. Parada, N. J., Pratt Jr., G. W.: Phys. Rev. Lett. *22,* 180 (1969)
76. Parada, N. J.: Phys. Rev. *B3,* 2042 (1971)
77. Hemstreet, L. A.: Phys. Rev. *B12,* 1212 (1975)
78. Allgaier, R. S., Scanlon, W. W.: Phys. Rev. *111,* 1029 (1958)
79. Stiles, P. J., Burstein, E., Langenberg, D. N.: J. Appl. Phys. Suppl. *32,* 2174 (1961)
80. Bauer, G., et al.: J. Appl. Phys. *47,* 1721 (1976)
81. Lischka, K., et al.: 14. Int. Conf. Phys. Semicond., Edinburgh 1978
82. Heinrich, H., et al.: Inst. Phys. Ser. *23,* 264 (1975)
83. Maier, H., Daniel, D. R., Preier, H.: J. Cryst. Growth *35,* 121 (1976)
84. Daniel, D. R., Maier, H., Preier, H.: J. Cryst. Growth *38,* 145 (1977)
85. Sato, Y., Fujimoto, M., Kobayashi, A.: Japan. J. Appl. Phys. *2,* 683 (1963)
86. Brebrick, R. F.: J. Electron. Mater *6,* 659 (1977)
87. Logothetis, E. M., Holloway, H.: J. Nonmetals *2,* 1 (1973)
88. Strauss, A. J.: J. Electron. Mater. *2,* 553 (1973)
89. Koval'chik, T. L., Maslakovets, Yu. P.: Sov. Phys. Techn. Phys. *1,* 2337 (1956)
90. Antcliffe, G. A., Wrobel, J. S.: Appl. Phys. Lett. *17,* 290 (1970)
91. Crocker, A. J.: J. Phys. Chem. Sol. *28,* 1903 (1967)
92. Bastard, G., Nozières, D.: Phys. Rev. *B13,* 2560 (1976)
93. Elliott, C. T., et al.: J. Phys. Chem. Sol. *33,* 1527 (1972)
94. Elliott, C. T., et al.: Phys. Rev. *B5,* 2985 (1972)
95. Finck, C., et al.: Proc. XI Int. Conf. Phys. Semicond., p. 944, Warsaw: PWN-Polish Scient. Publ. 1972
96. Dornhaus, R., et al.: Sol. State Comm. *17,* 837 (1975)
97. Zaitov, T. A., Stafeev, V. I., Khodakov, G. S.: Sov. Phys. Sol. State *14,* 2628 (1973)
98. Johnson, E. S., Schmit, J. L.: J. Electron. Mat. *6,* 25 (1977)
99. Scott, W., Stelzer, E. L., Hager, R. J.: J. Appl. Phys. *47,* 1408 (1976)
100. Gotacki, Z., et al.: J. Cryst. Growth *46,* 293 (1979)
101. Lawson, W. D.: J. Appl. Phys. *22,* 1444 (1951)

102. Lawson, W. D.: J. Appl. Phys. *23,* 495 (1952)
103. Strauss, A. J., Harman, T. C.: J. Electron. Mat. *2,* 71 (1973)
104. Gutknecht, K. H., Preier, H., Hesse, J.: Mat. Res. Bull. *10,* 277 (1975)
105. Butler, J. F., Harman, T. C.: J. Electrochem. Soc. *115,* 67 C (1968)
106. Dionne, G., Woolley, J. C.: J. Electrochem. Soc. *119,* 784 (1972)
107. Fano, V., Pergolari, R., Zanotti, L.: J. Mat. Sci. *14,* 535 (1979)
108. Mateika, D.: J. Crystal Growth *9,* 249 (1971)
109. Hiscocks, S. E. R., West, P. D.: J. Mat. Sci *3,* 76 (1968)
110. Burke, J. R., Jensen, J. D., Houston, B.: J. Electrochem. Soc. *120,* 431 (1973)
111. Kim, K. M., Witt, A. F., Gatos, H. C.: J. Electrochem. Soc. *119,* 1218 (1972)
112. Jäger, H., Seipp, E.: Verh. DPG. *1,* 86 (1978)
113. Bansaragtschin, B., Lehmann, G., Link, R.: Phys. Stat. Sol. (a) *37,* K 13 (1976)
114. Kimura, H.: J. Electron. Mat. *1,* 166 (1972)
115. Butler, J. F., Harman, T. C.: Appl. Phys. Lett. *12,* 347 (1968)
116. Prior, A. C.: J. Electrochem. Soc. *108,* 82 (1961)
117. Harman, T. C., Mc Vittie, J. P.: J. Electron. Mat. *3,* 843 (1974)
118. Maier, H., et al.: J. Mat. Sci. *13,* 297 (1978)
119. Maier, H., Viehmann, R.: to be publ.
120. Kasai, I., et al.: J. Cryst. Growth *23,* 201 (1974)
121. Preier, H., Herkert, R., Pfeiffer, H.: J. Cryst. Growth *22,* 153 (1974)
122. Lo, W.: J. Electron. Mat. *6,* 39 (1977)
123. Knittel, A. Z.: J. Cryst. Growth *26,* 33 (1974)
124. Parker, S. G., Pinnell, J. E., Johnson, R. E.: J. Electron. Mat. *3,* 731 (1974)
125. Pandey, R. K.: Sol. State Commun. *15,* 449 (1974)
126. Tamari, N., Shtrikman, H.: J. Cryst. Growth *43,* 378 (1978)
127. Parker, S. G.: J. Electron. Mat. *5,* 497 (1976)
128. Bradford, A., Wentworth, E.: Infrared Phys. *15,* 303 (1975)
129. Parker, S. G., Guldi, R. L.: J. Electrochem. Soc. *121,* 90 C (1974)
130. Shinohara, K., et al.: Japan. J. Appl. Phys. *11,* 273 (1972)
131. Bartlett, B. E., Deans, J., Ellen, P. C.: J. Mat. Sci. *4,* 266 (1969)
132. Gillham, C. J., Farrar, R. A.: J. Mat. Sci. *12,* 1994 (1977)
133. Delves, R. T.: Brit. J. Appl. Phys. *16,* 343 (1965)
134. Bartlett, B. E., et al.: J. Cryst. Growth *46,* 623 (1979), ibid. *47,* 341 (1979)
135. Bye, K. L.: J. Mat. Sci. *14,* 619 (1979)
136. Takase, Y.: Jap. J. Appl. Phys. *13,* 539 (1974)
137. Triboulet, R.: Rev. Phys. Appl. *12,* 123 (1977)
138. Dziuba, E. Z.: J. Electrochem. Soc. *116,* 104 (1969)
139. Triboulet, R., Triboulet, D., Didier, G.: J. Cryst. Growth *38,* 82 (1977)
140. Ueda, R., et al.: J. Cryst. Growth *13/14,* 668 (1972)
141. Harman, T. C.: J. Electron. Mat. *1,* 230 (1972)
142. Fiorito, G., Gasparrini, G., Passoni, D.: J. Electrochem. Soc. *125,* 315 (1978)
143. Swink, L. N., Brau, M. J.: Met. Trans. *1,* 629 (1970)
144. Stelzer, E. L., Schmit, J. L., Tufte, O. N.: IEEE Trans. El. Dev. *16,* 880 (1969)
145. Dietl, J., Jarosch, J.: US-Pat. 4.011.074, 1977
146. Brau, M. J.: US-Pat. 3.656.844, 1972
147. Maier, H., Ziegler, J.: to be publ.
148. Steininger, J.: J. Cryst. Growth *37,* 107 (1977)
149. Aust, K. T. in: Art and Science of Growing Crystals, (ed. Gilman, J.), p. 452, New York, London: Wiley 1963
150. Brau, M. J., Reynolds, R. A.: US-Pat. 3.849.205, 1974
151. Schmit, J. L., Stelzer, E. L.: J. Electron. Mat. *7,* 65 (1978)
152. Bartoli, F. J., et al.: Sol. State Commun. *25,* 963 (1978)
153. Small, M. B., Crossley, I.: J. Cryst. Growth *27,* 35 (1974)
154. Zschauer, K. H. in: Festkörperprobleme, Vol. 15 (ed. Queisser, H. J.), Braunschweig: Pergamon-Vieweg, 1975
155. Longo, J. T., et al.: J. Cryst. Growth *15,* 107 (1972)

156. Tomasetta, L. R., Fonstad, C. G.: Appl. Phys. Lett. *24*, 567 (1974)
157. Groves, S. H.: 18th Conf. Electron. Mat., Salt Lake City 1976
158. Nelson, H.: RCA Rev. *24*, 603 (1963)
159. Thompson, A. G., Wagner, H. W.: Phys. Stat. Sol. (a) *5*, 439 (1971)
160. Wang, C. C., Hampton, S. R.: Sol. State Electron. *18*, 121 (1975)
161. Tomasetta, L. R., Fonstad, C. G.: Mat. Res. Bull. *9*, 799 (1974)
162. Tamari, N., Shtrikman, H.: J. Appl. Phys., to be publ.
163. Crawford, M. G., Groves, W. O.: Proc. IEEE *6*, 862 (1973)
164. Hiscocks, S. E. R.: J. Cryst. Growth *17*, 222 (1972)
165. Rolls, W. H., Eddolls, D. V.: Infrared Phys. *13*, 143 (1973)
166. Eddolls, D. V.: Infrared Phys. *16*, 47 (1976)
167. Rogers, C. J., Daniel, D. R.: private communication
168. Bellavance, D. W., Johnson, M. R.: J. Electron. Mat. *5*, 363 (1976)
169. Ploog, K. in: Crystals, Growth, Properties and Applications, Vol. 3, Berlin, Heidelberg, New York: Springer 1980
170. Esaki, L., Chang, L. L.: Thin Sol. Films *36*, 285 (1976)
171. Cho, A. Y., Casey Jr., H. C.: Appl. Phys. Lett. *25*, 288 (1974)
172. Hohnke, D. K., Kaiser, S. W.: J. Appl. Phys. *45*, 892 (1974)
173. Tao, T. F., Wang, C. C.: J. Appl. Phys. *43*, 1313 (1972)
174. Hohnke, D. K., Holloway, H.: Appl. Phys. Lett. *24*, 633 (1974)
175. Hohnke, D. K., et al.: Appl. Phys. Lett. *29*, 98 (1976)
176. Holloway, H.: J. Nonmetals *1*, 347 (1973)
177. Smith, D. L., Pickhardt, V. Y.: J. Electron. Mat. *5*, 247 (1976)
178. Chan, W. S.: Infrared Phys. *14*, 177 (1974)
179. Walpole, J. N., et al.: Appl. Phys. Lett. *28*, 552 (1976)
180. Walpole, J. N., et al.: Appl. Phys. Lett. *29*, 307 (1976)
181. Lopez-Otero, A.: Thin Sol. Films *49*, 3 (1978)
182. Mc Lane, G. F., Sleger, K. J.: J. Electron. Mat. *4*, 465 (1975)
183. Duh, K., Preier, H.: J. Mat. Sci. *10*, 1360 (1975)
184. Bleicher, M., et al.: J. Mat. Sci. *12*, 317 (1977)
185. Preier, H., et al.: Appl. Phys. *12*, 277 (1977)
186. Schoolar, R. B., Jensen, J. D., Black, G. M.: Appl. Phys. Lett. *31*, 620 (1977)
187. Lopez-Otero, A.: Appl. Phys. Lett. *26*, 470 (1975)
188. Lopez-Otero, A., Haas, D. L.: Thin Sol. Films *23*, 1 (1974)
189. Jakobus, T., Hornung, J.: J. Cryst. Growth *45*, 224 (1978)
190. Lopez-Otero, A., Haas, L. D.: Thin Sol. Films *32*, 35 (1976)
191. Lowney, J. R., Cammack, D. R.: Mat. Res. Bull. *9*, 1639 (1974)
192. Kasai, I., Bassett, D. W., Hornung, J.: J. Appl. Phys. *47*, 3167 (1976)
193. Haas, L. D., Maier, H.: to be publ.
194. Bis, R. F., Farabough, E. N., Muth, E. P.: J. Appl. Phys. *47*, 736 (1976)
195. Schoolar, R. B., Jensen, J. D.: Appl. Phys. Lett. *31*, 536 (1977)
196. Maier, H., Bleicher, M.: Germ. Govt. Rept. T/RF 31/RF 310/51090 (1976)
197. Schaber, H., et al.: Sol. State Commun. *26*, 7 (1978)
198. Corsi, C.: Proc. IEEE *63*, 14 (1975)
199. Corsi, C.: J. Appl. Phys. *45*, 3467 (1974)
200. Corsi, C., et al.: Thin Sol. Films *33*, 135 (1976)
201. Schmit, J. L., Bowers, J. E.: Appl. Phys. Lett. *35*, 457 (1979)
202. Lanir, M., Wang, C. C., Vanderwyck, A. H. B.: Appl. Phys. Lett. *34*, 50 (1979)
203. Cohen-Solal, G., et al.: Compt. Rend. *261*, 931 (1965)
204. Marfaing, Y., Cohen-Solal, G., Bailly, F.: J. Phys. Chem. Sol. Suppl. *1*, 549 (1967)
205. Almasi, G. S., Smith, A. C.: J. Appl. Phys. *39*, 233 (1968)
206. Tufte, O. N., Stelzer, E. L.: J. Appl. Phys. *40*, 4559 (1969)
207. Pawlikowski, J. M.: Thin Sol. Films *44*, 241 (1977)
208. Saraie, J., et al.: Jap. J. Appl. Phys. *12*, 1259 (1973)
209. Svob, L., et al.: J. Appl. Phys. *46*, 4251 (1975) ibid. *46*, 4244 (1975)
210. Vohl, P., Wolfe, C. M.: J. Electron. Mat. *7*, 659 (1978)

211. Rolls, W., Lee, R., Eddington, R. J.: Sol. State Electron. *13*, 75 (1970)
212. Kennedy, C. A., Linden, K. J., Soderman, D. A.: Proc. IEEE *63*, 27 (1975)
213. Silberg, E., Zemel, A.: J. Electron. Mat. *8*, 99 (1979)
214. Wang, C. C., Lorenzo, J. S.: Infrared Phys. *17*, 83 (1977)
215. Johnson, M. R., Chapman, R. A., Wrobel, J. S.: Infrared Phys. *15*, 317 (1975)
216. Donnelly, J. P., et al.: J. Nonmetals *1*, 123 (1973)
217. Jakobus, T., et al.: Rev. Phys. Appl. *13*, 753 (1978)
218. Ameurlaine, J., et al.: Int. Electron. Devices Meeting, Washington. DC, p. 430, 1978
219. Spears, D. L.: Infrared Phys. *17*, 5 (1977)
220. Marine, J., Motte, C.: Appl. Phys. Lett. *23*, 450 (1973)
221. Fiorito, G., Gasparrini, G., Svelto, F.: Infrared Phys. *15*, 287 (1975)
222. Sood, A. K., Tredwell, T. J.: Int. Electron. Devices Meeting, Washington DC., p. 434, 1978
223. Ryssel, H.: private communication
224. Holloway, H.: J. Appl. Phys. *50*, 1386 (1979)
225. Holloway, H.: Thin Sol. Films *58*, 73 (1979)
226. Baars, J., Bassett, D., Schulz, M.: Phys. Stat. Sol. (a), *49*, 483 (1978)
227. Pawlikowski, J. M., et al.: Acta Physica Polonica *A 49*, 563 (1976)
228. Andrews, A. M., et al.: Appl. Phys. Lett. *26*, 438 (1975)
229. Wang, C. C., et al.: Sol. State Electron. *21*, 625 (1978)
230. Steckl, A. J.: Infrared Phys. *16*, 65 (1976)
231. Longo, J. T., et al.: IEEE Trans. Electron. Dev. *25*, 213 (1978)
232. Broudy, R., et al.: S. P. I. E. Proc., Vol. 132, 1978
233. Sirieix, M., et al.: Infrared Phys. *18*, 233 (1978)
234. Chapman, R. A., et al.: Appl. Phys. Lett. *32*, 434 (1978)
235. Buss, D. D., et al.: Int. Electron. Devices Meet., Washington, DC p. 496, 1978
236. Hesse, J., Preier, H. in: Festkörperprobleme, Vol. 15, (ed. by Queisser, H. J.), p. 229, Braunschweig: Pergamon-Vieweg 1975
237. Lo, W.: Appl. Phys. Lett. *28*, 154 (1976)
238. Grisar, R., et al.: IEEE Device Res. Conf., Boulder 1979
239. Walpole, J. N., et al.: Appl. Phys. Lett. *28*, 552 (1976)
240. Preier, H., et al.: Appl. Phys. Lett. *28*, 669 (1976)
241. Preier, H., et al.: J. Appl. Phys. *47*, 5476 (1976)
242. Carpenter, R., et al.: Infrared Phys. *18*, 193 (1978)
243. Kosogov, O. V., et al.: Sov. Phys. Semicond. *11*, 1113 (1977)
244. Groves, S. H., Nill, K. W., Strauss, A. J.: Appl. Phys. Lett. *25*, 331 (1974)
245. Ralston, R. W., et al.: J. Phys. *45*, 1323 (1974)
246. Karemset, D., Fonstad, C. G.: Appl. Phys. Lett. *34*, 432 (1979)
247. Tomasetta, L. R., Fonstad, C. G.: IEEE Quantum Electron. *11*, 384 (1975)
248. Anderson, W. W.: IEEE, Quantum Electron. *13*, 532 (1977)
249. Hewes, C. R., Adler, M. S., Senturia, S. D.: J. Appl. Phys. *44*, 1327 (1973)

Received November 12, 1979

Author Index Volumes 1–4

The volume numbers are printed in italics

A. F. Williams

A Theoretical Approach to Inorganic Chemistry

1979. 144 figures, 17 tables. XII, 316 pages
ISBN 3-540-09073-8

This book outlines the application of simple quantum mechanics to the study of inorganic chemistry, and shows its potential for systematizing and understanding the structure, physical properties, and reactivities of inorganic compounds. The considerable strides made in inorganic chemistry in recent years necessitate the establishment of a theoretical framework if the student is to acquire a sound knowledge of the subject. A wide range of topics is covered, and the reader is encouraged to look for further extensions of the theories discussed. The book emphasizes the importance of the critical application of theory and, although it is chiefly concerned with molecular orbital theory, other approaches are discussed. This text is intended for students in the latter half of their undergraduate studies. (235 references)

From the Foreword by Prof. C. K. Jørgensen

"...Dr. Alan Williams has acquired a considerable experience in work with transition metal complexes at the Universities of Cambridge and Geneva. In this book he has tried to avoid the variety of ephemeral and often contradictory rationalisations encountered in this field, and has made a careful comparison of modern opinions about chemical bonding. In my opinion this effort is fruitful for all students and active scientists in the field of inorganic chemistry. The distant relations to group theory, atomic spectroscopy and epistemology are brought into daylight.
...The interdisciplinary approach of the book shows up in the careful consideration given to many experimental techniques such as vibrational (infra-red and Raman), electronic (visible and ultraviolet), Mössbauer, magnetic resonance, and photoelectron spectra, with data for gaseous and solid samples as well as selected facts about solution chemistry. The book could not have been written a few years ago, and is likely to remain a highly information survey of modern inorganic chemistry and chemical physics."

Springer-Verlag
Berlin
Heidelberg
New York

Inorganic Chemistry Concepts

Editors:
M. Becke
C. K. Jørgensen
M. F. Lappert
S. J. Lippard
J. L. Margrave
K. Niedenzu
R. W. Parry
H. Yamatera

Springer-Verlag
Berlin
Heidelberg
New York

Volume 1
R. Reisfeld, C. K. Jørgensen

Lasers and Excited States of Rare Earths

1977. 9 figures, 26 tables. VIII, 226 pages
ISBN 3-540-08324-3

Contents:
Analogies and Differences Between Monatomic Entities and Condensed Matter. – Rare-Earth Lasers. – Chemical Bonding and Lanthanide Spectra. – Energy Transfer. – Applications and Suggestions.

Volume 2
R. L. Carlin, A. J. van Duyneveldt

Magnetic Properties of Transition Metal Compounds

1977. 149 figures, 7 tables. XV, 264 pages
ISBN 3-540-08584-X

Contents:
Paramagnetism: The Curie Law. – Thermodynamics and Relaxation. – Paramagnetism: Zero-Field Splittings. – Dimers and Clusters. – Long-Range Order. – Short-Range Order. – Special Topics: Spin-Flop, Metamagnetism, Ferrimagnetism and Canting. – Selected Examples.

Volume 3
P. Gütlich, R. Link, A. Trautwein

Mössbauer Spectroscopy and Transition Metal Chemistry

1978. 160 figures, 1 folding plate, 19 tables. X, 280 pages
ISBN 3-540-08671-4

Contents:
Basic Physical Concepts. – Hyperfine Interactions. – Experimental. – Mathematical Evaluation of Mössbauer Spectra. – Interpretation of Mössbauer Parameters of Iron Compounds. – Mössbauer-Active Transition Metals Other than Iron. – Some Special Applications.

MIX
Papier aus verantwortungsvollen Quellen
Paper from responsible sources
FSC® C105338

If you have any concerns about our products,
you can contact us on
ProductSafety@springernature.com

In case Publisher is established outside the EU,
the EU authorized representative is:
Springer Nature Customer Service Center GmbH
Europaplatz 3, 69115 Heidelberg, Germany

Printed by Libri Plureos GmbH
in Hamburg, Germany